Optimization Techniques with Fortran

Optimization Techniques with Fortran

JAMES L. KUESTER
Chemical Engineering
Arizona State University
Tempe, Arizona

JOE H. MIZE
Industrial Engineering & Management
Oklahoma State University
Stillwater, Oklahoma

McGRAW-HILL BOOK COMPANY
New York   St. Louis   San Francisco   Düsseldorf   London   Mexico   Panama   Sydney   Toronto
Johannesburg   Kuala Lumpur   Montreal   New Delhi   Rio de Janeiro   Singapore

OPTIMIZATION TECHNIQUES WITH FORTRAN

Copyright © 1973 by McGraw-Hill, Inc. All rights reserved.
Printed in the United States of America. No part of this
publication may be reproduced, stored in a retrieval system,
or transmitted, in any form or by any means, electronic,
mechanical, photocopying, recording, or otherwise, without
the prior written permission of the publisher.

07-035606-8

234567890WHWH7987654

Library of Congress Cataloging in Publication Data

Kuester, James L
   Optimization techniques with Fortran.

   Bibliography: p.
   1. Electronic data processing—Mathematical
optimization. 2. Electronic data processing—
Programming (Mathematics) 3. FORTRAN (Computer
program language) I. Mize, Joe H., joint author.
II. Title.
QA402.5.K84   519.7   73-6824
ISBN 0-07-035606-8

CONTENTS

Preface

Chapter 1. <u>Introduction</u>   1

PART ONE: SPECIAL PURPOSE METHODS

Chapter 2. <u>Linear Programming</u>   9

   I. Simplex (SIMPLX Algorithm)   10
  II. Revised Simplex (REVSIM Algorithm)   27
 III. Mixed Integer Programming (MINT Algorithm)   66
  IV. Zero-One Programming (ZONE Algorithm)   91

Chapter 3. <u>Quadratic Programming</u>   105

   I. Wolfe (WOLFE Algorithm)   106
  II. Quadratic Differential (QUADIF Algorithm)   120

Chapter 4. <u>Geometric Programming</u>   135

   I. Blau (GOMTRY Algorithm)   136

Chapter 5. <u>Dynamic Programming</u>   155

   I. Discrete (DP Algorithm)   156
  II. Continuous Dynamic Programming (DYNAM Algorithm)   183

Chapter 6. <u>Least Squares Objective Functions</u>   203

   I. Linear Regression (LINREG Algorithm)   205
  II. Gauss Newton (BARD Algorithm)   218
 III. Marquardt (BSOLVE Algorithm)   240
  IV. Powell (SSQMIN Algorithm)   251

PART TWO: SEARCH METHODS

Chapter 7. <u>Single Variable Unconstrained Methods</u>   275

   I. Coggins (COGGIN Algorithm)   276

Chapter 8. <u>Single Variable Constrained Methods</u>   285

   I. Fibonacci (FIBON Algorithm)   286

Chapter 9.   Multivariable Unconstrained Methods   297

    I.   Nelder and Mead (NELDER Algorithm)   298
   II.   Hooke and Jeeves (HOOKE Algorithm)   309
  III.   Rosenbrock (ROSENB Algorithm)   320
   IV.   Powell (BOTM Algorithm)   331
    V.   Fletcher and Reeves (FMCG Algorithm)   344
   VI.   Fletcher and Powell (FMFP Algorithm)   355

Chapter 10. Multivariable Constrained Methods   367

    I.   Box (COMPLEX Algorithm)   368
   II.   Constrained Rosenbrock (HILL Algorithm)   386
  III.   Rosen (PROJG Algorithm)   399
   IV.   Fiacco and McCormick (SUMT Algorithm)   412
    V.   Constrained Fletcher-Powell (CONMIN Algorithm)   464

BIBLIOGRAPHY   497

PREFACE

Optimization methods have been described adequately in several recent texts. (See for example, references 9, 5, 29, 47, 10.) Virtually all useful optimization algorithms have been designed for implementation on a digital computer. Computer programs have been written for many optimization techniques, but these programs are not generally included in the texts, and they are rather difficult to obtain. Consequently, most courses in optimization are taught using unrealistically small examples for which manual calculations can be performed.

The purpose of this book is to provide a central source of FORTRAN coded algorithms for a broad spectrum of optimization techniques. The book is intended as a supplementary text for courses in optimization theory, mathematical programming, operations research, and quantitative methods. It is hoped that the collection of programs provided will greatly facilitate the teaching of these courses.

In addition to its use as a supplementary text, the book should also be very useful for researchers and practicing engineers and management scientists who are involved in the application of optimization techniques. Sufficient information is provided about each method to allow the book to be used as a handy reference.

Copies of all programs listed in this book may be obtained from the authors.

The authors wish to thank each person and company who gave permission to include their computer algorithms. Without their cooperation, the creation of this central code source would not have been possible.

The authors would also like to acknowledge the very capable assistance of William Baumann, Merril Esher, Paul Inglish and Joe McKee in the preparation of this material. Finally, the authors are very grateful to Mrs. Margaret Estes for a very excellent job of typing and to Mrs. Betty Mize, who spent many hours assisting with the final manuscript.

James L. Kuester
Joe H. Mize

Chapter 1

INTRODUCTION

Presented in this book are a number of optimization techniques designed for a wide variety of optimization problems. The general problem under consideration is as follows,

<u>Objective Function</u>   Optimize $F(X_1, X_2, \ldots, X_N)$

<u>Constraints</u>   Subject to $G_k(X_1, X_2, \ldots, X_N) \leq 0$
                                                            $k = 1, 2, \ldots, M.$

The problem may be posed as either a maximization or minimization problem. If constraints exist, they may be of the equality or inequality form. The methods presented are restricted to a finite number of independent variables (also called control variables, forcing functions, decision variables, etc.). Thus the optimal profile type of problems in the calculus of variations category which involve an infinite set of independent variables are not included. Many system models in this category, however, are solved by numerical difference techniques thus transforming the infinite independent variable set into a finite set suitable for the optimization techniques described in this book.

In the finite category, optimization techniques may be classified in a variety of ways. A general distinction can be made between the classical or indirect methods amenable to analytical solutions and the mathematical programming and search methods which normally require a digital computer for finding a numerical solution to most realistic problems. The classical procedures, based on the condition that the first derivatives of the objective function with respect to the independent variables must vanish at the optimum, are restricted to very few real world problems and thus will not be considered in this text. Thus only numerical solution procedures, translated to computer algorithms, are presented. In this book, these techniques are grouped into two broad categories. The first category is called <u>SPECIAL PURPOSE METHODS</u>

and makes up <u>PART ONE</u> of the book. This group includes several linear programming algorithms, quadratic programming, geometric programming, dynamic programming and least squares objective functions. Each of these techniques is designed to be used with a specific problem structure.

<u>PART TWO</u> of the book consists of a variety of <u>SEARCH METHODS</u>. We have chosen to divide these methods into four groups, according to the type of problem they are intended to solve: (1) single variable unconstrained, (2) single variable constrained, (3) multivariable unconstrained and (4) multivariable constrained.

Each method presented in this book may be classified according to several characteristics beyond the broad categories described above. Table 1 is included to illustrate several of the important ways by which these methods may be classified.

All techniques are written in the same format, explained below:

A. <u>Purpose</u>

    States the general type of problem that this technique is intended to solve.

B. <u>Method</u>

    Describes the computational algorithm and presents a general logic diagram for the technique.

C. <u>Program Description</u>

  1) <u>Usage</u>:

    Describes main program and how it relates to all subroutines.

  2) <u>Subroutines Required</u>:

    Description of each subroutine, including the parameter list.

  3) <u>Description of Parameters</u>:

    Defines pertinent parameters, arrays and variables.

  4) <u>DIMENSION Requirements</u>:

    Describes all required changes to dimensioned arrays for the particular problem being solved.

  5) <u>Input Formats</u>:

    Describes in detail how to keypunch data cards.

  6) <u>Output</u>:

    Describes type of output, including controls for printing intermediate solutions.

  7) <u>Summary of User Requirements</u>:

    Self-explanatory.

TABLE 1

OPTIMIZATION TECHNIQUE CLASSIFICATIONS

Technique	Implicit Problems	Explicit Problems	Single Variable	Multivariable	Unconstrained	Equality Constraints	Inequality Constraints	Linear Objective Function	Quadratic Objective Function	Least Squares Objective Function	Geometric Programming Objective Function	General Nonlinear Objective Function	Linear Constraints	Geometric Programming Constraints	General Nonlinear Constraints	Derivatives Required (Anal. or Num.)	Integer Solutions	0-1 Integer Solutions	General Solutions
2.I. Simplex		X		X		X	X	X					X						X
2.II. Revised Simplex		X		X		X	X	X					X						X
2.III. Integer Programming		X		X		X	X	X					X				X		
2.IV. Zero-One Programming		X		X		X	X	X					X					X	
3.I. Wolfe		X		X		X	X		X				X						X
3.II. Differential		X		X		X	X		X				X						X
4.I. Blau		X		X		X					X			X					X
5.I. Discrete D.P.		X		X		X						X			X				X
5.II. Continuous D.P.	X	X		X		X						X			X				X
6.I. Linear Regression		X		X	X					X									X
6.II. Gauss-Newton	X	X		X			X			X			X			X			X
6.III. Marquardt	X	X		X			X			X			X			X			X
6.IV. Powell	X	X		X	X					X						X			X
7.I. Coggin	X	X	X		X							X							X
8.I. Fibonacci	X	X	X				X					X	X						X
9.I. Nelder-Mead	X	X		X	X							X							X
9.II. Hooke-Jeeves	X	X		X	X							X							X
9.III. Rosenbrock	X	X		X	X							X							X
9.IV. Powell	X	X		X	X							X							X
9.V. Fletcher-Reeves	X	X		X	X							X				X			X
9.VI. Fletcher-Powell	X	X		X	X							X				X			X
10.I. Box	X	X		X			X					X			X				X
10.II. Constrained Rosenbrock	X	X		X			X					X			X				X
10.III. Rosen	X	X		X			X					X	X		X				X
10.IV. Fiacco-McCormick	X	X		X		X	X					X			X	X			X
10.V. Constrained Fletcher-Powell	X	X		X		X	X					X			X	X			X

D. *Test Problem*

Presents an example problem, shows inputs and results. Indicates number of iterations and amount of computer time required.

E. *Program Listings and Example Output*

Complete listings of FORTRAN programs, subroutines and functions.

Some of the algorithms are formulated as maximization techniques while others are formulated as minimization problems. The translation between the two is achieved by changing the sign of the objective function,

$$\text{Maximize } \{F(X_1, X_2, \ldots, X_N)\} = \text{Minimize } \{-F(X_1, X_2, \ldots, X_N)\}.$$

The test problems solved in this book are all of the explicit type, i.e., the objective function and constraints can be directly calculated from simple algebraic functions. As indicated in Table 1, however, many of the techniques can handle both explicit and implicit problem types. Typical implicit problems would include those where the objective function and constraints may be evaluated from graphs, tables, numerical solutions to complex systems of equations, direct experimental measurements, etc. The computation time required for a function evaluation typically becomes a critical factor in implicit problems. Thus the time required to handle the optimization logic is usually negligible compared with that used in evaluating the function.

The multivariable test problems also involve relatively few independent variables and constraints (where applicable). The actual number of variables and equations a given technique can be expected to realistically handle will depend on the complexity and structure of the system model. Guidelines have been presented in several references (19, 8, 5). In general, the techniques in the explicit, special structure categories (linear programming, quadratic programming, geometric programming, linear least squares) would be expected to handle large problems consisting of perhaps several thousand equations and variables. The general nonlinear methods however are practical only for much smaller problems, with typical limits on the equations and variables being less than one hundred. Decomposition techniques, such as dynamic programming, may be appropriate for restructuring large problems into a sequence of smaller optimization problems.

No claim is made that the methods presented in this book are in any sense better than other methods. Indeed, one of the purposes of the book is to present a variety of methods to illustrate the diverse types of optimization techniques currently available. A method that works well on one problem may perform very poorly on another problem of the same general type. Only after much experience using all the methods can one judge which method would be "better" for a particular problem.

All computer programs presented in this book are written in ASA FORTRAN IV, with the exception of one subroutine in Section 2.II, which is written in IBM 360 Assembler language. Example problems were run on CDC 6400, IBM 360/65 and H-G 425 computers.

PART ONE:

SPECIAL PURPOSE METHODS

Chapter 2

LINEAR PROGRAMMING

The optimization techniques presented in this chapter are concerned with solving problems which are entirely linear. That is, these methods optimize a linear, algebraic objective function, subject to a number of linear, algebraic constraints.

The classical linear programming problem has the following form:

Optimize $\quad F = C_1 X_1 + C_2 X_2 + \ldots + C_N X_N$

Subject to $\quad A_{i1} X_1 + A_{i2} X_2 + \ldots + A_{iN} X_N \leq, =, \geq B_i$

$$i = 1, 2, \ldots, M$$

$$X_1, X_2, \ldots, X_N \geq 0$$

where A_{ij}, B_i and C_j are given constants and X_j are the decision variables. In particular, we are seeking the values of the X_j which will optimize (maximize or minimize) the objective function, F.

If the decision variables are permitted to vary continuously, then the L.P. problem is in standard form and may be solved using either the simplex algorithm or the revised simplex algorithm. Both of these methods are presented in this chapter.

If the decision variables may assume only integer values, a special variation of the method must be used. This method is called "integer programming." A more restricted case is when the decision variables may assume only the two integers, zero and one. The method required for this case is called "zero-one programming."

The four optimization methods presented in this chapter are:
    I. Simplex (SIMPLX Algorithm)
   II. Revised Simplex (REVSIM Algorithm)
  III. Mixed Integer Programming (MINT Algorithm)
  IV. Zero-One Programming (ZONE Algorithm)

I. **SIMPLEX (SIMPLX ALGORITHM)***

A. Purpose

This procedure finds the maximum of a multivariable, linear function subject to linear constraints:

Maximize $\quad F = C_1 X_1 + C_2 X_2 + \ldots + C_N X_N$

Subject to $\quad A_{i1} X_1 + A_{i2} X_2 + \ldots + A_{iN} X_N \leq, =, \geq B_i \quad i = 1, 2, \ldots, M$

$\quad\quad\quad\quad\quad X_1, X_2, \ldots, X_N \geq 0$

where the A_{ij}, B_i, and C_j are given constants and the X_j are the decision variables.

B. Method

The procedure is based on the original simplex algorithm developed by Dantzig (12), using the "contracted tableau" form of the simplex method. The method is an iterative technique that tends to find the global optimum of the objective function within the feasible region established by the constraints. The algorithm proceeds as follows:

1) An original tableau is generated representing the objective function and constraints. The RHS (Right Hand Side) values of the constraints are entered into a separate vector and must all be positive values.

2) Positive slack variables are generated automatically for $\leq$ type constraints and are placed in the initial basis. Negative slack variables are generated automatically for $\geq$ type constraints and do not appear in the initial basis. Artificial variables are generated automatically for $\geq$ and = type constraints and are placed in the initial basis.

3) The first phase of the method attempts to eliminate artificial variables without regard for the objective function. If successful, this phase produces an initial basic-feasible solution.

---

*The FORTRAN program contained in this section is based on SYSTEM/360 FORTRAN Linear Programming System, described on page 242 of "Catalog of Programs for IBM System/360 Models 25 and Above," GC 20-1619-8; program number 360D-15.2.006. Used by permission of International Business Machines Corporation.

4) Determine whether this solution is optimal. Eliminate the basic variables from the objective function and check the sign of the coefficient of each non-basic variable. If any of these coefficients is positive, optimality has not been reached and the solution continues at Step 5). Otherwise, the solution procedure terminates and all results are printed. [Step 8)].

5) Determine the entering basic variable by selecting the non-basic variable which would increase F by the greatest amount. Select the non-basic variable whose coefficient in the current objective function is largest. Let k denote the subscript of the entering basic variable.

6) Determine the leaving basic variable by selecting that basic variable which reaches zero first as the entering basic variable [from Step 5)] is increased. Consider all ratios B_i/A_{ik} greater than zero and select the variable associated with the smallest such ratio as the leaving basic variable.

7) Determine the new basic feasible solution. Elementary row operations are employed to solve for the basic variables in terms of the non-basic variables. The entire tableau is so transformed, including the objective row. Return to Step 4).

8) Codes are included to indicate whether:
   a) an optimal solution was found,
   b) no finite solution exists (constraints unbounded),
   c) no basic-feasible solution exists.

A flow diagram illustrating the above procedure is shown in Figure 2.I.

C. Program Description

1) Usage:

The program consists of a single main program. The problem is solved entirely in core. The output includes an iteration log (showing value of objective function on each iteration), the optimal solution and cost ranges, and reduced costs of slack and structural variables not in the optimal solution.

2) Subroutines Required:

None

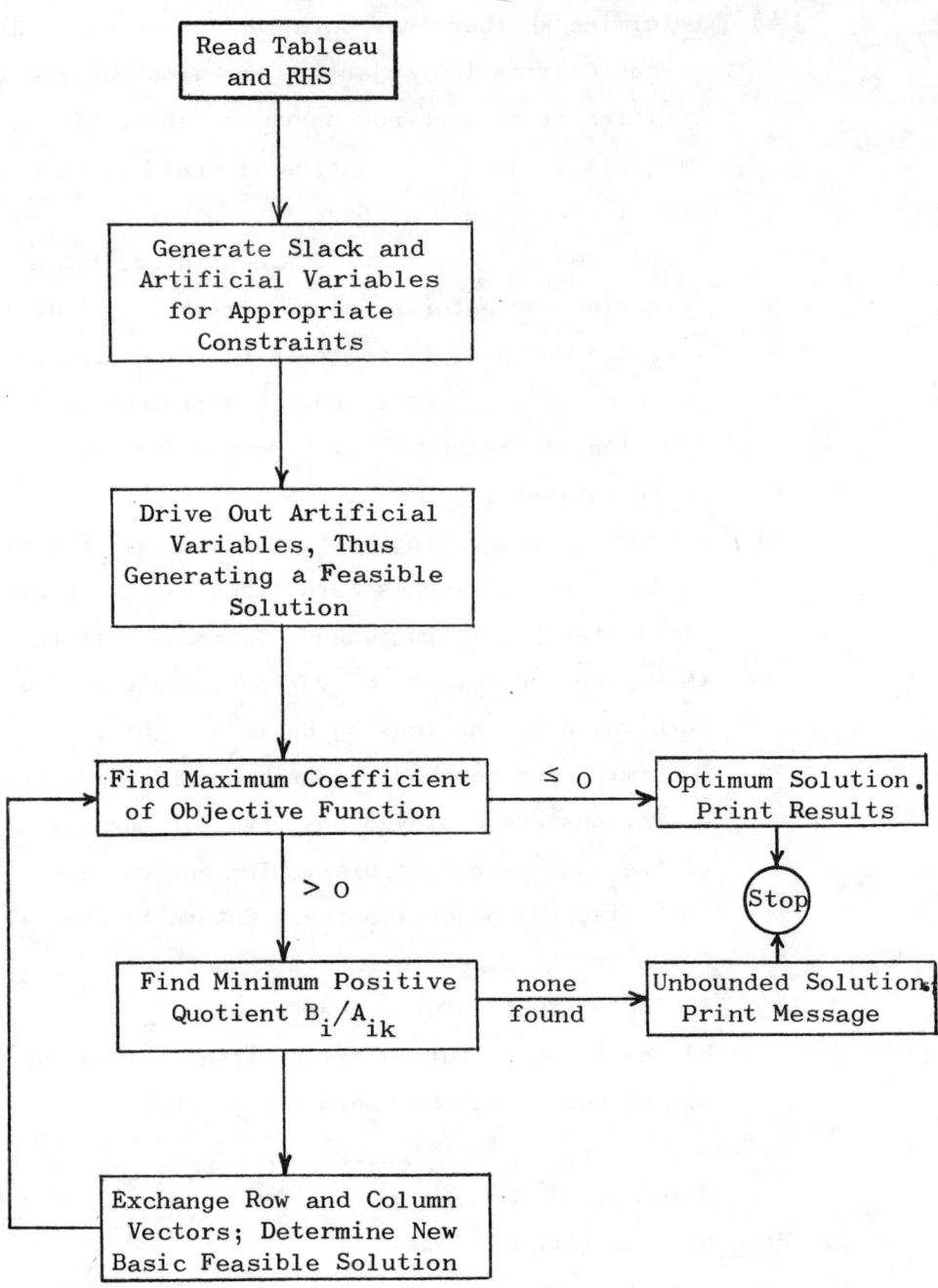

Figure 2.1. Simplex (SIMPLX ALGORITHM) Logic Diagram

3) <u>Description of Parameters:</u>

M1	Maximum number of rows in the problem to be solved, plus 1 for the objective function row
N1	Maximum number of structural variables (columns) in the problem, plus the number of ≥ constraints
B	M1 × N1 matrix representing LP tableau
RQ	M1 vector containing Right Hand Side (RHS) values of the constraints
CDID	Card group header: Punched ROW ID for row identification header Punched MATRIX for matrix header Punched FIRST B for RHS header Punched EOF for problem delimiter
LGE	Identifies type of constraint: + (12 punch) for ≤ - (11 punch) for ≥ 0 (or blank) for =
RNM1	First four characters of row (constraint) name
RNM2	Fifth (and final) character of row (constraint) name
CLNM1	First four characters of column name
CLNM2	Fifth (and final) character of column name
SYMB	Sign indicator: blank for positive - (11 punch) for negative
VALUE	Numerical value of matrix element, or RHS value
NI	Card reader unit number - Define in main program
NO	Printer unit number - Define in main program

4) <u>DIMENSION Requirements:</u>

The DIMENSION statement in the main program should be modified according to the requirements of each particular problem. The parameters included in the following DIMENSION statement conform to the Input Parameter definitions, above:

INTEGER ..., IBN1(M1), IBN2(M1), NBN1(M1), NBN2(M1)

REAL ..., BP(M1), RQ(M1), B(M1, N1), PI(N1), NBP(N1), XPI(N1)

5) **Input Formats:**

The LP problem is formulated in the following tabular arrangement:

Row Name	Type Constraint	Column Name				RHS Value
		Var. 1	Var. 2	...	Var. N1	
Obj. Func.	=	C_1	C_2	...	C_{N1}	
Constraint 1	$\leq, =, \geq$	$A_{1,1}$	$A_{1,2}$	...	$A_{1,N1}$	B_1
Constraint 2	$\leq, =, \geq$	$A_{2,1}$	$A_{2,2}$	...	$A_{2,N2}$	B_2
⋮						
Constraint M1	$\leq, =, \geq$	$A_{M1,1}$	$A_{M1,2}$	...	$A_{M1,N1}$	B_{M1}

Each row, including the Objective Function row, is given a unique name not exceeding 5 characters. Each column is given a unique name not exceeding 5 characters. If the problem is to minimize the objective function, the C_j's would be entered as negative values. All RHS values must be positive. Given the above formulation, the data is punched in four segments:

1. Problem Name - one card for each problem.
2. Row Name and Type Constraint - one card for each row.
3. Matrix Element - one card for each non-zero element in the matrix; contains Column Name, Row Name, and Element Value.
4. RHS Value - one card for each constraint.

The data deck for each problem is made up as follows:

CARD TYPE	FORMAT	CONTENTS
1	(2I5)	M1, N1
2	(IX, '39 blanks')	Alphameric Problem Name
3	(A2)	CDID (Punched ROW ID for row ID header)

4	(A2,9X,A1,1X,A4,A1)	CDID, LGE, RNM1, RNM2
		CDID is left blank.
		LGE (column 12) is punched + for $\leq$
		            − for $\geq$
		            0 for =
		RNM1, RNM2 (columns 14-18) contain 5 character row name

    (One Type 4 card for each row.)

5	(A2)	CDID (Punched MATRIX for matrix header)

6	(A2,5X,A4,A1,1X,A4, A1,A1,F11.6)	CDID, CLNM1, CLNM2, RNM1, RNM2, SYMB, VALUE CDID is left blank. CLNM1, CLNM2 (columns 8-12) contain 5 character column name RNM1, RNM2 (columns 14-18) contain 5 character row name SYMB (column 19) contains an 11-punch for negative value VALUE (columns 20-30) contains numerical value of matrix element, including decimal.
	(One Type 6 card for each non-zero matrix element.)	
7	(A2)	CDID (Punched FIRST B for RHS header)
8	(A2,11X,A4,A1,F12.6)	CDID, RNM1, RNM2, Value CDID is left blank RNM1, RNM2 (columns 14-18) contain 5 character row name Value (columns 20-30) contains RHS of constraint value; must be positive and include decimal point
	(One Type 8 card for each constraint.)	
9	(A2)	CDID (Punched EOF for problem delimiter)

Any number of problems can be solved on one computer run by providing a complete set of data as shown above, for each problem.

See Section D, <u>Test Problem</u>, for an example of the data format.

6) <u>Output</u>:

The problem name and certain statistical information (number of rows, number of columns, etc.) are printed. For each iteration, the iteration number, the variables entering and leaving solution, and the present value of the objective function are printed.*

If the program successfully completes Phase 1 (i.e., all artificial variables are eliminated), the message, "SOLUTION FEASIBLE", is printed. If the problem is infeasible, this message will not be printed and the optimal solution report (described below) will include at least one basis variable with a blank name and a unit profit of 0.

---

*In the program as presented, the complete tableau is not printed on each iteration. The user could easily modify the program to do this, if he desired.

Following the "SOLUTION FEASIBLE" message for feasible problems, the iteration log continues as described above. When an optimal solution has been reached, the total number of iterations, the optimal value of the objective function, and the optimal solution are printed. The optimal solution table contains the variable name, the amount of that variable in the optimal solution, the original unit contribution of that variable to the objective function, and the lower and upper cost range values. These numbers indicate the amount any one of the basis cost coefficients can change, provided that no other cost or constraint changes, without affecting the final optimal solution.

7) Summary of User Requirements:
   a) Formulate problem in tabular form, giving names to rows and columns of matrix. Assure that all RHS values are positive.
   b) Specify values for M1, N1, NI, and NO.
   c) Adjust DIMENSION values in INTEGER and REAL statements in main program. Also, adjust FORMAT statements, as required.

D. Test Problem

The following problem is taken from reference (55). The calculations were performed on an IBM 360/65.

It is desired to blend the following ingredients in the least expensive way, such that the stated restrictions are satisfied:

Ingredient	Analysis			Cost per Pound
	Protein	Fat	Fibre	
Barley	0.115	0.02	0.06	$ 0.025
Corn	0.086	0.038	0.025	0.028
Screenings	0.13	0.03	0.07	0.022
Salt				0.011

Restrictions

1) Total weight of blend = 2000 lb.
2) Total protein $\geq$ 200 lb.
3) Total fat $\geq$ 54 lb.
4) Total fibre $\leq$ 90 lb.

5) Amount of corn between 400 and 1000 lb.

6) Weight of salt used = 5 lb.

This problem can be represented in tabular form, as follows:

Row Name	Constraint		BARLX	CORNX	SCRNX	SALTX	RHS Value
PROFT	=	0	-0.025	-0.028	-0.022	-0.011	
WGHTE	=	0	1.0	1.0	1.0	1.0	2000
PROTL	$\geq$	-	0.115	0.086	0.13		200
FATLO	$\geq$	-	0.02	0.038	0.03		54
FIBRH	$\leq$	+	0.06	0.025	0.07		90
CORNL	$\leq$	-		1.0			400
CORNH	$\geq$	+		1.0			1000
SALTE	=	0				1.0	5

Note that the cost coefficients are entered as negative numbers; since this program maximizes the objective function, we must reverse the signs in order to minimize costs. Remember that only non-zero matrix elements should be punched. The punched data is shown in Table 2.I. The solution is shown in the next section, following the program listing. Execution time on the 360/65 was 1 second.

TABLE 2.I

DATA FOR SAMPLE PROBLEM

```
                         Column No.
Card              1111111111222222222233333
 No.     12345678901234567890123456789012345

  1           8       7
  2         SAMPLE PROBLEM    NUMBER 1
  3         ROW ID
  4                   O PROFT
  5                   O WGHTE
  6                   - PROTL
  7                   - FATLO
  8                   + FIBRH
  9                   - CORNL
 10                   + CORNH
 11                   O SALTE
 12         MATRIX
 13                 BARLX PROFT-   .025
 14                 BARLX WGHTE   1.000
 15                 BARLX PROTL    .115
 16                 BARLX FATLO    .020
 17                 BARLX FIBRH    .060
 18                 CORNX PROFT-   .028
 19                 CORNX WGHTE   1.000
 20                 CORNX PROTL    .086
 21                 CORNX FATLO    .038
 22                 CORNX FIBRH    .025
 23                 CORNX CORNL   1.000
 24                 CORNX CORNH   1.000
 25                 SCRNX PROFT-   .022
 26                 SCRNX WGHTE   1.000
 27                 SCRNX PROTL    .130
 28                 SCRNX FATLO    .030
 29                 SCRNX FIBRH    .070
 30                 SALTX PROFT-   .011
 31                 SALTX WGHTE   1.000
 32                 SALTX SALTE   1.000
 33         FIRST B
 34                       WGHTE 2000.000
 35                       PROTL  200.000
 36                       FATLO   54.000
 37                       FIBRH   90.000
 38                       CORNL  400.000
 39                       CORNH 1000.000
 40                       SALTE    5.000
 41         EOF
```

E. Program Listings and Example Output

```
      INTEGER   RNM1,RNM2,CLNM1,CLNM2,BLNK,NEG,POS,SYMB,CDID,RO,MA,FI,EO,
     1          IBN1(40),IBN2(40),NBN1(40),NBN2(40)
      REAL      PIVOT,LST,XNBP,FN,CJBAR,X,VALUE,BP(40),RQ(40),B(40,40),
     1          PI(40),NBP(40),XPI(40)
C
      DATA RO,MA,POS,NEG,FI,EO,BLNK/2HRO,2HMA,1H+,1H-,2HFI,2HEO,4H    /
C
C
C     INPUT PROGRAM
C
54323 CONTINUE
      M=0
      N=1
      ISW = 0
      NROWS = 0
      NGE = 0
      NLE = 0
      NEQ = 0
      NEL = 0
      NRHS = 0
      NCOLS = 0
      NI = 5
      NO = 6
      READ(NI,1) M1,N1
    1 FORMAT(2I5)
      IF(M1.EQ.0) STOP
C
C     CLEAR MATRIX TO ZERO
C
      DO 12 I = 1, M1
      DO 12 J = 1, N1
   12 B(I,J) = 0.
C     READ PROBLEM TITLE
      READ(NI,800)
      WRITE(NO,8151)
      WRITE(NO,800)
      WRITE(NO,801)
 8151 FORMAT('1')
  800 FORMAT(1X,'                                                    ')
  801 FORMAT(//)
C
C     READ FIRST CARD - SHOULD BE ROWID
C
      READ(NI,2) CDID
    2 FORMAT (A2)
      IF (CDID-RO)3,680,3
    3 WRITE(NO,3333)
 3333 FORMAT (//'   ROWID CARD MISSING  '//)
 3334 STOP
C
C     READ AND STORE ROWID CARDS
C     INCLUDING DUMMY READ
C     FOR OBJECTIVE ROW NAME
```

```
C            GENERATE POS AND NEG SLACKS AS REQUIRED
C
  680 READ(NI,681)
  681 FORMAT (80X)
  101 READ(NI,102) CDID,LGE,RNM1,RNM2
  102 FORMAT(A2,9X,A1,1X,A4,A1)
      IF (CDID-MA)103,504,103
  504 CONTINUE
      GO TO 104
  103 M = M + 1
      NROWS = NROWS+1
      IF (LGE-POS)105,106,105
  105 IF (LGE-NEG)107,108,107
  106 IBN1(M)=RNM1
      IBN2(M)=RNM2
      NLE = NLE+1
      BP(M) = 0.
      GO TO 101
  108 IBN1(M) = RNM1
      IBN2(M) = RNM2
      NGE = NGE+1
      BP(M) = -1.0
      B(M,N) = -1.0
  401 NBN1(N) = RNM1
      NBN2(N) = RNM2
      NBP(N) = 0.
      N = N+1
      GO TO 101
  107 IBN1(M) = RNM1
      IBN2(M) = RNM2
      NEQ = NEQ+1
      BP(M) = -2.0
      GO TO 101
C
C            READ AND STORE FIRST MATRIX ELEMENT
C
  104 READ(NI,195) CDID,CLNM1,CLNM2,RNM1,RNM2,SYMB,VALUE
  195 FORMAT(A2,5X,A4,A1,1X,A4,A1,A1,F11.6)
      GO TO 119
  109 IF(NBN1(N) - CLNM1)111,600,111
  600 IF(NBN2(N)-CLNM2)111,601,111
  601 CONTINUE
  112 DO 113 I=1,M
      IF(IBN1(I)-RNM1)113,602,113
  602 IF(IBN2(I)-RNM2)113,603,113
  113 CONTINUE
      WRITE(NO,8113)
 8113 FORMAT (//'    INCORRECT INGREDIENT CARD '//)
      STOP
  603 CONTINUE
  114 IF (SYMB-NEG) 116,115,116
  115 B(I,N) = -VALUE
      GO TO 117
  116 B(I,N) =  VALUE
C
C            READ AND STORE MATRIX ELEMENTS
C
  117 READ(NI,195) CDID,CLNM1,CLNM2,RNM1,RNM2,SYMB,VALUE
```

```
            NEL = NEL+1
            GO TO 109
    111 N = N+1
            NCOLS = NCOLS+1
            IF (CDID-FI) 119,190,119
    119 NBN1(N) = CLNM1
            NBN2(N) = CLNM2
    201 IF (SYMB-NEG)202,203,202
    202 NBP(N) =   VALUE
            GO TO 117
    203 NBP(N) = -VALUE
            GO TO 117
C
C           READ AND STORE RHS ELEMENTS
C
    190 DO 191 I=1,M
    191 RQ(I) = 0.
            GO TO 120
    120 READ(NI,121) CDID,RNM1,RNM2,VALUE
    121 FORMAT(A2,11X,A4,A1,F12.6)
            IF (CDID-EO) 122,193,122
    122 DO 124 I=1,M
            IF(IBN1(I)-RNM1)124,610,124
    610 IF(IBN2(I)-RNM2)124,611,124
    124 CONTINUE
            WRITE(NO,8124) RNM1,RNM2
   8124 FORMAT(///'  NO ROW NAME FOR  ',A4,A1///)
            STOP
    611 CONTINUE
    125 RQ(I) = VALUE
            NRHS = NRHS+1
            GO TO 120
    193 N = N-1
            WRITE(NO,551) NROWS,NCOLS,NLE,NGE,NEQ,NRHS,NEL
    551 FORMAT(' ROWS ',I2,',    COLS ',I2,',    LE ROWS ',I2,',    GE ROWS '
        1,I2,/',   E ROWS ',I2,',    NONZERO RHS''S ',I2,',    NONZERO MATRIX
        1ELEMENTS   ',I5)
C
C      BLANK OUT ARTIFICIAL NAMES
C
            DO 10 I=1,M
            IF(BP(I)+1.0)19,11,10
     11 IBN1(I) = BLNK
            IBN2(I) = BLNK
            GO TO 10
     19 BP(I) = -1.0
            IBN1(I) = BLNK
            IBN2(I) = BLNK
     10 CONTINUE
C
C      ACCUMULATE COUNT OF INFEASIBILITIES
C
            NINF =0
            DO 6000 I=1,M
            IF(BP(I))6001,6000,6000
   6001 NINF = NINF+1
   6000 CONTINUE
C
```

```
C         GENERATE INDICATORS FOR MINIMIZATION OF INFEASIBILITY
C
      DO 6101 J=1,N
      XPI(J) =0.
      DO 6101 I=1,M
      IF(BP(I))6102,6101,6101
 6102 XPI(J) = XPI(J)-B(I,J)
 6101 CONTINUE
      DO 6002 I=1,M
 6002 BP(I) = 0.
      IPHASE = 1
C
C     MAIN ROUTINE
C
 9201 WRITE(NO,9202)
 9202 FORMAT ('0 ITERATION     VAR IN       VAR OUT           OBJ FN',/)
      IT = 0
54325 CONTINUE
C
C        CALCULATE SHADOW PRICES
C
      DO 194 J=1,N
      PI(J) = -NBP(J)
      DO 194 I=1,M
  194 PI(J) = PI(J) + BP(I)*B(I,J)
C
C        SELECT BEST NONBASIS VECTOR
C
 9101 LST = -.0000001
      KCOL = 0
      GO TO (751,552),IPHASE
  751 IF(NINF)54321,54321,552
  552 CONTINUE
      DO 9102 J=1,N
C
C        IGNORE ARTIFICIAL VARIABLES
C
      IF(NBN1(J)-BLNK+NBN2(J)-BLNK)651,9102,651
  651 CONTINUE
      GO TO (6003,6004),IPHASE
 6003 IF(XPI(J)-LST)6005,6006,6006
 6005 KCOL=J
      LST = XPI(J)
      GO TO 9102
 6004 CONTINUE
      IF(PI(J)-LST)9103,9102,9102
 9103 KCOL = J
      LST = PI(J)
 6006 CONTINUE
 9102 CONTINUE
      IF (KCOL)54321,54321,9104
C
C        DETERMINE KEYROW
C
 9104 KROW = 0
      CJBAR = LST
      LST = 1.0E20
      DO 9105 I=1,M
```

```
         IF(B(I,KCOL))9105,9105,9106
 9106 RATIO = RQ(I)/B(I,KCOL)
         IF (RATIO-LST)9107,9105,9105
 9107 LST = RATIO
         KROW=I
 9105 CONTINUE
         IF(KROW)9112,9112,9114
 9112 WRITE(NO,9113) NBN1(KCOL),NBN2(KCOL)
 9113 FORMAT(' VARIABLE ',A4,A1,'   UNBOUNDED ')
         GO TO 54323
 9114 CONTINUE
C
C      TRANSFORM
C
C         DIVIDE BY PIVOT
         PIVOT = B(KROW,KCOL)
         DO 9108 J=1,N
 9108 B(KROW,J) = B(KROW,J)/PIVOT
         RQ(KROW) = RQ(KROW)/PIVOT
         DO 9109 I=1,M
         IF(I-KROW)9110,9109,9110
 9110 RQ(I) = RQ(I) - RQ(KROW)*B(I,KCOL)
         DO 4444 J = 1, N
         IF(J-KCOL)9111,4444,9111
 9111 B(I,J) = B(I,J) - B(KROW,J)*B(I,KCOL)
 4444 CONTINUE
 9109 CONTINUE
         DO 9300 I=1,M
 9300 B(I,KCOL) = -B(I,KCOL)/PIVOT
         B(KROW,KCOL) = 1.0/PIVOT
C
C         INTERCHANGE BASIS AND NONBASIS VARIABLES
C
         RNM1 = NBN1(KCOL)
         RNM2 = NBN2(KCOL)
         NBN1(KCOL) = IBN1(KROW)
         NBN2(KCOL) = IBN2(KROW)
         IBN1(KROW) = RNM1
         IBN2(KROW) = RNM2
         LST = NBP(KCOL)
         NBP(KCOL) = BP(KROW)
         BP(KROW) = LST
         IT = IT + 1
         IF(NBN1(KCOL)-BLNK+NBN2(KCOL)-BLNK)6201,6200,6201
 6200 NINF = NINF-1
 6201 CONTINUE
C
C         COMPUTE OBJECTIVE FUNCTION
C
         FN = 0.
         DO 9301 I=1,M
 9301 FN = FN + BP(I)*RQ(I)
         GO TO (7000,7001),IPHASE
 7000 SAVE = PI(KCOL)
         DO 7003 J=1,N
         PI(J) = PI(J) - SAVE*B(KROW,J)
         XPI(J) = XPI(J) - CJBAR*B(KROW,J)
 7003 CONTINUE
```

```
              PI(KCOL) = -SAVE/PIVOT
              XPI(KCOL) = -CJBAR/PIVOT
              GO TO 7004
 7001  CONTINUE
              DO 9302 J=1,N
 9302  PI(J) = PI(J) - CJBAR*B(KROW,J)
              PI(KCOL) = -CJBAR/PIVOT
 7004  CONTINUE
C              CHECK FOR ESSENTIAL ZERO
              DO 6111 I=1,M
              DO 6111 J=1,N
              X=B(I,J)
              IF(ABS(X)-.0000001)6112,6112,6111
 6112  B(I,J) = 0.
 6111  CONTINUE
C
C
C     LOG ITERATION
C
              WRITE(NO,9120)IT,IBN1(KROW),IBN2(KROW),NBN1(KCOL),NBN2(KCOL),FN
 9120  FORMAT(I9,7X,A4,A1,8X,A4,A1,3X,F13.3)
              GO TO 9101
C
C
54321  CONTINUE
              IF(IPHASE-1)8000,8000,54322
 8000  IPHASE = 2
              IF(NINF)8003,8003,8004
 8004  WRITE(NO,8005)
 8005  FORMAT('0 SOLUTION INFEASIBLE',/)
              GO TO 54322
 8003  CONTINUE
              WRITE(NO,8002)
 8002  FORMAT('0   SOLUTION FEASIBLE ',/)
              GO TO 54325
54322  CONTINUE
C
C     OUTPUT ROUTINE
C
              WRITE(NO,301) IT,FN
  301  FORMAT('1','   ITERATION',I5,'   OBJ FN ',F15.3/)
              WRITE(NO,302)
  302  FORMAT(3X,'BASIS VAR',17X,'AMOUNT',6X,'UNIT PROFIT',6X,'LOW  '
     1,6X,'HIGH',/)
              DO 3033 I=1,M
C
C     COST RANGING
C
              VALUE = 1.0E20
              LST = 1.0E20
              DO 12300 J=1,N
              IF(NBN1(J)-BLNK+NBN2(J)-BLNK)12305,12300,12305
12305  CONTINUE
              IF(B(I,J))12301,12300,12302
12302  X=PI(J)/B(I,J)
              IF(X-LST)12303,12300,12300
12303  LST=X
              GO TO 12300
```

```
12301 X=-PI(J)/B(I,J)
      IF(X-VALUE)12304,12300,12300
12304 VALUE = X
12300 CONTINUE
      LST = BP(I) - LST
      VALUE = BP(I) + VALUE
 3033 WRITE(NO,304) IBN1(I),IBN2(I),RQ(I),BP(I),LST,VALUE
  304 FORMAT(7X,A4,A1,7X,F16.6,3X,F11.6,3X,F11.6,3X,F11.6)
      WRITE(NO,305)
  305 FORMAT('1    VARIABLE           REDUCED COST'//)
      DO 309 J=1,N
      IF(NBN1(J)-BLNK+NBN2(J)-BLNK)311,309,311
  311 WRITE(NO,310) NBN1(J),NBN2(J),PI(J)
  310 FORMAT(' ',4X,A4,A1,10X,F12.5)
  309 CONTINUE
C
C     RETURN TO READ NEXT PROBLEM
C
      GO TO 54323
      END
```

SAMPLE PROBLEM   NUMBER 1

ROWS 7,   COLS 4,   LE ROWS 0,   GE ROWS 3
  E ROWS 4,   NONZERO RHS'S 7,   NONZERO MATRIX ELEMENTS   20

ITERATION	VAR IN	VAR OUT	OBJ FN
1	CORNX		-11.200
2	CORNL		-28.000
3	SALTX		-28.055
4	SCRNX		-39.788
5	FATLO		-47.347
6	PROTL		-48.484
7	BARLX		-51.340

SOLUTION FEASIBLE

ITERATION   7   OBJ FN   -51.340

BASIS VAR	AMOUNT	UNIT PROFIT
BARLX	464.998291	-0.025000
FATLO	9.200027	0.0
SCRNX	530.001709	-0.022000
PROTL	8.374975	0.0
CORNX	1000.000000	-0.028000
CORNL	600.000000	0.0
SALTX	5.000000	-0.011000

II. REVISED SIMPLEX (REVSIM ALGORITHM)*

A. Purpose

This procedure finds the minimum of a multivariable, linear function subject to linear constraints:

Minimize $\quad F = C_1 X_1 + C_2 X_2 + \ldots + C_N X_N$

Subject to $\quad A_{i1} X_1 + A_{i2} X_2 + \ldots + A_{iN} X_N \leq, =, \geq B_i$

$$i = 1, 2, \ldots, M$$

$$X_1, X_2, \ldots, X_N \geq 0$$

where the A_{ij}, B_i, and C_j are given constants and the X_j are the decision variables.

B. Method

This method accomplishes the same results as the original simplex method (see Section 2.I), but in a manner which is more efficient for implementation on a digital computer. Specifically, it computes and retains only the information essential for the current stage, and this information is carried along in a more compact form.

Since the revised simplex method makes explicit use of matrix manipulations, it is convenient to describe the problem in matrix notation (28, p. 508):

Minimize $\quad X_o = \underline{C}\,\underline{X}$

Subject to $\quad \underline{A}\,\underline{X} \leq \underline{B} \quad$ and $\quad \underline{X} \geq \underline{0}$

where $\underline{C}$ is the row vector $\underline{C} = [C_1, C_2, \ldots, C_N]$, $\underline{X}$, $\underline{B}$, and $\underline{0}$ are column vectors

$$\underline{X} = \begin{bmatrix} X_1 \\ X_2 \\ \vdots \\ X_N \end{bmatrix}, \quad \underline{B} = \begin{bmatrix} B_1 \\ B_2 \\ \vdots \\ B_N \end{bmatrix}, \quad \underline{0} = \begin{bmatrix} 0 \\ 0 \\ \vdots \\ 0 \end{bmatrix}$$

---

*The FORTRAN program contained in this section is based on <u>MFOR 360 Linear Programming Code</u>, described on page 243 of "Catalog of Programs for IBM System/360 Models 25 and Above," GC 20-1619-8; program number 360 D-15.2.007. Used by permission of International Business Machines Corporation.

and $\underline{A}$ is a M×N matrix,

$$\underline{A} = \begin{bmatrix} A_{11} & A_{12} & \cdots & A_{1N} \\ A_{21} & A_{22} & \cdots & A_{2N} \\ \vdots & \vdots & & \vdots \\ A_{M1} & A_{M2} & \cdots & A_{MN} \end{bmatrix}$$

A column vector is introduced for the slack variables,

$$\underline{X}_s \quad \begin{bmatrix} X_{N+1} \\ X_{N+2} \\ \vdots \\ X_{N+M} \end{bmatrix}$$

The constraints thus become

$$[\underline{A},\underline{I}] \begin{bmatrix} \underline{X} \\ \underline{X}_s \end{bmatrix} = \underline{B} \quad \text{and} \quad \begin{bmatrix} \underline{X} \\ \underline{X}_s \end{bmatrix} \geq \underline{0}$$

It is shown in many texts, such as Hillier and Lieberman (28, pp. 508-516), that the entire set of equations for a linear programming problem can be transformed into the following arrangement:

$$\begin{bmatrix} 1 & | & \underline{C}_P \underline{P}^{-1} \underline{A} - \underline{C} & | & \underline{C}_P \underline{P}^{-1} \\ \underline{0} & | & \underline{P}^{-1} \underline{A} & | & \underline{P}^{-1} \end{bmatrix} \begin{bmatrix} X_o \\ \underline{X} \\ \underline{X}_s \end{bmatrix} = \begin{bmatrix} \underline{C}_P \underline{P}^{-1} \underline{B} \\ \underline{P}^{-1} \underline{B} \end{bmatrix}$$

where $\underline{P}$ is a M×M basis matrix, $\underline{C}_P$ is a row vector of cost coefficients for the basis variables, and all other terms are as previously defined. The significant feature of this arrangement, and the key to the revised simplex method, is that any value in the current set of equations can be calculated directly by performing the appropriate matrix operation in the above expression.

The computational algorithm proceeds as follows:

1) An original tableau is generated representing the objective function and constraints. Slack and artificial variables are entered explicitly into the problem formulation.

2) For each non-basic variable, compute $Z_j - C_j = \underline{C}_P \underline{P}^{-1} \underline{A}_j - C_j$, where $\underline{A}_j$ is the jth column of $\underline{A}$. For $(N+1) \leq j \leq (N+M)$, $C_j = 0$, $A_{jj} = 1$, and $A_{ij} = 0$ for $i \neq j$, the above expression reduces to $Z_j = \underline{C}_P (\underline{P}^{-1})_{j-N}$, where the column vector $(\underline{P}^{-1})_i$ is

$$(\underline{P}^{-1})_i = \begin{bmatrix} (\underline{P}^{-1})_{1i} \\ (\underline{P}^{-1})_{2i} \\ \vdots \\ (\underline{P}^{-1})_{Mi} \end{bmatrix}$$

If $Z_j - C_j$ is non-positive for all N non-basic variables, then an optimal solution has been found. Otherwise, select as the entering basic variable, that variable that would decrease Z at the fastest rate. Denote this variable by k.

3) Determine the leaving basic variable by considering all positive B'_i / A'_{ik} and selecting that variable having the smallest such value. A'_{ik} is computed as follows:

$$\begin{bmatrix} A'_{ik} \\ A'_{2k} \\ \vdots \\ A'_{Mk} \end{bmatrix} = \underline{P}^{-1} \underline{A}_k$$

which reduces to $(\underline{P}^{-1})_{k-N}$ when $k > N$. The B_i are computed as follows:

$$\begin{bmatrix} B'_1 \\ B'_2 \\ \vdots \\ B'_M \end{bmatrix} = \underline{P}^{-1} \underline{B}$$

Denote the leaving variable by r.

4) Update the basic data for the next iteration. This is done by finding the new $\underline{P}^{-1}$:

$$(\underline{P}^{-1}_{new})_{ij} = \begin{cases} (\underline{P}^{-1}_{old})_{ij} - \dfrac{A'_{ik}}{A'_{rk}} (\underline{P}^{-1}_{old})_{rj}, & i \neq r \\ \dfrac{1}{A'_{rk}} (\underline{P}^{-1}_{old})_{rk}, & i = r \end{cases}$$

Return to Step 2).

C. Program Description*

1) Usage:

The program consists of a main program, twenty three FORTRAN subroutines, and one IBM 360 Assembler language subroutine. The problem is solved entirely in core.

The main program initializes certain constants, calls the major input routine, INP, and also calls the routine TAP for reading in the coefficient matrix.

2) Subroutines Required:

SUBROUTINE BOS checks for feasibility; calculates prices; finds non-basis column with minimum reduced cost; tests for termination; determines entering column; chooses pivot row; updates the inverse and solution vector.

SUBROUTINE CAP(KK,K) enters the right hand sides of constraints.

SUBROUTINE DEL(J,D) calculates reduced cost for a given column.

SUBROUTINE ERR calculates errors on each row and, if present solution is feasible, on each column. On rows, the error is defined as the amount by which the equation $\underline{A}\,\underline{X} = \underline{B}$ is not satisfied. On columns corresponding to variables in the basis, the error is the difference of the reduced cost from zero.

SUBROUTINE GET sets up the price vector.

SUBROUTINE HOT prints out the column dependent quantities; i.e., the name, value, true cost and reduced cost of each variable.

---

*The program as available from IBM is capable of considerably more flexibility than is included in this description. Included here are only those features pertinent to the solution of ordinary linear programming problems. For details of the full capabilities of this program, see reference (54).

SUBROUTINE INP(KIN) the major input routine.

SUBROUTINE JMY(J) generates one transformed matrix column.

SUBROUTINE KRT applies the transformation previously created to the price vector through multiplication.

SUBROUTINE LOT prints out the row dependent quantities; i.e., the row name, the right hand side, the associated price, and either the row error or the corresponding entry of the last column to be transformed.

SUBROUTINE NEW pivots in singletons or slacks, depending on a previously set parameter.

SUBROUTINE MIN(JM) finds the minimum reduced cost among the non-basic variables.

SUBROUTINE OUP prints control cards and associated data.

SUBROUTINE ROW(IR) selects the pivot row.

SUBROUTINE PIV(IR) adds a transformation to the inverse, using the product form of the inverse.

SUBROUTINE SOT used to print out a "short output," including the name of the objective row, the number of infeasibilities, the names of the matrix and right hand side, the iteration number, the number of steps and the number of pivots that have been performed, the present value of the objective function, the sum of the infeasibilities, the value of the determinant, the minimum reduced cost among the non-basic variables, the name of the column just brought into the basis, the one just removed, and the name of the pivot row.

SUBROUTINE TAP(K) enters the matrix of coefficients.

SUBROUTINE UOUT(K) the output control routine.

SUBROUTINE VER performs reinversion of the basis.

SUBROUTINE XCK(K) checks present solution vector, setting to zero any element falling inside a given tolerance limit. Also checks for feasibility.

SUBROUTINE YXB generates a new solution vector from the right hand side by applying to it the transformations that have been created.

FUNCTION TIME(I) provides a method for measuring relative time.

FUNCTION SS(I) controls whether optimal reinversion should be performed.

UTILITY PACKAGE written in IBM 360 Assembler language, this routine determines the type of control card just read in and is a supporting routine for FUNCTION TIME.

3) <u>Description of Parameters</u>:

NI	Card reader unit number; define in main program
NO	Printer unit number; define in main program
M	Number of rows, including objective function
N	Number of columns, including slack variables
A	Matrix entries
KN	Column name vector
NR	Row name vector
B	Right hand sides vector
S	Maximum number of non-zero matrix entries
T	Maximum number of transformations
L	Maximum number of transformation entries

4) <u>DIMENSION Requirements</u>:

The COMMON statement in the main program and subroutines should be modified according to the requirements of each particular problem. The parameters included in the following statements conform to the <u>Input Parameter</u> definitions, above:

COMMON M, MA, MB, MC, MD, ME, MF, N, KM(50), KP(8,2), Z(20), T(10), KBCD(30)

COMMON A(S), IA, KL, KN(N), NR(M), B(M), JH(M), KB(N), P(M), X(M), Y(M), IP(T), LE, IE, E(L)

INTEGER*2 IA(S), KL(N), LE(T), IE(L)

(If half-words are not available on a particular computer, IA, KL, LE, and IE should be dimensioned S, N, T, and L, respectively.)

5) <u>Input Formats</u>:

Data input is accomplished through three types of control card sets: Type 0, Type 1 and Type N. Type 0 control cards are for specific purposes and have no data cards following them. There is one Type 1 control card, PRMD, followed by one associated card containing other information (explained below). Type N cards are followed by an arbitrary number of data cards. The data cards must have blanks in columns 1-4 and the data for the Type N cards ends with an END card, punched in columns 1-3. All control cards must be punched left justified.

TYPE 0 CONTROL CARDS

BEGN   The first card of each run; initializes various constants to nominal values.

ROW   The row name appearing in columns 5-8 will be the name of the objective row.

SOLV   This causes the current problem to be executed.

TIME   This causes the time in seconds since the last resetting of the internal times to be printed.

STOP   Indicates the end of all the control cards; stops the job.

END   No action is caused. It terminates a data set for Type N control cards.

TYPE 1 CONTROL CARD

(Not used in version presented here.)

TYPE N CONTROL CARDS

MTRX   If a name is to be associated with this matrix, it should be punched in columns 5-8 of the MTRX card. Each data card contains one matrix entry. The format is:

Col. 1-4 Blank
Col. 5-8 Column Name
Col. 9-12 Row Name
Col.13-24 Numerical entry; should punch decimal point; otherwise it is assumed to be between columns 18 and 19.

All entries with the same column name must be adjacent in the deck. No restriction is made on the ordering of row names. Zero entries may be omitted.

RHS   If a name is to be associated with the right hand side, it should be placed in columns 5-8 on the RHS card. Each data card contains one right hand side entry in the following format:

Col. 1-4 Blank
Col. 5-8 Ignored
Col. 9-12 Row Name
Col.13-24 Right hand side entry. If not punched, decimal point is assumed between columns 18 and 19.

No restriction is made on the ordering of the row names.

TOLR Each data card contains the information needed for the input of one floating point constant in the following format:

Col. 1-4 Blank
Col. 5-8 Constant Name
Col. 9-16 Numerical value in E8.1 format

BEGN sets these quantities to the following nominal values:

COL. 5-8	MEANING	NOMINAL VALUE
PIV	Minimum element size considered for pivoting	1.0E-05
REJ	Pivot rejection tolerance	1.0E-03
COST	Reduced cost considered negative below this value	-1.0E-03
RSET	Tolerance for setting X to zero	1.0E-05
ENTR	Minimum element size in new transformation	1.0E-07
TMLT	Minimum element size considered for applying transformation	1.0E-10
MIX	Mixed printing ratio	0.0
TIMX	Maximum time since last re-setting of internal timer, in seconds. Job will be terminated if this time limit is exceeded.	1.0E+03

FREQ Each data card contains the information to input one fixed point constant in the following format:

Col. 1-4 Blank
Col. 5-8 Name of Constant
Col. 9-16 Numerical value (right-justified integer)

BEGN sets these quantities to nominal values of 10,000.

COL. 5-8	MEANING
OUTP	Frequency of output
INVT	Reinversion frequency
CTOF	Maximum number of interactions allowed.

NOTE: An END card (punched in columns 1-3) should follow each group of Type N cards described above.

6) Output:

The output was described in section C.2), above, in SUBROUTINES SOT, LOT, and HOT.

7) Summary of User Requirements:

a) Adjust array sizes in COMMON statements. Also, adjust KM(6), KM(20), KM(21), KM(36), and KM(40) in main program. These must not exceed the array sizes specified in COMMON statement; i.e.,

$$KM(6) \leq S$$
$$KM(20) \leq M$$
$$KM(21) \leq N$$
$$KM(36) \leq T-1$$
$$KM(40) \leq L-1$$

b) Specify NI and NO in main program; adjust FORMAT statements, if required.

c) Formulate problem in tabular form, giving names to all rows and columns of tableau.

D. <u>Test Problem</u>

Calculations for the following problem were performed on an IBM 360/65 computer:

Maximize: $F = 3X_1 + 5X_2 + 4X_3$

Subject to: $2X_1 + 3X_2 \leq 8$
$2X_2 + 5X_3 \leq 10$
$3X_1 + 2X_2 + 4X_3 \leq 16$

The tableau is set up as follows:

	COL1	COL2	COL3	SLK1	SLK2	SLK3
ROW1	2.0	3.0	0.0	1.0	0.0	0.0
ROW2	0.0	2.0	5.0	0.0	1.0	0.0
ROW3	3.0	2.0	4.0	0.0	0.0	1.0
COST	-3.0	-5.0	-4.0	0.0	0.0	0.0

We will limit the maximum number of iterations to 25 and will set the pivot rejection tolerance to 1.0E-03. The data is punched as shown in Table 2.II. The objective function is called "COST," the right hand side vector is called "REQ," and the matrix is called "-A-." "TIME" control cards are inserted to print out total execution time prior to solution and total execution time following the completion of the solution. Only non-zero matrix elements are punched. The objective function coefficients are entered as negative values since this is a maximization problem and the algorithm is designed for minimization. The sign of

TABLE 2.II

DATA FOR SAMPLE PROBLEM

```
                              Column No.
  Card                   111111111122222
  No.      123456789012345678901234

   1       BEGN
   2       FREQ
   3           CTOF      25
   4       END
   5       TØLR
   6           REG       1.E-4
   7       END
   8       RØW CØST
   9       RHS REQ
  10           RØW1      8.4
  11           RØW2     10.0
  12           RØW3     16.0
  13       END
  14       MTRX-A-
  15           CØL1RØW1  2.0
  16           CØL1RØW3  3.0
  17           CØL1CØST -3.0
  18           CØL2RØW1  3.0
  19           CØL2RØW2  2.0
  20           CØL2RØW3  2.0
  21           CØL2CØST -5.0
  22           CØL3RØW2  5.0
  23           CØL3RØW3  4.0
  24           CØL3CØST -4.0
  25           SLK1RØW1  1.0
  26           SLK2RØW2  1.0
  27           SLK3RØW3  1.0
  28       END
  29       TIME
  30       SØLV
  31       TIME
  32       STØP
```

the final answer will be reversed from the true sign.

Algorithm Answers: $F = 18.93$
$X_1 = 2.54$
$X_2 = 0.98$
$X_3 = 1.61$

Central Processor Time: 1 second

The listing and output for this problem are contained in the following section.

E. Program Listings and Example Output

```
CFIN    THE MFOR MAIN ROUTINE
        COMMON NI, NO
        COMMON M,MA,MB,MC,MD,ME,MF,N,KM(50),KP(8,2),Z(20),T(10),KBCD(30)
        COMMON A(200),IA,KL,KN(20),NR(10),B(10),JH(10),KB(20),
       1 P(10),X(10),Y(10),IP(50),LE,IE,E(1000)
        INTEGER*2 IA(200),KL(20),LE(50),IE(1000)
C
        NI = 5
        NO = 6
123     FORMAT(1H1)
10      CONTINUE
        T(1)=TIME(0)
        DO 3 IOK=1,50
3       KM(IOK)=0
        WRITE(NO,123)
        M=0
        N=0
        MA=0
        MB=0
        MC=0
        MD=0
        ME=0
        MF=0
        DO 7 IO=1,10
        Z(IO)=0.
7       T(IO)=0.
        DO 8 IO=11,20
8       Z(IO)=0.
        KM(48)=1077952576
        KM(38)= -1027225643
        KM(39)= -488384809
        KM(33)= -975578654
        KM(1) = KM(48)
        KM(3) = KM(48)
        KM(5) = 10000
        KM(7)=1
        KM(8) = 10000
        KM(9) = 10000
        KM(10)= KM(48)
        KM(13) = KM(48)
C       HERE IS WHERE WE SET THE VALUES FOR MAX STORAGE ALLOWABLE IN
C               DIMENSIONED ARRAYS IN COMMON (MUST AGREE WITH OR BE LESS
C                  THAN SIZE IN COMMON STATEMENT)
        KM(6) = 200
        KM(20) = 10
        KM(21) = 20
        KM(36) = 49
        KM(40) = 999
        KP(1,1)=0
        KP(2,1)=0
        KP(3,1)=9
        KP(4,1)=9
        KP(5,1)=9
```

```
          KP(6,1)=6
          KP(7,1)=9
          KP(8,1)=9
          Z(1)  =    1.E-5
          Z(2)  =    1.E-5
          Z(3)  =   -1.E-3
          Z(5)  =    1.E-10
          Z(6)  =    1.E-7
          Z(7)  =    1.E-3
          Z(9)  =    1.
          Z(11)=1.0E+03
          K = 0
C    30    CALL        INP( K,A,B,E.IA,IE,IP,JH,KB,KL,KN,LE,NR,P,X,Y)
30         CALL INP(K)
           IF (K)  20, 10, 20
      20   CALL TAP(K)
           KM(49)=1
           GO TO 30
             END
```

```
CHOT      CALLED BY QOT            OUTPUT COLUMN DEPENDENT QUANTITIES (RS M1)
          SUBROUTINE HOT
          COMMON NI, NO
          COMMON M,MA,MB,MC,MD,ME,MF,N,KM(50),KP(8,2),Z(20),T(10),KBCD(30)
          COMMON A(200),IA,KL,KN(20),NR(10),B(10),JH(10),KB(20),
         1 P(10),X(10),Y(10),IP(50),LE,IE,E(1000)
          INTEGER*2 IA(200),KL(20),LE(50),IE(1000)
C
  501    WRITE(NO,502)
    700  DO 602  J = 1,N
         CALL DEL(J,D)
         COST = 0.
         K1 =KL(J)
         K2=KL(J+1)-1
         DO 05 K = K1,K2
         IF(IA(K)-MC)05,06,05
     05 CONTINUE
         GO TO 07
     06 COST = A(K)
     07    K1 = KB(J)
C510    WRITE(NO,504) KN(J),X(KI), COST, D, NR(KI)
  510    WRITE(NO,504) KN(J),X(KI), COST, D
    602     CONTINUE
    603 RETURN
    502 FORMAT(1H0   / 7H    NAME,9X,5HVALUE,17X,4HTRUE,4X,6H/ COST/,5X,7HRED
        1UCED,/1X)
C       1UCED,11X,3HROW/1X)
  504   FORMAT(3X,A4,F18.6,F22.6,F20.6,8X)
 C504   FORMAT(3X,A4,F18.6,F22.6,F20.6,8X,A4)
            END
```

```
CBOS        CALLED BY INP    (RS MFOR)      MASTER COMPUTING ROUTINE
       SUBROUTINE BOS
       COMMON NI, NO
       COMMON M,MA,MB,MC,MD,ME,MF,N,KM(50),KP(8,2),Z(20),T(10),KBCD(30)
       COMMON A(200),IA,KL,KN(20),NR(10),B(10),JH(10),KB(20),
     1 P(10),X(10),Y(10),IP(50),LE,IE,E(1000)
       INTEGER*2 IA(200),KL(20),LE(50),IE(1000)
C
       KK1 = KM(2) +1
       KK2 = KM(9)
       DO 100   KK0 = KK1,KK2
C                                   INVERSION FREQUENCY
       IF (KM(28) - KM(08) )  05, 03, 05
   03  KM(35) = KM(35) +1
04     CALL VER
       KERR=1
       GO TO 300
C                                   CHECK CHANGE OF PHASE
05     CALL XCK(K)
       IF (KM(4) - K)  06,14,10
C                                   GONE INFEASIBLE
 6     KM(4)=1
       WRITE(NO,900) KM(2)
900    FORMAT(30H *BECAME INFEASIBLE, ITERATION,I4)
       IF (KM(28))  04, 14, 04
C                                   END PHASE ONE
  10   KM(4) = 0
       WRITE(NO,901) KM(2),KM(42)
901    FORMAT(23H *FEASIBLE ON ITERATION,I4,1H,,I6,6H STEPS)
       CALL UOUT(2)
       Z(4) = 0.
14     CALL GET
C                                   ITERATION COUNTER
       KM(2) =    KM(2)   +1
       KM(37) = 0
15     CALL MIN(JM)
C                                   TEST END OF PHASE
       IF ( Z(13) - Z(3) )   23, 16, 16
  16   IF (Z(4))  12, 18, 12
  18   IF (KM(37))  21, 19, 21
  19   IF (KM(4)) 40, 30, 40
  21   JM = KM(37)
       GO TO 36
  22   LSUB = 0
       JM = KM(27)
       GO TO 24
  23   LSUB = KM(7)
  24   KM(12) = KN(JM)
       CALL JMY(JM)
       IF ( KM(7) )  230, 232, 230
 230   IF (LSUB) 232, 231, 232
 231   TEST = Y(MC)
       IF(KM(4))   233, 237, 233
 233   TEST = Z(4) * TEST
       CALL XCK(K)
       DO 235 I = MF,M
       IF (X(I))  236, 238, 238
 236   TEST = TEST + Y(I)
```

```
              GO TO 235
  238     IF (JH(I)) 235, 239, 235
  239     TEST = TEST - Y(I)
  235     CONTINUE
  237  IF (TEST - Z(3) )    232, 452, 452
  232  CALL ROW(IR)
C                                   TEST PIVOT
       IF (IR) 29, 28, 29
   28  IF (Z(4)) 12, 50, 12
C                                   PIVOT FOUND
   29  Z(17)= Y(IR)
       IF (ABS(Z(17))-Z(7)) 31,31,36
C                                   PIVOT REJECTED
   31  WRITE(NO,902) KN(JM),NR(IR),Z(17),KM(2)
       KB(JM) =-1
       KM(37) =    JM
       KM(47) = KM(37)
       IF (LSUB) 22, 15, 22
   36  KM(13) =  KM(12)
       KM(11) = NR(IR)
       JOLD   = JH(IR)
       KM(10) =  KN(JOLD)
       IF ( KP(1,1) ) 361, 362, 361
  361  CALL UOUT(1)
  362  KM(42) = KM(42) +1
       KB(JM) = IR
       JH(IR) = JM
       IF (JOLD)  44,45,44
   44  KB(JOLD) = 0
       GO TO 451
   45  KM(17)=KM(17)-1
  451  CALL PIV(IR)
C                             TEST FOR RUNNING OUT OF SPACE
       IF (KM(24))  47,  52,  47
   52  IF (LSUB) 22, 452, 22
   12  Z(4) = 0.0
       WRITE(NO,903) KM(2)
  903  FORMAT(40H MIXED PRICING RATIO ZEROED BY ITERATION,I6)
C                                   INVERSION COUNTER
  452    KM(28) = KM(28) +1
C                                   QOT(6) FREQUENCY
   48  KM(25) = KM(25) +1
       IF ( KM(25) - KM(05) ) 53, 51, 53
   51  CALL UOUT(6)
C                                   TEST FOR TIME INVERSION
   53  IF(SS(LOP)) 49,54,49
C   TEST FOR EXCEEDING THE TIME LIMIT AS SET IN Z(11)
   54  IF(TIME(1)-Z(11))  100,100,08
   47     KM(35) =1
   49     KM(35)= KM(35) +1
   43     KM(28) = KM(08)
  100  CONTINUE
       K=8
       WRITE(NO,904)
       GO TO 105
    8  K=8
       WRITE(NO,905)
       GO TO 105
```

```
     30   K=3
          WRITE(NO,906)
          GO TO 105
     40   K=4
          WRITE(NO,907)
          GO TO 105
     50   K=5
          WRITE(NO,908)
    105   KM(14)=1
          T(1)=TIME(1)
          WRITE(NO,909) T(1)
    909   FORMAT(13H THE TIME IS ,F10.2, 9H SECONDS.)
          CALL UOUT (K)
     81   KERR=2
    300   CALL ERR
          GO TO (5,99),KERR
       99 RETURN
    902   FORMAT(6H (COL ,A4,6H, ROW ,A4,2H)=,F12.8,12H REJECTD,ITN,I4)
    904   FORMAT(26H =ITERATION LIMIT EXCEEDED)
    905   FORMAT(27H =JOB PULLED, TOO MUCH TIME)
    906   FORMAT(18H =OPTIMAL SOLUTION)
    907   FORMAT(22H =NO FEASIBLE SOLUTION)
    908   FORMAT(19H =INFINITE SOLUTION)
          END
```

```
CGET      CALLED BY BOS, ERR              GET PRICES
          SUBROUTINE GET
          COMMON NI, NO
          COMMON M,MA,MB,MC,MD,ME,MF,N,KM(50),KP(8,2),Z(20),T(10),KBCD(30)
          COMMON A(200),IA,KL,KN(20),NR(10),B(10),JH(10),KB(20),
         1 P(10),X(10),Y(10),IP(50),LE,IE,E(1000)
          INTEGER*2 IA(200),KL(20),LE(50),IE(1000)
C
          DO 19 I = 1, M
       19 P(I) = 0.
          IF (KM(4))   21,22,21
       22 P(MC) = 1.
          GO TO 11
       21 P(MC) = Z(4)
          DO 25 I = MF,M
          IF (X(I)) 26,27,27
       26 P(I)=1.
          GO TO 25
       27 IF (JH(I)) 25,29,25
       29 P(I) = -1.0
       25 CONTINUE
       11 IF (ME)   10,99,10
     10   CALL KRT
       99 RETURN
          END
```

```
CCAP    CALLED BY INP    (RS MFOR)      RHS, ALTERA, ALTERB, OBJECT, BASIS
        SUBROUTINE CAP(KK,K)
        COMMON NI, NO
        COMMON M,MA,MB,MC,MD,ME,MF,N,KM(50),KP(8,2),Z(20),T(10),KBCD(30)
        COMMON A(200),IA,KL,KN(20),NR(10),B(10),JH(10),KB(20),
       1 P(10),X(10),Y(10),IP(50),LE,IE,E(1000)
        INTEGER*2 IA(200),KL(20),LE(50),IE(1000)
        KERR=0
        KM(4)=1
        GO TO (01,02,03,05,05,06),KK
C KK=01    RHS,FIRST    RIGHT-HAND SIDES
    01 IF (M)    51, 41, 51
    41  WRITE(NO,909)
   909  FORMAT(30H THIS PROBLEM HAS NO OBJECTIVE)
        MC=1
        M=1
        MF=2
        GO TO 05
    51 DO 11 I=1,M
    11    B(I) = 0.
C KK=05    ALTERB      RIGHT-HAND SIDES WITHOUT ZEROING
   05   READ(NI,15) K,KBCD(3),JJ,AA
        IF (K - KM(48)) 99,25,99
    25 DO 35 J=1,M
           IF (NR(J)-JJ) 35,45,35
    35 CONTINUE
        IF(KK-1)55,21,55
    21   IF(KM(49))55,31,55
    31   M=M+1
         J = M
         NR(J) = JJ
    65 B(J) =AA
        GO TO 05
    55   WRITE(NO,910) JJ
   910   FORMAT(19H ILLEGAL ROW(COL), ,A4,10H, IN ENTRY)
         KERR=1
         GO TO 5
    45 IF(KK-4) 65,4,65
C KK=04    ALTERA         CHANGE MATRIX ENTRY
    04 DO 14 II =1,N
           IF (KN(II) - KBCD(3) ) 14,24,14
    14 CONTINUE
    54 JJ = KBCD(3)
        GO TO 55
    24 NOO = KL(II)
        NOT = KL(II+1)-1
        DO 34 IJ = NOO,NOT
        IF(IA(IJ)-J) 34,44,34
    34 CONTINUE
        GO TO 55
    44 A(IJ) = AA
        GO TO 5
C KK=06    OBJECT      LIST OF OBJECTIVE ROWS    IGNORE IF M NE MC
    06 IF (M-MC) 26,32,26
    26   WRITE(NO,911)
   911   FORMAT(25H OBJECTIVES READ TOO LATE)
         GO TO 97
    16   M=M+1
```

```
         MC = M
         MF=M+1
         NR(M) = KBCD(I)
         GO TO 52
C KK=02         BASIS           LIST OF BASIS HEADINGS
   02 DO 22 I=1,N
   22 KB(I) = 0
      IF (KM(22) )    32, 03, 32
32    READ(NI,12) K,(KBCD(I),I=3,13)
      IF (K - KM(48)) 98,42, 98
   42 DO 52 I = 3,13
         IF(KBCD(I) -KM(48)) 62,52,62
  62   IF(KK-3)72,73,16
   73    DO 83 II=1,M
            IF ( NR(II) - KBCD(I) )    83,93,83
   83    CONTINUE
         GO TO 92
   93    JH(II) = 0
         GO TO 52
   72    DO 82 JJ=1,N
            IF ( KN(JJ) - KBCD(I) )  82,102,82
   82    CONTINUE
92    CONTINUE
      WRITE(NO,96) KBCD(I)
96    FORMAT(14H ILLEGAL NAME,,A4)
         GO TO 52
  102  KB(JJ)=1
   52 CONTINUE
      GO TO  32
C KK=03         ARTROW          LIST OF ARTIFICIAL ROWS
   03 DO 13 I = MF,M
   13 JH(I) = 12345
      KM(22)=1
      GO TO 32
C RETURN FOR 01 04 05
   99 IF (KERR) 97,98,97
97    KOPOL=8
      CALL INP (KOPOL)
   98 RETURN
C15   FORMAT(A4,4X,A4,2X,A4,F12.6)
15    FORMAT(3A4,F12.6)
12    FORMAT(12A4)
         END
```

CERR   CALLED BY BOS, INP   CALCULATES ERRORS   (ADAPTED FROM RO M2*)

```
      SUBROUTINE ERR
      COMMON NI, NO
      COMMON M,MA,MB,MC,MD,ME,MF,N,KM(50),KP(8,2),Z(20),T(10),KBCD(30)
      COMMON A(200),IA,KL,KN(20),NR(10),B(10),JH(10),KB(20),
     1 P(10),X(10),Y(10),IP(50),LE,IE,E(1000)
      INTEGER*2 IA(200),KL(20),LE(50),IE(1000)
C         THIS VERSION OF ERR DOUBLE SPACES WHEN IT IS DONE AND
C         RESETS THE EJECT PARAMETER  KM(14)
```

```
C
      IM=1
      KM(12)=KM(33)
C                              STORE AX-B AT Y
         DO 01 I = 1,M
   01 Y(I) =-B(I)
         DO 02 I = 1,M
      JB = JH(I)
       IF (JB) 03,04,03
   04 Y(I) = Y(I) + X(I)
      GO TO 02
   03 IZ = KL(JB)
      IB=KL(JB+1)-1
         DO 05  L = IZ,IB
      IT=IA(L)
   05 Y(IT) = Y(IT) + X(I)*A(L)
   02    CONTINUE
C                              FIND SUM AND MAXIMUM OF ERRORS
   80 T(1) = 0.
      T(2) = 0.
         DO 81 I = 1,M
      T(1)=T(1)+ABS(Y(I))
      IF(ABS(T(2))-ABS(Y(I))) 82,81,81
   82 T(2) = Y(I)
      IM = I
   81    CONTINUE
C                              OUTPUT ERRORS
   83    WRITE(NO,912) NR(IM), T(2), T(1)
  912    FORMAT(18H MAX ERROR ON ROW ,A4,1H=,E13.5,6H, SUM=,E12.5)
   06 IF (KM(4) + KM(24) )      99,07,99
C                              COMPUTE COLUMN ERRORS
   07    CALL GET
      T(1) = 0.0
      T(2) = 0.0
         DO 11 I = 1,M
      JHI = JH(I)
      IF (JHI) 09,11,09
   09    CALL DEL(JHI,DAL)
      T(1)=T(1)+ABS(DAL)
      IF(ABS(T(2))-ABS (DAL)) 12,11,11
   12    T(2) = DAL
         IM = I
   11 CONTINUE
   84    LB=JH(IM)
         WRITE(NO,913) KN(LB), T(2), T(1)
  913    FORMAT(18H MAX ERROR ON COL ,A4,1H=,E13.5,6H, SUM=,E12.5)
   99    WRITE(NO,914)
  914    FORMAT(1H0)
         KM(14) = 0
         RETURN
         END
```

```
C INP CALLED BY FIN (RS MFOR) INPUT SYMBOLIC CONTROL CARDS
      SUBROUTINE INP(KIN)
      COMMON NI, NO
      COMMON M,MA,MB,MC,MD,ME,MF,N,KM(50),KP(8,2),Z(20),T(10),KBCD(30)
      COMMON A(200),IA,KL,KN(20),NR(10),B(10),JH(10),KB(20),
     1 P(10),X(10),Y(10),IP(50),LE,IE,E(1000)
      INTEGER*2 IA(200),KL(20),LE(50),IE(1000)
C
      K = KIN
      IF(K)992,419,992
992   IF(K-8)4,421,4
419   KTIE=1
421   GO TO (401,402),KTIE
402   WRITE(NO,923)
923   FORMAT(17H SKIP TO NEXT RUN/1H1)
403   READ(NI,420) K
413   IF(K-KM(38))4033,02,4033
4033  IF(K-KM(39))403,990,403
02    KK=12
      GO TO 120
401   KTIE=2
C                 READ + WRITE CONTROL
1     READ(NI,900) K,(KBCD(I),I=3,21)
5     WRITE(NO,930) K,(KBCD(I),I=3,21)
4     CALL SEARCH(K,KK,2)
03    GO TO              10, 20, 30, 40, 50, 60, 70, 80, 90, 01, 110,1
     120,130,140,150,170,160,180,190,200, 40,220,230,240,250,260,270,280
     2 ,290,300,310,320,330,340,500,990,1),KK
990   WRITE(NO,991)
991   FORMAT(20H THE END HAS COME.       /)
      STOP
C                               READ INTEGRAL CONSTANTS KM
10    READ(NI,902) K, KBCD(3),KFIX
      IF (K - KM(48))  05,11, 05
11    CALL SEARCH(KBCD(3),KK,1)
      IF(KK-3)12,12,10
12    WRITE(NO,924) KBCD(3),KFIX
      IF(KK-2)13,14,15
13    KM(5) = KFIX
      GO TO 10
14    KM(8) = KFIX
      GO TO 10
15    KM(9) = KFIX
      GO TO 10
C                               READ FLOATING CONSTANTS Z
20    READ(NI,903) K,KBCD(3),T(1)
      IF (K - KM(48) )  05, 21, 05
21    CALL SEARCH(KBCD(3),KK,3)
      IF(KK-8)25,26,25
26    Z(11)=ABS(T(1))
      KK=11
      GO TO 24
25    IF(KK-7)22,22,20
22    Z(KK)=ABS(T(1))
      IF(KK-3)24,23,24
23    Z(3) = - Z(3)
24    WRITE(NO,925) KBCD(3),Z(KK)
      GO TO 20
```

```
C                             OUTPUT CONTROL
 30      READ(NI,904) (KP(I,1),I=1,8)
         GO TO 01
C                             INPUT MATRIX (RETURN TO MAIN ROUTINE)
  050    KIN = 1
         KM(1)=KBCD(3)
  999    RETURN
C                             START PHASE ONE
 60      KNEW=2
 599     KM(43)=1
 600     CALL NEW
         GO TO (1,61),KNEW
 61      KVER=2
 649     KM(23)=1
 650     CALL VER
         GO TO (1,70),KVER
C                             COMPUTE
 70      KBOS=1
 700     CALL BOS
         GO TO (1,351),KBOS
C                             FORM BASIS INVERSE
 80      KVER=1
         GO TO 650
C                             INSERT SINGLETONS IN BASIS
 90      KNEW=1
         GO TO 599
C                             CRASH
 110     KM(23)=1
         GO TO 80
C                             BEGIN NEW RUN
  120    KIN = 0
         GO TO 999
C                             CHECK SOLUTION
 130     CALL ERR
         GO TO 001
C                             READ BASIS HEADINGS
 140     KK=2
         GO TO 211
C                             READ ARTIFICIAL ROWS
 150     KK=3
         GO TO 211
C                             SPECIAL OUTPUT
 160     CALL UOUT(6)
         GO TO 001
C                             SUBOPTIMIZE
 170     KM(7)=1
         GO TO 1
C                             PUNCH BASIS HEADINGS
 180     CALL OUP
         GO TO 001
C                             SET INVERSION FREQUENCY
 190     KM(8)=M/10+6
         GO TO 001
C                             NO SUBOPTIMIZATION
 200     KM(7)=0
         GO TO 1
C                             READ RHS
 40      KK=1
```

```
              KM(3)=KBCD(3)
211           CALL CAP(KK,K)
              GO TO 4
C             NO CORE ROUTINE HAS BEEN WRITTEN YET SO DUMMY CALL -----
230           GO TO 1
C                                       EXTRA ROUTINES
240           CALL EXT1
              GO TO 001
250           CALL EXT2
              GO TO 001
260           CALL EXT3
              GO TO 001
270           CALL EXT4
              GO TO 001
C                                       ALTER A
 280          KK=4
              GO TO 211
C                                       ALTER B
 290          KK=5
              GO TO 211
C                                       GENERATE NEW X:S FROM RHS
300           CALL YXB
              GO TO 1
C                                       CORE CLOCK
310           T(1)=TIME(1)
              WRITE(NO,926) T(1)
926           FORMAT(16H THE TIME IS NOW, F10.2 ,8H SECONDS)
              GO TO 01
C                                       READ OBJECTIVE NAMES
 320          KK=6
              GO TO 211
C                                       READ OBJECTIVE ROW
  330   IF(M) 331,331,332
  331   MC=1
        MF=2
        M=1
        NR(1)=KBCD(3)
        GO TO 01
  332   MM=MF-1
        DO 333 I = 1,MM
        IF (KBCD(3) - NR(I) )  333, 334, 333
  333 CONTINUE
335     WRITE(NO,927) KBCD(3)
927     FORMAT(16H OBJECTIVE ROW, 2X,A4,30H, IS NOT IN LIST OF OBJECTIVES)
        GO TO 402
  334 MC = I
        GO TO 1
C                                       SLACKS
 340    KNEW=1
        GO TO 600
C                                       GO FOR ALL OBJECTIVES
220     MM=MF-1
        DO 351 I = 1,MM
           MC = I
        KBOS=2
           GO TO 700
  351 CONTINUE
        GO TO 1
```

```
C                        BLANK CARD  -  ERROR
500     WRITE(NO,928)
928     FORMAT(35H CONTROL CARD HAS BLANK FIRST FIELD)
        GO TO 402
420     FORMAT(A4)
900     FORMAT(20A4)
902     FORMAT(2A4,I8)
903     FORMAT(2A4,E8.1)
  904   FORMAT(8I1)
908     FORMAT(2I4)
924     FORMAT(9X,A4,I8)
925     FORMAT(9X,A4,E12.4)
930     FORMAT(1H ,20A4)
        END




CJMY    CALLED BY BOS,VER
        SUBROUTINE JMY (J)
        COMMON NI, NO
        COMMON M,MA,MB,MC,MD,ME,MF,N,KM(50),KP(8,2),Z(20),T(10),KBCD(30)
        COMMON A(200),IA,KL,KN(20),NR(10),B(10),JH(10),KB(20),
      1 P(10),X(10),Y(10),IP(50),LE,IE,E(1000)
        INTEGER*2 IA(200),KL(20),LE(50),IE(1000)
        KM(12)=KN(J)
        DO 8 I=1,M
        Y(I)=0.0
8       CONTINUE
        LUMP=KL(J)
        LUMP1=KL(J+1)-1
        DO 14 K=LUMP,LUMP1
        IMP=IA(K)
        Y(IMP)=A(K)
14      CONTINUE
        IF(ME) 16,28,16
16      DO 27 I=1,ME
        D=Y(IP(I))
        Y(IP(I))=0.0
        IF(ABS(D)-Z(5)) 27,27,21
21      LUMP2=LE(I)
        LUMP3=LE(I+1)-1
        DO 26  K=LUMP2,LUMP3
        II=IE(K)
        Y(II)=Y(II)+E(K)*D
26      CONTINUE
27      CONTINUE
28      RETURN
        END
```

```
CKRT      CALLED BY GET     APPLY REVERSE TRANSFORM TO P.
          SUBROUTINE KRT
          COMMON NI, NO
          COMMON M,MA,MB,MC,MD,ME,MF,N,KM(50),KP(8,2),Z(20),T(10),KBCD(30)
          COMMON A(200),IA,KL,KN(20),NR(10),B(10),JH(10),KB(20),
         1 P(10),X(10),Y(10),IP(50),LE,IE,E(1000)
          INTEGER*2 IA(200),KL(20),LE(50),IE(1000)
          DO 12 K6=1,ME
          K1=ME+1-K6
          T(2)=0.0
          LU3=LE(K1)
          LU4=LE(K1+1)-1
          DO 9 K3=LU3,LU4
          IMP=IE(K3)
          T(2)=T(2)+P(IMP)*E(K3)
9         CONTINUE
          P(IP(K1))=T(2)
12        CONTINUE
          RETURN
              END




CMIN       CALLED BY BOS  FIND COLUMNS WITH MINIMUM REDUCED COST
           SUBROUTINE MIN(JM)
           COMMON NI, NO
           COMMON M,MA,MB,MC,MD,ME,MF,N,KM(50),KP(8,2),Z(20),T(10),KBCD(30)
           COMMON A(200),IA,KL,KN(20),NR(10),B(10),JH(10),KB(20),
          1 P(10),X(10),Y(10),IP(50),LE,IE,E(1000)
           INTEGER*2 IA(200),KL(20),LE(50),IE(1000)
           IF ( KM(47) - KM(37) )   06, 12, 06
     06 DO 10 I=1,N
           IF(KB(I))08,10,10
     08 KB(I)=0
     10 CONTINUE
           KM(47) = 0
     12 JM = 0
           Z(13) = 1.E+20
           Z(14) = 1.E+20
     32 DO 36 J=1,N
           IF(KB(J))   36,33, 36
  33       CALL DEL (J,D)
           IF(D-Z(13))  14, 46, 46
     14    Z(14) = Z(13)
           Z(13) = D
           KM(27) = JM
           JM = J
           GO TO 36
     46    IF (D - Z(14)     )  15, 36, 36
     15    Z(14) = D
           KM(27) = J
     36 CONTINUE
           RETURN
               END
```

```
CNEW      CALLED BY INP     INSTALL SINGLETONS OR SLACKS      RS MFOR
          SUBROUTINE NEW
          COMMON NI, NO
          COMMON M,MA,MB,MC,MD,ME,MF,N,KM(50),KP(8,2),Z(20),T(10),KBCD(30)
          COMMON A(200),IA,KL,KN(20),NR(10),B(10),JH(10),KB(20),
         1 P(10),X(10),Y(10),IP(50),LE,IE,E(1000)
          INTEGER*2 IA(200),KL(20),LE(50),IE(1000)
C               IF KM(43)=0 WE INSTALL SLACKS
C               IF KM(43)=1 WE INSTALL SINGLETONS
          LE(1)=1
          ME = 0
          KM(17)=M-MF+1
          KM(19) = 0
          KM(31) = 0
          KM(44) = 0
          Z(9) = 1.0
          DO 3001  I = 1,M
            X(I) = B(I)
            JH(I) = 0
 3001   CONTINUE
          DO 3020  J = 1,N
            KB(J) = 0
            KTA = KL(J)
          KTB=KL(J+1)-1
            KQ = 0
            IR = 0
            DO 3010  L = KTA,KTB
          LQ=IA(L)
          IF(A(L))3011,3010,3011
 3011       IF (LQ-MF)  3012, 3013, 3013
 3012 KQ=1
            GO TO 3010
 3013       IF (IR)   3020, 3014, 3020
 3014       IR = LQ
          LR = L
 3010     CONTINUE
          IF (IR)  3021, 3020, 3021
 3021     IF (JH(IR))  3020, 3019, 3020
 3019     IF (A(LR) - 1.0)  3024, 3022, 3024
 3024 KQ=1
 3022     IF (KQ - KM(43))  3004, 3023, 3020
 3023     IF (KQ)    3004, 3005, 3004
 3004     IF ( A(LR)*B(IR))  3020, 3005, 3005
 3005     KB(J) = IR
          JH(IR) = J
          KM(17)=KM(17)-1
          IF (KQ)  3007, 3006, 3007
 3006 KM(44)=KM(44)+1
          GO TO 3020
 3007     KM(11) = NR(IR)
          KM(12) = KN(J)
          KM(13) = KM(12)
          DO 3016 I=1,M
            Y(I) = 0.0
 3016     CONTINUE
          DO 3015  L = KTA,KTB
          LQ=IA(L)
            Y(LQ) = A(L)
```

```
      3015    CONTINUE
              CALL PIV(IR)
              IF (KM(24)) 3198,3020,3198
      3198    KM(35)=3
              CALL VER
      3020    CONTINUE
              WRITE(NO,915) KM(44),ME
      915     FORMAT(12H ::INSTALLED,I6,8H SLACKS,,I6,11H NON-SLACKS)
              KM(43) = 0
              RETURN
              END




COUP      CALLED BY INP, QOT       PUNCH BASIS CARDS     (RS M1 AND RS MFOR)
          SUBROUTINE OUP
          COMMON NI, NO
          COMMON M,MA,MB,MC,MD,ME,MF,N,KM(50),KP(8,2),Z(20),T(10),KBCD(30)
          COMMON A(200),IA,KL,KN(20),NR(10),B(10),JH(10),KB(20),
         1 P(10),X(10),Y(10),IP(50),LE,IE,E(1000)
          INTEGER*2 IA(200),KL(20),LE(50),IE(1000)
          WRITE(NO,1) KM(1),KM(3),KM(2)
          NTAL = 0
          DO 2 I=MF,M
          IF (JH(I))   3,2,3
    3     NTAL=NTAL+1
          JHI = JH(I)
          KBCD(NTAL) = KN(JHI)
          IF (NTAL - 11 )        02, 04, 02
    4     WRITE(NO,5) (KBCD(IQQ),IQQ=1,11)
          NTAL = 0
    2     CONTINUE
          IF (NTAL) 6,15,6
    6     WRITE(NO,5) (KBCD(IQQ),IQQ=1,NTAL)
   15     NTAL = 0
          WRITE(NO,7)
          DO 8 I= MF,M
          IF (JH(I))   8,10,8
   10     NTAL=NTAL+1
          KBCD(NTAL) = NR(I)
          IF ( NTAL - 11 )        08, 09, 08
    9     WRITE(NO,5) (KBCD(IQQ),IQQ=1,11)
          NTAL = 0
    8     CONTINUE
          IF (NTAL)  12,13,12
   12     WRITE(NO,5) (KBCD(IQQ),IQQ=1,NTAL)
   13     WRITE(NO,14)
          RETURN
    1     FORMAT(5H BASS,2X,A4,2X,A4,6H ITERN,I6)
   C1     FORMAT( 4HBASS,2X,A4,2X,A4,6H ITERN,I6)
    5     FORMAT(4X,11A4)
    7     FORMAT(4H END/5H AROW      )
   C7     FORMAT(4HEND /4HAROW)
   14     FORMAT(4H END)
  C14     FORMAT (3HEND)
          END
```

```
CPIV    CALLED BY BOS,NEW,VER   PIVOT, I.E., ADD A TRANSFORMATION (RO M2*)
        SUBROUTINE PIV(IR)
        COMMON NI, NO
        COMMON M,MA,MB,MC,MD,ME,MF,N,KM(50),KP(8,2),Z(20),T(10),KBCD(30)
        COMMON A(200),IA,KL,KN(20),NR(10),B(10),JH(10),KB(20),
       1 P(10),X(10),Y(10),IP(50),LE,IE,E(1000)
        INTEGER*2 IA(200),KL(20),LE(50),IE(1000)
C
        KM(26)=KM(26)+1
        T(1) = Y(IR)
        T(3) = X(IR)/T(1)
        T(4)=ABS(Z(6)*T(1))
        X(IR) = 0.
        Y(IR) = -1.
        K = LE(ME+1)
        DO 01 I = 1,M
        IF(ABS(Y(I))-T(4)) 01,01,06
   06   X(I) = X(I) - T(3) * Y(I)
7       IF(K-KM(40)+M)8,8,11
11      KM(24)=1
   08   E(K) = - Y(I)/T(1)
        IE(K)=I
9       K=K+1
   01   CONTINUE
        Y(IR) = T(1)
        KM(19)=K-1
3       ME=ME+1
        Z(9) = Z(9) * T(1)
910     IF(ABS(Z(9))-1.0) 911,05,921
  911   Z(9) = Z(9) *10.
        KM(31)=KM(31)-1
        GO TO 910
921     IF(ABS(Z(9))-10.0) 05,922,922
  922   Z(9) = Z(9)/10.
        KM(31)=KM(31)+1
        GO TO 921
   05   IF (ME - KM(36) ) 18,18,98
   18   IP(ME) = IR
   19   LE(ME+1) = K
   99   RETURN
98      KM(24)=1
        RETURN
        END


CROW    CALLED BY BOS     COMPOSITE ROW SELECTION  ... RS MSUB VERSION
        SUBROUTINE ROW(IR)
        COMMON NI, NO
        COMMON M,MA,MB,MC,MD,ME,MF,N,KM(50),KP(8,2),Z(20),T(10),KBCD(30)
        COMMON A(200),IA,KL,KN(20),NR(10),B(10),JH(10),KB(20),
       1 P(10),X(10),Y(10),IP(50),LE,IE,E(1000)
        INTEGER*2 IA(200),KL(20),LE(50),IE(1000)
C
C AMONG EQS. WITH X=0, FIND MAX ABS(Y) AMONG ARTIFICIALS, OR, IF NONE,
C   GET MAX POSITIVE Y(I) AMONG REALS.
```

```
1000  IR=0
      AA = 0.0
      IAB=0
      DO 1050 I = MF, M
      IF ( X(I) ) 1050, 1041, 1050
1041  YI=ABS(Y(I))
      IF ( YI - Z(1) )     1050, 1050, 1042
1042  IF ( JH(I) ) 1043, 1044, 1043
1043  IF(IAB) 1050,1048,1050
1048  IF ( Y(I) ) 1050, 1050, 1045
1044  IF(IAB) 1045,1046,1045
1045  IF ( YI   - AA  )  1050, 1050, 1047
1046  IAB=1
1047  AA = YI
      IR=I
1050  CONTINUE
      Z(8) = 0.
      IF(IR)1099,1001,1099
1001  AA = 1.0E+20
C                FIND MIN. PIVOT AMONG POSITIVE EQUATIONS
      DO 1010 IT = MF , M
      IF ( Y(IT) - Z(1) ) 1010, 1010, 1002
1002  IF ( X(IT) ) 1010, 1010, 1003
1003  XY = X(IT) / Y(IT)
      IF ( XY - AA ) 1004, 1005, 1010
1005  IF ( JH(IT)) 1010, 1004, 1010
1004  AA = XY
      Z(8) = XY
      IR=IT
1010  CONTINUE
      IF (KM(30)) 1016, 1099, 1016
C  FIND PIVOT AMONG NEGATIVE EQUATIONS, IN WHICH X/Y IS LESS THAN THE
C  MINIMUM X/Y IN THE POSITIVE EQUATIONS, THAT HAS THE LARGEST ABSF(Y)
1016  BB = - Z(1)
      DO 1030 I = MF , M
      IF (X(I))   1012, 1030, 1030
1012  IF ( Y(I)  -  BB ) 1022, 1030, 1030
1022  IF ( Y(I) * AA - X(I) ) 1024, 1024, 1030
1024  BB = Y(I)
      IR=I
      Z(8) = X(I) / BB
1030  CONTINUE
1099  RETURN
      END




CTAP    CALLED BY FIN        INPUT MATRIX
      SUBROUTINE TAP (K)
      COMMON NI, NO
      COMMON M,MA,MB,MC,MD,ME,MF,N,KM(50),KP(8,2),Z(20),T(10),KBCD(30)
      COMMON A(200),IA,KL,KN(20),NR(10),B(10),JH(10),KB(20),
     1 P(10),X(10),Y(10),IP(50),LE,IE,E(1000)
      INTEGER*2 IA(200),KL(20),LE(50),IE(1000)
C
```

```
            IF  (KM(49))   08, 09, 08
08      WRITE(NO,916)
916     FORMAT(30H MATRIX READ TWICE IN SAME RUN)
        GO TO 42
9       N=0
        KM(18) = 0
        IF (M )    01, 201, 01
201     WRITE(NO,917)
917     FORMAT(30H THIS PROBLEM HAS NO OBJECTIVE)
        M=1
        MC=1
        MF=2
01      READ(NI,90) K,J,I,AA
        IF (K - KM(48)) 51, 02, 51
C                  LOOK FOR ROW NAME
   02   DO 03   II=1,M
        IF (NR(II) - I)  03,04,03
   03   CONTINUE
        M=M+1
        IF(M-KM(20)) 11,11,101
   11   II = M
        NR(M) = I
C                  BUMP MATRIX ENTRY COUNT
  4     KM(18)=KM(18)+1
        IF(KM(18)-KM(6)) 10,10,102
   10   KM18   = KM(18)
C                  LOOK FOR COLUMN NAME
        IF (N)  06,06,07
   07   IF (KN(N) - J )   06, 13, 06
  6     N=N+1
        IF(N-KM(21)) 12,12,103
   12   KL(N) = KM18
        KN(N) = J
C                  FILE AN  ENTRY
13      IA(KM18)=II
        A(KM18) = AA
        GO TO 01
   51   KL(N+1)=KM18+1
        KWOT=1
   30   CONTINUE
        WRITE(NO,918) M,N,KM(18)
918     FORMAT(17H       PROBLEM HAS I4,6H ROWS,I4,14H COLUMNS, AND I5,16H M
     1ATRIX ENTRIES. )
        KM(17)=M-MF+1
        GO TO (41,42),KWOT
42      KOPOL=8
        CALL INP (KOPOL)
   41   KBCD(3)=J
        RETURN
101     WRITE(NO,920)
        GO TO 25
102     WRITE(NO,921)
        GO TO 25
103     WRITE(NO,922)
   25   KWOT=2
        GO TO 30
  920   FORMAT(35H PROBLEM TOO BIG, TOO MANY ROWS.     )
  921   FORMAT(35H PROBLEM TOO BIG, TOO MANY ENTRIES. )
```

```
  922     FORMAT(35H PROBLEM TOO BIG, TOO MANY COLUMNS. )
C 90     FORMAT(A4,4X,A4,2X,A4,F12.6)
   90     FORMAT(3A4,F12.6)
          END
```

```
CVER      CALLED BY BOS, INP      INVERSION AND FEASIBLE CRASHING ROUTINE
          SUBROUTINE VER
          COMMON NI, NO
          COMMON M,MA,MB,MC,MD,ME,MF,N,KM(50),KP(8,2),Z(20),T(10),KBCD(30)
          COMMON A(200),IA,KL,KN(20),NR(10),B(10),JH(10),KB(20),
         1 P(10),X(10),Y(10),IP(50),LE,IE,E(1000)
          INTEGER*2 IA(200),KL(20),LE(50),IE(1000)
C                   INITIATE
          KM(32)  = KM(02)
          Z(15)=TIME(1)
          WRITE(NO,920) KM(2),ME,KM(19),Z(15),KM(35)
  920     FORMAT(19H REINVERTING AFTER I5,13HTH ITERATION.I5,22H TRANSFORMAT
         1IONS WITH I5,15H ENTRIES, TIME=,F12.2,5H TYPE,I2)
          KBAD = 0
          KM(24) = 0
          KM(28) = 0
          IF (KM(23))   1102, 1101, 1102
 1102 DO 1105  J =1,N
          IF (KB(J) )  1106,1106,1105
 1106     KB(J) = -12345
 1105 CONTINUE
          IF (LE(1) )    1107, 1101, 1107
 1101 ME = 0
          KM(17)=M-MF+1
          KM(19) = 0
          KM(31) = 0
          KM(44) = 0
          Z(9) = 1.
          LE(1) =1
          DO 13   I = 1,M
   13 X(I) = B(I)
          IF (KM(23) ) 1107, 1109, 1107
 1109     DO 01   I = MF,M
          IF (JH(I)) 11,01,11
   11 JH(I) = 12345
   01 CONTINUE
          DO 1131 J = 1,N
          IF (KB(J)) 1132, 1131, 1133
 1132     KB(J) = 0
          GO TO 1131
 1133     KB(J) = -12345
 1131 CONTINUE
 1107 CONTINUE
C                  FORM INVERSE
   02 IF (KM(17) ) 1142,  20, 1142
 1142 KM(45) = 12345
          DO 1150 JJ = 1,N
          IF (KB(JJ))   1151 , 1150, 1150
```

```
     1151   KM(46) = KL(JJ+1) - KL(JJ)
            IF (KM(46) - KM(45) )   1152, 1150, 1150
     1152   KM(45) = KM(46)
            J = JJ
     1150 CONTINUE
            IF (KM(45) - 12345 )   1161, 20,      1161
     1161 KB(J) = 0
            KM(12) = KN(J)
     03     CALL JMY(J)
C                          CHOOSE PIVOT
            IR = 0
            IF (KM(23))   1121,1122,1121
     1122   T(2) = Z(1)
            DO 1123  I = MF, M
            IF (JH(I) - 12345) 1123, 1124, 1123
     1124 IF(ABS(Y(I))-T(2)) 1123,1123,1125
     1125 T(2)=ABS(Y(I))
            IR = I
     1123 CONTINUE
     1128   IF (IR)   08, 07, 08
C                          TROUBLE CHECK
     7    KBAD=KBAD+1
          GO TO 2
C                          PIVOT
     08   JH(IR) = J
          KB(J)  = IR
          KM(17) =KM(17)-1
          IF(KM(45)-1)62,1155,62
     1155 JJ = KL(J)
          IF ( A(JJ) - 1.0 ) 62,   1141,    62
     1141 KM(44)=KM(44)+1
          GO TO 2
     62   CALL PIV(IR)
          IF ( KM(24) )   20 , 02 , 20
C                   RESET ARTIFICIALS
     20   DO 09 I = 1,M
          IF (JH(I)-12345) 09,12,09
     12   JH(I) = 0
     09   CONTINUE
          DO 1171 J = 1,N
          IF (KB(J))   1172,1171,1171
     1172   KB(J) = 0
     1171 CONTINUE
          Z(16)=TIME(1)
          WRITE(NO,921) KM(44),KBAD,ME,KM(19),Z(16)
     921  FORMAT(21H *INVERSION COMPLETED,I5,8H SLACKS,I5,11H POOR COLS,I5,
         122H TRANSFORMATIONS WITH I5,15H ENTRIES, TIME= F12.2/1X)
          KM(23) = 0
          KM(35) = 0
          IF (KM(24)) 24,99,24
     99 RETURN
     24   WRITE(NO,922)
     922  FORMAT(29H =TOO MANY ENTRIES IN INVERSE)
          KM(14)=1
          KM(19)=LE(ME)-1
          ME=ME-1
          CALL UOUT(7)
     1170 KOPOL=8
```

```
      CALL INP (KOPOL)
C                              FEASIBLE CRASHING
1121  AA = 1.0E+20
      CC = 0.
      DO 2100 I = MF,M
         IF (JH(I))  2101,2100,2101
2101     IF (Y(I) - Z(1))  2100, 2100, 2103
2103     IF (X(I) + Z(2))  2100, 2102, 2102
2102     IF (X(I) - Z(2) )  2110, 2110, 2104
2104  AA=AMIN1(AA,X(I)/Y(I))

2100  CONTINUE
2111  DO 2120 I = MF,M
         IF (JH(I))  2120, 2121, 2120
2121  IF(ABS(Y(I))-Z(7))2120,2120,2122
2122  IF(ABS(X(I))-Z(2))2130,2130,2123
2123     BB = X(I)/Y(I)
         IF (BB)   2120, 2120, 2124
2124     IF ( BB - AA )   2130, 2130, 2120
2130  IF(ABS(Y(I))-CC) 2120,2120,2125
2125  CC=ABS(Y(I))
      IR = I
2120  CONTINUE
      GO TO 1128
2110  AA = 0.
      GO TO 2111
      END
```

```
CXCK    CHECK SOLUTION FOR FEASIBILITY      CALLED BY BOS
      SUBROUTINE XCK(K)
      COMMON NI, NO
      COMMON M,MA,MB,MC,MD,ME,MF,N,KM(50),KP(8,2),Z(20),T(10),KBCD(30)
      COMMON A(200),IA,KL,KN(20),NR(10),B(10),JH(10),KB(20),
     1 P(10),X(10),Y(10),IP(50),LE,IE,E(1000)
      INTEGER*2 IA(200),KL(20),LE(50),IE(1000)
C
C         RESET X AND CHECK FOR INFEASIBILITIES
      KM(30) = 0
      K = 0
         DO 01 I = MF,M
      IF(ABS(X(I))-Z(2)) 02,02,03
  02  X(I) = 0.0
      GO TO 01
  03  IF (X(I)) 07,01,05
  05  IF(JH(I)) 01,06,01
   7  KM(30)=1
   6  K=1
  01     CONTINUE
  99  RETURN
      END
```

CYXB   GENERATES NEW X FROM NEW RHS AND OLD TRANSFORMATIONS. CD. BY INP
         SUBROUTINE YXB
         COMMON NI, NO
         COMMON M,MA,MB,MC,MD,ME,MF,N,KM(50),KP(8,2),Z(20),T(10),KBCD(30)
         COMMON A(200),IA,KL,KN(20),NR(10),B(10),JH(10),KB(20),
        1 P(10),X(10),Y(10),IP(50),LE,IE,E(1000)
         INTEGER*2 IA(200),KL(20),LE(50),IE(1000)
         DO 01 I=1,M
    01   X(I) = B(I)
         IF (ME) 02, 99, 02
    02   DO 10 KAM = 1,ME
           K1 = IP(KAM)
           D = X(K1)
           X(K1) = 0.0
         IF(ABS(D)-Z(J)) 10,10,04
    04   MTA = LE(KAM)
         MTB=LE(KAM+1)-1
            DO 03   K4 = MTA,MTB
         K5=IE(K4)
    03   X(K5) = X(K5) + E(K4)*D
    10   CONTINUE
    99   RETURN
         END

CTIME THE FUNCTION TIME USED TO INTERROGATE THE TIMER
         FUNCTION TIME(I)
C   ARGUMENT OF 0 INDICATES WE WANT TO REINITIALIZE OUR RELATIVE
C   TIMER, WHILE AN ARGUMENT WHICH IS NON-ZERO INDICATES THAT WE
C   WANT TO FIND THE TIME SINCE THE LAST REINITIALIZATION
         IF(I) 2,1,2
    1   IT = ITIME(L)
    2   TIME=(ITIME(L)-IT)/100.0
         RETURN
         END

CSS   CONTROLS OPTIMAL REINVERSION   CALLED BY BOS
         FUNCTION SS(I)
         COMMON NI, NO
         COMMON M,MA,MB,MC,MD,ME,MF,N,KM(50),KP(8,2),Z(20),T(10),KBCD(30)
         COMMON A(200),IA,KL,KN(20),NR(10),B(10),JH(10),KB(20),
        1 P(10),X(10),Y(10),IP(50),LE,IE,E(1000)
         INTEGER*2 IA(200),KL(20),LE(50),IE(1000)
         DIMENSION XTIME(5)
         SS=0.
         DO 61 J=1,4
    61   XTIME(J)=XTIME(J+1)
         XTIME(5)=TIME(1)
         IF(XTIME(5))62,99,62
    62   XD=KM(2)-KM(32)
         IF(XD-5.)99,63,63
    63   IF(4.*(XTIME(5)-Z(15))-(XTIME(5)-XTIME(1))*XD) 98,99,99
    98   SS=1.
    99   RETURN
         END

```
CUOUT ONCE WAS NAMED QOT      CALLED BY BOS,INP,VER
      SUBROUTINE UOUT(K)
      COMMON NI, NO
      COMMON M,MA,MB,MC,MD,ME,MF,N,KM(50),KP(8,2),Z(20),T(10),KBCD(30)
      COMMON A(200),IA,KL,KN(20),NR(10),B(10),JH(10),KB(20),
     1 P(10),X(10),Y(10),IP(50),LE,IE,E(1000)
      INTEGER*2 IA(200),KL(20),LE(50),IE(1000)
C                                SET CONDITION NUMBER
      KM(16) = K
C                                RESET QOT(6) COUNTER
      IF(KM(16)- 6) 22,21,22
   21 KM(25) = 0
C                                TEST IF PUNCHING IS WANTED
   22 KD = KP(K,1)
      IF(KD-5)3,2,2
    2 KD=KD-5
      CALL OUP
C                                EJECT PAGE BEFORE LONGER OUTPUT
    3 IF(KD-1)99,4,5
   05 IF (KM(14))    09,08,09
    8 KEJ=1
   80 WRITE(NO,90)
   89 GO TO (9,14),KEJ
C                                CONTROL 1.  PAGE WILL NOT BE RESTORED
   04 KM(14) = 0
   09 CALL SOT
      IF(MOD(KD,2))11,10,11
   10 CALL LOT
   11 IF (KD - 2 ) 99, 13, 12
   12 CALL HOT
C                                EJECT PAGE
   13 KEJ=2
      GO TO 80
C                                NOTE PAGE EJECTED
   14 KM(14)=1
   99 RETURN
   90 FORMAT(1H1)
      END
```

```
CSOT     CALLED BY QOT        SHORT OUTPUT
         SUBROUTINE SOT
         COMMON NI, NO
         COMMON M,MA,MB,MC,MD,ME,MF,N,KM(50),KP(8,2),Z(20),T(10),KBCD(30)
         COMMON A(200),IA,KL,KN(20),NR(10),B(10),JH(10),KB(20),
        1 P(10),X(10),Y(10),IP(50),LE,IE,E(1000)
         INTEGER*2 IA(200),KL(20),LE(50),IE(1000)
C
         KM(29) = 0
         T(2) = 0.0
C                                      COMPUTE INFEASIBILITIES
         DO 01  I = MF,M
           IF (X(I))     03, 01, 04
   03      T(2) = T(2) - X(I)
           GO TO 06
   04      IF (JH(I))    01, 05, 01
   05      T(2) = T(2) + X(I)
    6      KM(29)=KM(29)+1
   01    CONTINUE
         T(3) = - X(MC)
  101    WRITE(NO,102) KM(16),NR(MC),KM(29),KM(1),KM(3),KM(2),KM(42),
        1KM(26),T(3),T(2),Z(9),KM(31),Z(13),KM(13),KM(10),KM(11)
   202 RETURN
C
  102    FORMAT(2H (,I1,42H) MATRIX R.H.S. ITER STEPS PIVS OBJECTIVE ,A4,2X
        1,I6,59H INFEAS    DETERMINANT  MIN. R/COST   NEW COL   OLD COL  PIV ROW
        2/6X,A4,3XA4,I5,I6,I5,F17.5,F13.3,F9.5,1HE,I3,2XF12.5,3(5X,A4),/1X)
         END




CLOT     CALLED BY QOT           OUTPUT ROW DEPENDENT QUANTITIES
         SUBROUTINE LOT
         COMMON NI, NO
         COMMON M,MA,MB,MC,MD,ME,MF,N,KM(50),KP(8,2),Z(20),T(10),KBCD(30)
         COMMON A(200),IA,KL,KN(20),NR(10),B(10),JH(10),KB(20),
        1 P(10),X(10),Y(10),IP(50),LE,IE,E(1000)
         INTEGER*2 IA(200),KL(20),LE(50),IE(1000)
C
  301    WRITE(NO,302) KM(12)
   200 DO 100 I=1,M
         L = JH(I)
  502    WRITE(NO,303) NR(I),B(I),P(I),Y(I)
 C502    WRITE(NO,303) KN(L),X(I),NR(I),B(I),P(I),Y(I)
   100 CONTINUE
    99 RETURN
  302    FORMAT(5X,3HROW,10X,3HRHS,14X,5HPRICE,12X,A4//)
  303    FORMAT(4X,A4,2F18.6,E15.6)
 C302    FORMAT(7H   NAME,9X,5HVALUE13X,3HROW10X,3HRHS14X,5HPRICE,12X,A4//)
 C303    FORMAT(3X,A4,F18.6,8X,A4,2F18.6,E15.6)
         END
```

```
CDEL    CALLED BY ERR, HOT, MIN        REDUCED COST OF ONE COLUMN (RO M2*)
        SUBROUTINE DEL (J,D)
        COMMON NI, NO
        COMMON M,MA,MB,MC,MD,ME,MF,N,KM(50),KP(8,2),Z(20),T(10),KBCD(30)
        COMMON A(200),IA,KL,KN(20),NR(10),B(10),JH(10),KB(20),
     1  P(10),X(10),Y(10),IP(50),LE,IE,E(1000)
        INTEGER*2 IA(200),KL(20),LE(50),IE(1000)
C
        DT = 0.
        MTA = KL(J)
        MTB=KL(J+1)-1
C
   01 DO 03   K = MTA,MTB
        IN=IA(K)
        IF ( P(IN) ) 02,03,02
   02    DT = DT + P(IN)*A(K)
   03    CONTINUE
C
        D = DT
        RETURN
        END
```

360 ASSEMBLER LANGUAGE ROUTINE

```
000000                                          1  SUBRTN  START 0
                                                2          ENTRY SEARCH
000000 E2C5C1D9C3C840                           3          DC    CL7'SEARCH '
000007 08                                       4          DC    X'8'
000008                                          5          USING *,15
000008 90E8 D00C                  00000C        6  SEARCH  STM   14,8,12(13)
00000C 5821 0000                  00000         7          L     2,0(1)
000010 5831 0004                  00004         8          L     3,4(1)
000014 5841 0008                  00008         9          L     4,8(1)
000018 5854 0000                  00000        10          L     5,0(4)
00001C 5950 F12C                  0013C        11          C     5,Z3
000020 4740 F02C                  0003C        12          BL    LIST1
000024 4720 F038                  00048        13          BH    LIST3
000028 4160 F094                  0009C        14  LIST2   LA    6,TWO
00002C 4170 C024                  00024        15          LA    7,36
000030 47F0 F040                  00048        16          B     CHECK
000034 4160 F068                  00070        17  LIST1   LA    6,ONE
000038 4170 C003                  00003        18          LA    7,3
00003C 47F0 F040                  00048        19          B     CHECK
000040 4160 F074                  0007C        20  LIST3   LA    6,THREE
000044 4170 C008                  00008        21          LA    7,8
000048 4180 0001                  00001        22  CHECK   LA    8,1
00004C D503 2000 6000 00000       00000        23  CHECK1  CLC   0(4,2),0(6)
000052 4780 F05A                  0006A        24          BE    YES
000056 5A80 F124                  0012C        25          A     8,Z1
00005A 5A6C F128                  00130        26          A     6,Z2
00005E 4670 F044                  0004C        27          BCT   7,CHECK1
000062 5083 0000                  00000        28  YES     ST    8,0(3)
000066 98E8 D01C                  0001C        29          LM    2,8,28(13)
00006A 92FF D00C                  0000C        30          MVI   12(13),X'FF'
00006E 07FE                                    31          BCR   15,14
000070 D6E4E3D7C9D5E5E3                       32  ONE     DC    C'OUTPINVTCTOF'
00007C D7C9E540D9E2C5E3                       33  THREE   DC    C'PIV RSETCUSTMIX '
00008C E3D4D3E3C5D5E3D9                       34          DC    C'TMLTENTRREJ TIMX'
00009C C6D9C5D8E3D6D3D9                       35  TWO     DC    C' FREQTOLRPRMDFRST'
0000AC D4E3D9E7E2D6D3E5                       36          DC    C'MTRXSOLVGO  INVT'
0000BC E2D5C7D3C5D5C440                       37          DC    C'SNGLEND CRSHBEGN'
0000CC C5D9D9E2C2C1E2E2                       38          DC    C'ERRSBASSAROWSUPT'
0000DC D6E4E3D7D7D5C3C8                       39          DC    C'OUTPPNCHSINVNSUP'
0000EC D9C8E240C7D6C7D6                       40          DC    C'RHS GOGOCOREEXT1'
0000FC C5E7E3F2C5E7E3F3                       41          DC    C'EXT2EXT3EXT4ALTA'
00010C C1D3E3C2D5C5E6E7                       42          DC    C'ALTBNEWXTIMEOBJ '
00011C D9D6E640E2D3C3D2                       43          DC    C'ROW SLCK    STOP'
00012C 00000001                               44  Z1      DC    F'1'
000130 00000004                               45  Z2      DC    F'4'
000134 00000002                               46  Z3      DC    F'2'
                                              47          ENTRY EXT1
00013E C5E7E3F140                             48          DC    CL5'EXT1 '
00013D 06                                     49          DC    X'6'
00013E                                        50          USING *,15
00013E 92FF D00C          0000C               51  EXT1    MVI   12(13),X'FF'
000142 07FE                                   52          BCR   15,14
                                              53          ENTRY EXT2
000144 C5E7E3F240                             54          DC    CL5'EXT2 '
000149 06                                     55          DC    X'6'
```

```
00014A                                      56            USING *,15
00014A 92FF D00C            0000C           57 EXT2       MVI   12(13),X'FF'
00014E 07FE                                 58            BCR   15,14
                                            59            ENTRY EXT3
000150 C5E7E3F340                           60            DC    CL5'EXT3 '
000155 06                                   61            DC    X'6'
000156                                      62            USING *,15
000156 92FF D00C            C000C           63 EXT3       MVI   12(13),X'FF'
00015A 07FE                                 64            BCR   15,14
                                            65            ENTRY EXT4
00015C C5E7E3F440                           66            DC    CL5'EXT4 '
000161 06                                   67            DC    X'6'
000162                                      68            USING *,15
000162 92FF D00C            0000C           69 EXT4       MVI   12(13),X'FF'
000166 07FE                                 70            BCR   15,14
                                            71            ENTRY ITIME
000168 C9E3C9D4C5                           72            DC    CL5'ITIME'
00016D 06                                   73            DC    X'6'
00016E                                      74            USING *,12
00016E 90EC C00C            0000C           75 ITIME      STM   14,12,12(13)
000172 18CF                                 76            LR    12,15
                                            77            TIME  BIN
000174 4110 0001            00001           78+           LA    1,1(0,0)
000178 0ACB                                 79+           SVC   11 ISSUE TIME SVC
00017A 982C D01C            0001C           80            LM    2,12,28(13)
00017E 92FF D00C            0000C           81            MVI   12(13),X'FF'
000182 07FE                                 82            BCR   15,14
                                            83            END
```

```
FREQ      CTOF        25
END       REJ   0.1000E-03
TOLK
END
ROW CCST
RHS REQ
MTRX-A-
          PROBLEM HAS     4 ROWS,    6 COLUMNS, AND    13 MATRIX ENTRIES.
TIME
THE TIME IS NOW    0.40 SECONDS
SOLV
::INSTALLED     3 SLACKS,    0 NON-SLACKS
REINVERTING AFTER  0TH ITERATION.    0 TRANSFORMATIONS WITH    0 ENTRIES, TIME=    0.42 TYPE 0
*INVERSION COMPLETED  3 SLACKS,    0 POOR COLS,    0 TRANSFORMATIONS WITH    0 ENTRIES, TIME=    0.44

*FEASIBLE ON ITERATION   0,    0 STEPS
=OPTIMAL SOLUTION
THE TIME IS      0.45 SECONDS.
BASS -A-  REQ  ITERN   3
     CCL2CCL3CCL1
END
AROW
END
(3) MATRIX R.H.S. ITER STEPS PIVS OBJECTIVE COST      0 INFEAS  DETERMINANT  MIN. R/COST   NEW COL  OLD COL  PIV ROW
     -A-   REQ   3    3   3             -18.92680     0.0      4.10000E  1    0.26829      COL1     SLK3     ROW3

ROW           RHS             PRICE              SLK2

COST          0.0             1.000000           0.585366E 00
ROW1          8.000000        1.097559           0.195122E 00
ROW2         10.000000        0.585366           0.121951E 00
ROW3         16.000000        0.268293          -0.292683E 00

NAME          VALUE            TRUE     /COST/          REDUCED

CCL1          2.536586        -3.000000                 -0.000004
CCL2          0.975610        -5.000000                 -0.000008
CCL3          1.609756        -4.000000                 -0.000001
SLK1          0.000000         0.0                       1.097559
SLK2          0.000000         0.0                       0.585366
SLK3          0.000000         0.0                       0.268293

MAX ERROR ON ROW COST= -0.24796E-04, SUM= 0.29564E-04
MAX ERROR ON COL COL2= -0.76294E-05, SUM= 0.12398E-04

TIME
THE TIME IS NOW    0.70 SECONDS
STOP
THE END HAS COME.
```

III. MIXED INTEGER PROGRAMMING (MINT ALGORITHM)*

A. Purpose

This program finds the minimum of a multivariable, linear function subject to linear constraints, in which some or all of the variables may be restricted to integer values:

Minimize $\quad F = C_1X_1 + C_2X_2 + \ldots + C_{N1}X_{N1} + C_{N1+1}Y_{N1+1} + \ldots + C_NY_N$

Subject to $\quad A_{ij}X_j + \bar{A}_{ik}Y_k \leq, =, \geq B_i \quad i = 1, \ldots, m$
$\quad\quad\quad\quad\quad\quad\quad\quad\quad\quad\quad\quad\quad\quad\quad j = 1, \ldots, N1$
$\quad\quad\quad\quad\quad\quad\quad\quad\quad\quad\quad\quad\quad\quad\quad k = N1+1, \ldots, N$

X_j are each integer and subject to an upper bound
$X_j, Y_k \geq 0$.

B. Method

The algorithm is based on the Land and Doig (32) method. A dual simplex algorithm is imbedded in the program to obtain the starting, continuous solution and evaluate each integer trial. The specified integer variables are tested one at a time in paired values to establish direction and value. The algorithm is as follows:

1) The algorithm employs a dual simplex linear programming algorithm (not product form) hereinafter referred to as the LP. The tableau is carried in compact Tucker (45) form: the initial number of rows equals the number of problem constraints plus one; the initial number of columns equals the number of true variables plus one. Whenever a zero-constrained slack variable becomes non-basic, it is removed from the problem, resulting in a reduction by one of the number of columns in the tableau. Zero-constrained slack variables arise from two sources: equality constraints in the initial tableau; constraining a basic integer variable to an integer value (see 4 below). The number of rows in the tableau remains constant throughout.

---

*The FORTRAN program contained in this section is based on <u>Branch and Bound Mixed Integer Programming</u>, described on page 242 of "Catalog of Programs for IBM System 360 Models 25 and Above," GC 20-1619-8; program number 360D-15.2.005. Used by permission of International Business Machines Corporation.

2) Carry out an LP on the initial tableau. Print the solution. Check to see if all integer variables are integer valued. If so, the problem is terminated; if not, set the initial tolerance for the problem. (Tolerance is defined as the value below which the objective function must stay in order for a continuation of the current sequence of integer-constrained interger variables to be considered as a candidate for the mixed integer solution. Note that the objective function value at the continuous solution represents an absolute lower bound for the mixed integer solution.) Set to 1 the index of the integer variable being constrained.

3) Choose from those integer variables which are non-basic in the current tableau the one with highest coefficient in the objective function (shadow price). (The program makes use of the fact that the shadow price represents an underestimate of the increase in the objective function associated with constraining the non-basic integer variable to 1.) If no non-basic integer variable exists, go to 4. Otherwise, store the current tableau and constrain the variable chosen to zero. This is done simply by removing the corresponding column from the tableau. (A non-basic variable is constrained to a non-zero integer value by adding the product to this value with each element in the corresponding column in the constant column of the tableau. The corresponding column is then removed from the tableau.) Go to 6.

4) Store the current tableau. Consider all integer variables X_i which are basic in the current tableau (there must be at least one) with value X_i^f. For each X_i determine the absolute difference between the increase in the objective function associated with the initial LP pivot step when X_i is constrained to $[X_i^f]$ and when X_i is constrained to $[X_i^f] + 1$. Choose as the integer variable to be constrained that X_i for which this difference is a maximum and constrain it to the value yielding the smaller increase. The actual constraining is accomplished by adding the integer value to the constant column of the row corresponding to the variable, and then stipulating that the row corresponds to a zero-constrained slack variable. Carry out an LP. If the objective function stays within the tolerance go to 6; otherwise go to 5.

5) If the current integer variable was constrained to $[X_i^f]$, record the fact that constraining it to values $[X_i^f] - k$ ($k = 1, 2, \ldots$) within its range need not be considered. Conversely, if X_i was set to $[X_i^f] + 1$, make note that values $[X_i^f] + 1 + k$ need not be considered. Go to 9.

6) Test the constrained variable index. If it is equal to N1, the number of integer variables in the problem, go to 9. Otherwise increase it by one and go to 3.

7) Decrease the constrained variable index by one and test it.

8) If it is zero go to 11. Otherwise go to 9.

9) Determine for the integer variable corresponding to the current value of the index whether its range has been exhausted (explicitly or implicitly) on neither, on one or on both sides of its current value. If it has been exhausted on both sides, go to 7. If the variable to be constrained has been exhausted on one side, constrain it to the unexhausted integer value closest to its current value in the proper direction. If the range is unexhausted on either side, determine in which direction to go using the method employed in 4, and proceed as for only that side open. (Note that the range of an integer variable which was non-basic when constrained is immediately exhausted from below.) Carry out an LP. If the objective function stays within the tolerance go to 6. Otherwise, note that the range of the current variable is exhausted in the direction in which its current value lies from its original value (see 5). Go to 9.

10) A better feasible mixed integer solution has been obtained. Print the solution. Replace the tolerance by the objective function value. Go to 8.

11) For the current tolerance, all ranges of all the integer variables have been exhausted. If at least one feasible mixed integer solution has been obtained, the last printed solution is an optimal solution to the mixed integer problem and the problem is terminated. Otherwise, the tolerance is increased, the continuous solution tableau is restored, the index of constrained integer variables is set to one, and control goes to 3.

If the program is terminated abnormally, the last printed feasible mixed integer solution (if any) is the best obtained. A flow diagram illustrating the above procedure is shown in Figure 2.III.

C. Program Description

1) Usage:

The program consists of a main program only. Program size, solution estimate, and tableau coefficients along with control parameters are read in. The objective function to be minimized is the first row of the tableau.

2) Subroutines Required:

None.

3) Description of Parameters:

ISIZE   Intermediate storage area = NZR1VR*(2*N-NZR1VR+1)/2 or as large as possible.

NMRUNS   Number of runs or problems to be solved

IOUT2   Print control for initial working tableau:
0 = No print
1 = Print tableau

IOUT3   Print control for continuous solution tableau:
0 = No print
1 = Print tableau

IPACK   Matrix format:
0 = Unpacked, read all coefficients
1 = Packed, read non-zero coefficients only

SOLMIN   Estimate of objective function if known, zero otherwise

PCTTOL   Tolerance as fraction of objective function for continuous solution (may be left at zero)

M   Total number of rows

N   Total number of columns equals sum of X and Y variables plus 1 for constraints

NM1   DO loop parameters: NM1 = N-1

NZR1VR   Number of integer variables

UPBND   Vector of integer variable's upper bounds; size = N - 1

IROW   Vector of constraint types; size = M - 1:
$+1 \geq b_i$
$0 = b_i$
$-1 \leq b_i$

ITEMP   Column of coefficients being read in row i including objective row

VAL   Coefficient value of columns specified by ITEMP for row i

ATAB   Initial working tableau, N X M array

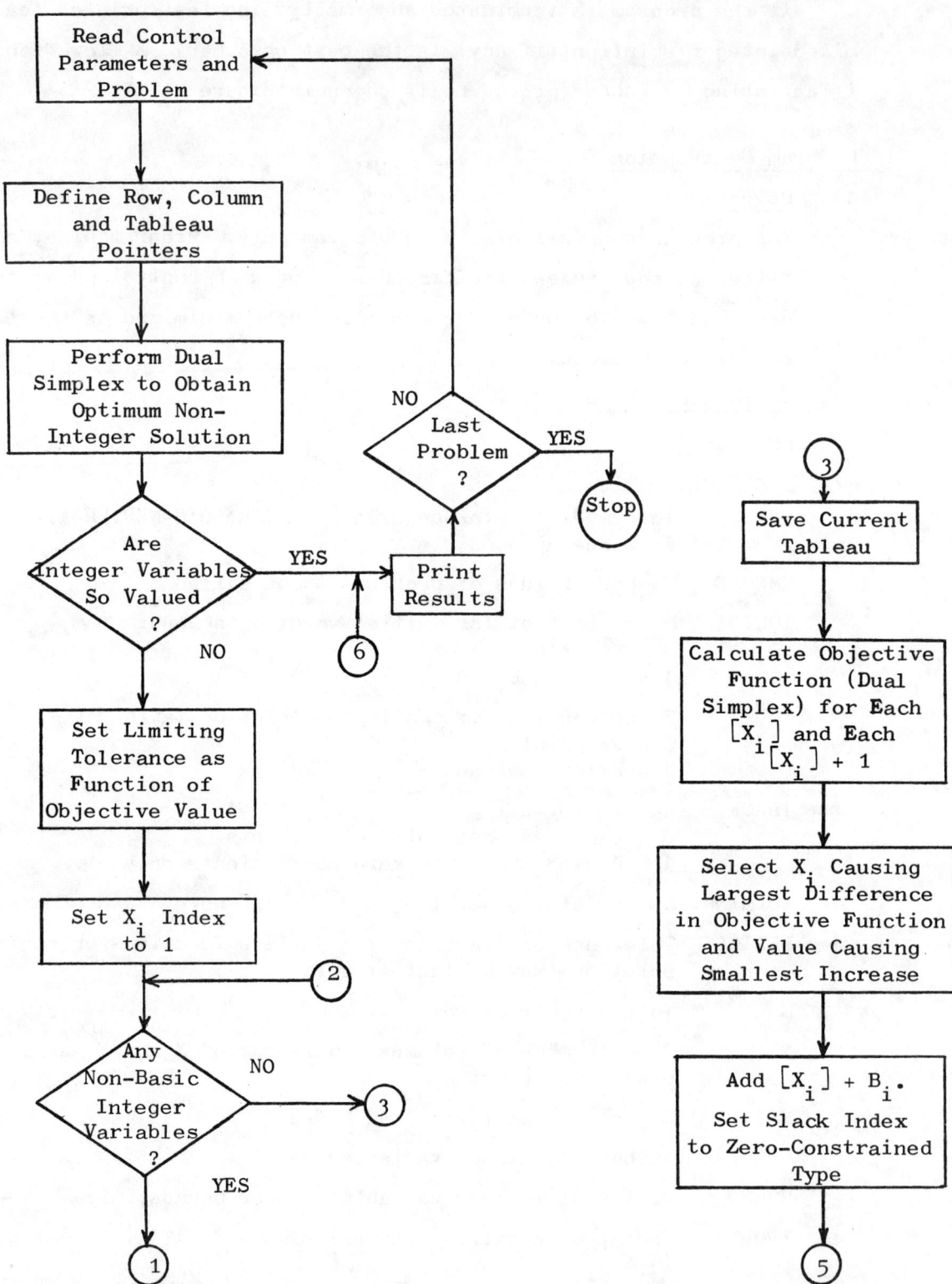

Figure 2.III. Mixed Integer Programming (MINT ALGORITHM) Logic Diagram

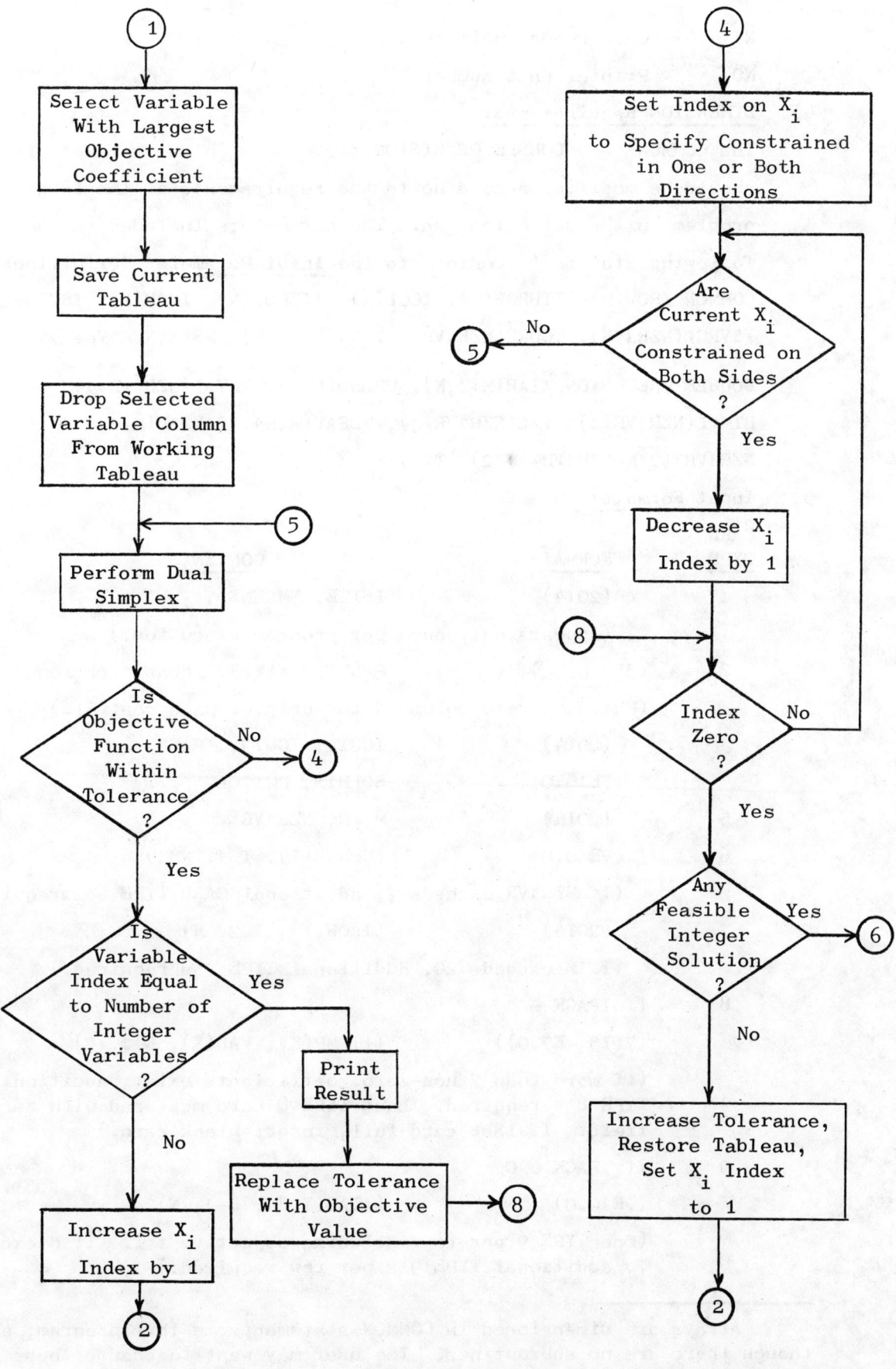

Figure 2.III. (Continued)

NI   Card reader unit number

NO   Printer unit number

4) DIMENSION Requirements:

The COMMON* and DOUBLE PRECISION statements in the main program should be modified according to the requirements of the largest problem in the set being run. The parameters included in the following statements conform to the Input Parameter definitions above:

COMMON IROW(M), ITBROW(M), ICOL(N), ITBCOL(N), IVAR(N), ISVROW(M,NZR1VR), ISVRCL(NZR1VR), ICORR(NZR1VR), ISVN(NZR1VR), KSVN(NZR1VR+1)

DOUBLE PRECISION ATAB(M+1,N), UPBND(N+1), TPVAL(NZR1VR+1), BTMVL(NZR1VR+1), VAL(NZR1VR+q), TBSAV(M,N), SAVTAB(M+1, NZR1VR*(2*N-NZR1VR+1)/2), T(N)

5) Input Formats:

CARD TYPE	FORMAT	CONTENTS
1	(20I4)	ISIZE, NMRUNS

(Appears only once per program execution.)

2	(55H)	Problem title, identification

(Put 1 in card column 1 for printer page control.)

3	(20I4)	IOUT2, IOUT3, IPACK
4	(7E10.0)	SOLMIN, PCTTOL
5	(20I4)	M, N, NZR1VR
6	(7E10.0)	(UPBND(I), I=1, NM1)

(If NZR1VR exceeds 7, additional CARD TYPE 6's required.)

7	(20I4)	(IROW(I), I=2, M)

(If M exceeds 20, additional TYPE 7's required.)

8	If IPACK = 1	
	(7(I3, E7.0))	(ITEMP(K), VAL(K), K=1, 7)

(If more than 7 non-zero coefficients exist, additional TYPE 8's required. Last TYPE 8 card must end with zero field. If last card full, insert blank card.)

9	If IPACK = 0	
	(7E10.0)	(ATAB(I,J), J=1, N)

(One TYPE 9 per row including objective fct. If N exceeds 7, additional TYPE 9's per row required.)

---

   * Arrays are dimensioned in COMMON statements in this program, even though there are no subroutines. The user may want to change these to DIMENSION statements. The program ran successfully in its present form on a CDC 6400 and an IBM 360/65.

6) Output:

The main program prints out the problem title supplied, print control parameters, problem size and number of integer variables, bounds on the integer variables, codes for the constraint types, and the matrix format type code as part of the initial data.

The coefficient tableau is printed as raw data for checking purposes.

If IOUT2 = 1, the initial working tableau (as input to the first dual simplex solution) is printed in the Tucker form as used.

If IOUT3 = 1, the tableau from the continuous solution is printed.

The objective function value and values of each variable are printed for the continuous solution and for each feasible integer solution along with the present iteration number.

Error messages are printed for abnormal terminations suggesting the reason and giving the iteration number.

7) Summary of User Requirements:
   a) Determine values for each problem set for SOLMIN, PCTTOL, M, N, NZR1VR, UPBND, IROW, NMRUNS, NI, and NO.
   b) Calculate intermediate storage area for ISIZE.
   c) Define code for matrix type input for each problem.
   d) Specify print control criteria for IOUT2, IOUT3.
   e) Adjust COMMON size statements as needed to hold largest problem or satisfy machine limits.
   f) Adjust FORMAT statements as necessary.

D. Test Problem

The following problem is from Wagner (46, p. 497). The calculations were performed on an IBM 360/65 computer.

Maximize: $60X_1 + 60X_2 + 40X_3 + 10X_4 + 20X_5 + 10X_6 + 3X_7$

Subject to: $3X_1 + 5X_2 + 4X_3 + 1X_4 + 4X_5 + 3X_6 + 1X_7 \leq 10$

all $X_j \geq 0$

(X_1, X_2, X_3, X_4) integer valued

$(X_1, X_2, X_3, X_4) \leq 10$

Parameters: ISIZE = 50; NMRUNS = 1; IOUT2 = 1; IOUT3 = 1; IPACK = 0; SOLMIN = 0.0; PCTTOL = 0.0; M = 8; N = 2; NZR1VR = 4

Algorithm Answers: F = 190.000
$X_1 = 3$
$X_4 = 1$
All other $X_j = 0$

Number of Iterations: 9

Central Processor Time: 2 seconds

The listing and output for this problem are contained in the following section.

E. **Program Listings and Example Output**

```
      DOUBLE PRECISION DABS
      DOUBLE PRECISION ATAB(8,11), UPBND(12), TPVAL(10), BTMVL(10),
     1VAL(10), TBSAV(8,11), SAVTAB(9,65), T(11)
      DOUBLE PRECISION SOLMIN, PCTTOL, TLRNCE, YVECT, ATAB11, AMAX,
     1RTIO, ALFA, ARTIO, ADELT, ZOPT, ATAB12, X1, AMAX2, AMAX3, ALW,
     2AUP, RTIO2, DIFF1, DIFF2, DIFF, SVALW, ANDCT4
      COMMON IROW (8), ITBROW (8)
      COMMON ICOL (11), ITBCOL (11), IVAR (11)
      COMMON ISVROW (8, 10)
      COMMON ISVRCL (10), ICORR (10), ISVN (10)
      COMMON KSVN (11)
C     ARRAY ITEMP USED FOR PACKED FORMAT DATA INPUT ONLY
      DIMENSION ITEMP (7)
      X1 = 1.0
   10 FORMAT (1H0, (7D10.3))
C     UNPACKED FORMAT NO. 11
   11 FORMAT ( 7D10.0     )
   12 FORMAT ( 1X, 8D13.7)
   13 FORMAT (1H0,24HPRINT CONTROL PARAMETERS)
   14 FORMAT (1H0,30HUPPER BOUND ON VARIABLE 1 TO N)
   15 FORMAT(20I4)
C     PACKED FORMAT NO. 16
   16 FORMAT ( 7(I3,D7.0))
   17 FORMAT (1H0,18HMATRIX FORMAT CODE)
   18 FORMAT (4H0I =, I4, 6I10)
   19 FORMAT (27H0STRUCTURAL VARIABLES: X(I))
   20 FORMAT (44H0ROWS X COLUMNS AND NO. OF INTEGER VARIABLES//
     1 I4,2H X, I3, 24X, I6)
   21 FORMAT (30H0CONSTRAINT TYPES IN ROW ORDER)
   22 FORMAT (52H0INPUT TABLEAU ECHO, CONSTRAINT VALUE LEFT.  BY ROW.)
   23 FORMAT (1H0,10D13.3/(1H , 10D13.3))
   24 FORMAT (1H0,13HITERATION NO.,I6)
   25 FORMAT ( 1H0,8D13.5/(1H , 8D13.5))
   26 FORMAT ( 1H , I6, 7I13)
   27 FORMAT(1H+114X,I5)
   29 FORMAT (18H0TOLERANCE SET AT E15.7,14H  AT ITERATION,I6)
   30 FORMAT(21H PROBLEM NOT FEASIBLE)
   35 FORMAT (21H0OBJECTIVE FUNCTION =, F15.7,14H  AT ITERATION,I6)
   40 FORMAT (29H0CONTINUOUS SOLUTION COMPLETE)
   42 FORMAT (38H0FINAL TABLEAU FOR CONTINUOUS SOLUTION)
   45 FORMAT(40H0CONTINUOUS SOLUTION IS INTEGER SOLUTION)
   46 FORMAT (1H0,30HNO INTEGER VARIABLES REQUESTED)
   50 FORMAT (23H0OPTIMALITY ESTABLISHED)
   55 FORMAT(33H0PROBLEM TOO BIG FOR MACHINE SIZE)
   60 FORMAT(55H
   65 FORMAT (30H0END OF PROBLEM, ITERATION NO., I6)
   70 FORMAT(26H0BRANCH POINT INCREASED TOI4)
   75 FORMAT(26H0BRANCH POINT DECREASED TOI4)
   78 FORMAT (24H0INITIAL WORKING TABLEAU)
      NI = 5
      NO = 6
      READ (NI, 15) ISIZE, NMRUNS
C     INITIALIZATION
```

```
      68 CONTINUE
         INDCT7=1
         KSVN(1)=1
         INDCTR=1
         ICNTR=0
         IOUT1 = 0
         I1ROW=1000
         ADELT = 5.0E-7
C***
C        READ AND WRITE PROBLEM IDENTIFICATION: PUT 1 IN COL. 1
         READ(NI,60)
         WRITE(NO,60)
C***
C        IOUT2 = INITIAL WORKING TABLEAU
C        IOUT3=CONTINUOUS SOLUTION TABLEAU
         READ (NI, 15) IOUT2, IOUT3, IPACK
         WRITE(NO,13)
         WRITE(NO,15) IOUT2, IOUT3
C***
C        SOLMIN UPPER BOUND ON OBJ. FUNCTION FOR INTEGER SOLUTION
C     PCTTOL=INPUT TOLERANCE AS FRACTION OF OBJECTIVE FUNCT. FOR CONT. SOL
C        SET EACH ZERO FOR UNKNOWN PROBLEM.
         READ ( NI, 11) SOLMIN, PCTTOL
C***
C        INPUT PARAMETERS  M = TOTAL NO. OF ROWS, N = TOTAL NO. OF COLS.
C        NZR1VR = NO. OF INTEGER VARIABLES
         READ(NI,15)M,N,NZR1VR
         WRITE(NO,20) M, N, NZR1VR
      73 DO 72 I=1,N
      72 T(I)=0.
         NM1=N-1
      74 IF(SOLMIN)786,787,786
C***
C        INPUT UPPER BOUND ON OBJECTIVE FUNCTION
     786 TLRNCE=SOLMIN
         PCTTOL=-1.
         GO TO 90
     787 ITOL=1
         SOLMIN = 1E35
         IF(PCTTOL)90,788,90
     788 PCTTOL=.1
C***
C     .  INPUT UPPER BOUNDS ON VARIABLES (ZERO MEANS NO UPPER BOUND)
      90 READ (NI,11) (UPBND(I), I = 1, NM1)
         WRITE(NO,14)
         WRITE(NO,10) (UPBND(I), I = 1,NM1)
         IROW(1)=0
C**      CONSTRAINT TYPES: ( +1, = 0, ' -1 .
         READ(NI,15)(IROW(I),I=2,M)
         WRITE (NO, 21)
         WRITE (NO, 15) (IROW(I), I = 2, M)
C**      MATRIX FORMAT:   PACKED = 1,   UNPACKED = 0  .
         WRITE(NO,17)
         WRITE(NO,15) IPACK
         IF(IPACK)901,949,901
     901 DO 92 I=1,M
         DO 903 J = 1, N
     903 ATAB(I,J) = 0.0
```

```
      904 READ(NI,16)(ITEMP(K),VAL(K),K=1,7)
          DO 907 L=1,7
          IF(ITEMP(L))905,92,905
      905 K = ITEMP(L)
      909 ATAB(I,K)=VAL(L)
      907 CONTINUE
          GO TO 904
       92 CONTINUE
          GO TO 9510
C***
C         READ MATRIX ELEMENTS
      949 DO 952 I=1,M
          READ ( NI, 11) (ATAB(I,J) , J = 1, N)
      952 CONTINUE
          IF ( M .LT. 2) GO TO 450
C***
C         PRINT INPUT TABLEAU FOR ERROR CHECK
     9510 WRITE(NO,22)
          DO 80 I = 1, M
          WRITE (NO, 23) (ATAB(I,J), J = 1, N)
       80 CONTINUE
     9520 DO 954 I=2,M
          IF(IROW(I))953,9521,9521
     9521 DO 9523 J=2,N
     9523 ATAB(I,J)=-ATAB(I,J)
          GO TO 954
      953 ATAB(I,1)=-ATAB(I,1)
      954 CONTINUE
      450 CONTINUE
      955 DO 98 I=2,N
          IF(UPBND(I-1))96,96,98
       96 UPBND (I-1) = 1E3
       98 CONTINUE
C***
C         COMPUTE NO. OF Y VECTORS
      981 YVECT=UPBND(1)+1.
          IF ( NZR1VR .LT. 2) GO TO 322
          DO 982 I=2,NZR1VR
      982 YVECT=YVECT*(UPBND(I)+1.)
      322 CONTINUE
C***
C         SET SOLUTION VECTOR OF VARIABLES EQUAL TO ZERO
C         AND SAVE ORIGINAL UPPER BOUNDS
      985 DO 99 I=2,N
       99 IVAR(I-1)=0
C***
C         INITIALIZE ROW AND COLUMN IDENTIFIERS,+K=VARIABLE NO. K,
C         ZERO = ZERO SLACK, -K = POSITIVE SLACK
          IF ( M .LT. 2) GO TO 451
          DO 102 I=2,M
          IF(IROW(I))100,102,100
      100 IROW(I)=1-I
      102 CONTINUE
      451 CONTINUE
          ATAB11=ATAB(1,1)
          ICOL(1) = 0
          DO 103 J=2,N
          IF(ATAB(1,J))1022,1025,1025
```

```
      1022 DO 1023 I=1,M
           ATAB(I,1)=ATAB(I,1)+ATAB(I,J)*UPBND(J-1)
      1023 ATAB(I,J)=-ATAB(I,J)
           ICOL(J)=1000+J-1
           GO TO 103
      1025 ICOL(J)=J-1
       103 CONTINUE
C***
C     OUTPUT INITIAL TABLEAU
           IF(IOUT2)104,254,104
       104 WRITE(NO,78)
           WRITE(NO,26)(ICOL(J),J=1,N)
           DO 110 I=1,M
           WRITE(NO,25)(ATAB(I,J),J=1,N)
       110 WRITE(NO,27)IROW(I)
           GO TO 254
C***
C     START DUAL LP
C     CHOOSE PIVOT ROW, MAXIMUM POSITIVE VALUE IN CONSTANT COLUMN
       112 AMAX = 0.0
           IF ( M .LT. 2) GO TO 452
           DO 120 I=2,M
           IF(ATAB(I,1))120,120,115
       115 IF(ATAB(I,1)-AMAX)120,120,117
       117 AMAX=ATAB(I,1)
           IPVR=I
       120 CONTINUE
       452 CONTINUE
C***
C     IF NO POSITIVE VALUE, LP FINISHED (PRIMAL FEASIBLE)
           IF(AMAX)265,265,130
C     CHOOSE PIVOT COLUMN, ALGEBRAICALLY MAXIMUM RATIO A(1,J)/A(PIVOTROW
C     FOR A (PIVOTROW,J) NEGATIVE. IF NO NEGATIVE A(PIVOTROW,J) PROBLEM
C     INFEASIBLE
       130 AMAX = -1E35
           IF(N-2)143,132,132
       132 IPVC=0
           DO 140 J=2,N
           IF(ATAB(IPVR,J))133,140,140
       133 RTIO=ATAB(1,J)/ATAB(IPVR,J)
           IF(RTIO-AMAX)140,137,135
       135 AMAX=RTIO
       136 IPVC=J
           GO TO 140
       137 IF(ATAB(IPVR,J)-ATAB(IPVR,IPVC))136,140,140
       140 CONTINUE
           IF( IPVC)150,143,150
       143 GO TO (145,435,542,610,665),INDCTR
       145 WRITE(NO,30)
           GO TO 1001
C***
C     CARRY OUT PIVOT STEP
       150 ALFA=ATAB(IPVR,IPVC)
C**        UPDATE TABLEAU
           DO 180 J=1,N
           IF(ATAB(IPVR,J))152,180,152
       152 IF(J-IPVC)153,180,153
       153 ARTIO=ATAB(IPVR,J)/ALFA
```

```
          DO 175 I=1,M
          IF(ATAB(I,IPVC))157,175,157
      157 IF(I-IPVR)160,175,160
      160 ATAB(I,J)=ATAB(I,J)-ARTIO*ATAB(I,IPVC)
          IF(DABS(ATAB(I,J))-ADELT) 165, 165, 175
      165 ATAB(I,J) = 0.0
      175 CONTINUE
      180 CONTINUE
          DO 190 J=1,N
      190 ATAB(IPVR,J)=ATAB(IPVR,J)/ALFA
C***
C         EXCHANGE ROW AND COLUMN IDENTIFIERS
          ISV=IROW(IPVR)
          IROW(IPVR)=ICOL(IPVC)
          IF(ISV)197,195,197
C***
C         IF PIVOT ROW WAS ZERO SLACK, SET MODIFIED PIVOT COLUMN ZERO.
      195 DO 196 I=1,M
      196 ATAB(I,IPVC)=ATAB(I,N)
          ICOL(IPVC)=ICOL(N)
          N=N-1
          GO TO 200
      197 DO 198 I=1,M
      198 ATAB(I,IPVC)=-ATAB(I,IPVC)/ALFA
          ICOL(IPVC)=ISV
          ATAB(IPVR,IPVC)=1./ALFA
C***
C         COUNT PIVOTS
      200 ICNTR=ICNTR+1
          IF(IROW(IPVR)+1000)210,205,210
      205 DO 207 J=1,N
      207 ATAB(IPVR,J)=ATAB(M,J)
          IROW(IPVR)=IROW(M)
          M=M-1
      210 IF(IOUT1)240,2505,240
C***
C         OUTPUT CURRENT TABLEAU
      240 WRITE (NO,24) ICNTR
          WRITE(NO,26)(ICOL(J),J=1,N)
          DO 250 K=1,M
          WRITE(NO,25)(ATAB(K,L),L=1,N)
      250 WRITE(NO,27)IROW(K)
     2505 GO TO (254,251,252,253,2535), INDCTR
C***
C         IF SEEKING INTEGER SOLUTION, TEST OBJECTIVE FUNCTION AGAINST CURRE
      251 IF(ATAB(1,1)-TLRNCE)254,435,435
      252 IF(ATAB(1,1)-TLRNCE)254,542,542
      253 IF(ATAB(1,1)-TLRNCE)254,610,610
     2535 IF(ATAB(1,1)-TLRNCE)254,665,665
C***
C         IF CONSTANT COLUMN OF ZERO SLACK ROW IS NEG., REVERSE SIGNS OF ENT
      254 IF ( M .LT. 2) GO TO 453
          DO 260 K = 2, M
          IF(IROW(K))260,255,260
      255 IF(ATAB(K,1))256,260,260
      256 DO 258 L=1,N
      258 ATAB(K,L)=-ATAB(K,L)
      260 CONTINUE
```

```
      453 CONTINUE
C         GO TO NEXT PIVOT STEP
          GO TO 112
      265 CONTINUE
C***
C         IF ANY BASIS VARIABLE EXCEEDS ITS UPPER BOUND, COMPLEMENT IT, AND
C         PIVOT ON CORRESPONDING ROW
          IF ( M .LT. 2) GO TO 454
          DO 275 I=2,M
          IF(IROW(I))275,275,266
      266 J=IROW(I)
          IF(J-1000)268,268,267
      267 J=J-1000
      268 IF(UPBND(J)+ATAB(I,1))269,275,275
      269 IF(ADELT+UPBND(J)+ATAB(I,1))270,274,274
      270 ATAB(I,1)=-ATAB(I,1)-UPBND(J)
          DO 271 K=2,N
      271 ATAB(I,K)=-ATAB(I,K)
          IPVR=I
          IF(J-IROW(I))272,273,272
      272 IROW(I)=J
          GO TO 130
      273 IROW(I)= IROW(I)+1000
          GO TO 130
      274 ATAB(I,1)=-UPBND(J)
      275 CONTINUE
      454 CONTINUE
C***
C         TRUE END OF LINEAR PROGRAMMING
C         SET SOLUTION VECTOR VALUES FOR BASIC VARIABLES
          IF ( M .LT. 2) GO TO 455
          DO 280 I=2,M
          IF(IROW(I))280,280,277
      277 IF(IROW(I)-1000)279,279,278
      278 J=IROW(I)-1000
          T(J)=UPBND(J)+ATAB(I,1)
          GO TO 280
      279 J=IROW(I)
          T(J)=-ATAB(I,1)
      280 CONTINUE
      455 CONTINUE
C***
C         SET SOLUTION VECTOR VALUES FOR NON-BASIC VARIABLES IN COMPLEMENTED
          DO 285 I=2,N
          IF(ICOL(I))285,285,282
      282 IF(ICOL(I)-1000)284,284,283
      283 J=ICOL(I)-1000
          T(J)=UPBND(J)
          GO TO 285
      284 J=ICOL(I)
          T(J)=0.
      285 CONTINUE
          GO TO (286,437,548,615,670),INDCTR
C***
C         FIRST TIME,WRITE CONTINUOUS SOLUTION TABLEAU
C         IF REQUESTED
      286 WRITE(NO,40)
          IF(IOUT3)287,291,287
```

```
      287 WRITE(NO,42)
          WRITE(NO,26)(ICOL(J),J=1,N)
      288 DO 290 I=1,M
          WRITE(NO,25)(ATAB(I,J),J=1,N)
      290 WRITE(NO,27)IROW(I)
      291 ZOPT =DABS( ATAB(1,1))
          WRITE (NO, 35) ZOPT, ICNTR
          WRITE (NO, 19)
          WRITE (NO,18) (I, I = 1, NM1)
          WRITE (NO, 10) (T(I), I = 1, NM1)
C***
C         COMPUTE ABSOLUTE TOLERANCE
          ATAB12=ATAB(1,1)
          ATAB11 =DABS (ATAB11 - ATAB(1,1))
          IF(PCTTOL)294,293,292
      292 TLRNCE=PCTTOL*ATAB11+ATAB12
          GO TO 294
      293 TLRNCE = 1E35
      294 CONTINUE
C***
C         DETERMINE WHETHER CONTINUOUS SOLUTION IS MIXED INTEGER SOLUTION
          IF ( M .LT. 2) GO TO 456
      301 DO 310 I=2,M
          IF(IROW(I))310,310,302
      302 IF(IROW(I)-1000)303,303,304
      303 IF(IROW(I)-NZR1VR)305,305,310
      304 IF(IROW(I)-1000-NZR1VR)305,305,310
      305 AJO1 = ATAB(I,1)
          AJO2 = ADELT
          AJO3 = X1
          IF(AMOD(-AJO1,AJO3)-AJO2) 310,310,306
      306 IF(1.0-AMOD(-AJO1,AJO3)-AJO2) 310,310,295
      310 CONTINUE
      456 CONTINUE
          IF ( NZR1VR) 307, 308, 307
      307 WRITE (NO,45)
          GO TO 998
      308 WRITE (NO,46)
          GO TO 998
C***
C DETERMINE WHETHER PROBLEM FITS IN MEMORY , AND IF SO  WHETHER TO SAVE
C     ALL INTERMEDIATE TABLEAUS OR ONLY SOME
      295 IF(N-NZR1VR)297,297,298
      297 ISVLOC=(N*(N+1))/2
          GO TO 299
      298 ISVLOC=(NZR1VR*(2*N-NZR1VR+1))/2
      299 IF(ISIZE-ISVLOC)3001,3001,300
      300 I1ROW=0
          GO TO 315
     3001 NONBSC=0
          DO 3006 J=2,N
          IF(ICOL(J))3006,3006,3002
     3002 IF(ICOL(J)-1000)3003,3004,3004
     3003 IF(ICOL(J)-NZR1VR)3005,3005,3006
     3004 IF(ICOL(J)-1000-NZR1VR)3005,3005,3006
     3005 NONBSC=NONBSC+1
     3006 CONTINUE
          IF(N-NZR1VR)3007,3007,3008
```

```
      3007 ISVLOC=N+((N-NONBSC)*(N-NONBSC+1))/2
           GO TO 3009
      3008 ISVLOC=N+((NZR1VR-NONBSC)*(N-NONBSC+N-NZR1VR+1))/2
      3009 IF(ISIZE-ISVLOC)3010,3010,315
      3010 WRITE(NO,55)
           GO TO 998
       315 CONTINUE
C***
C     BEGIN INTEGER PROGRAMMING
       400 I1=1
       402 AMAX = -X1
           KSVN(I1+1)=KSVN(I1)
C***
C     CHOOSE NEXT INTEGER VARIABLE TO BE CONSTRAINED
C     TRY NONBASIC VARIABLES FIRST, CHOOSING ONE WITH LARGEST SHAD PRICE
           DO 4085 I=2,N
           IF(ICOL(I))4085,4085,405
       405 IF(ICOL(I)-1000)406,407,407
       406 IF(ICOL(I)-NZR1VR)408,408,4085
       407 IF(ICOL(I)-1000-NZR1VR)408,408,4085
       408 IF(AMAX-ATAB(1,I))4082,4085,4085
      4082 ISVI=I
           AMAX=ATAB(1,I)
      4085 CONTINUE
C***
C     IF NONE LEFT, TRY BASIC VARIABLES
           IF ( AMAX + X1) 4087, 420, 4087
C***
C     VARIABLE CHOSEN
      4087 IVAR(I1)=ICOL(ISVI)
           BTMVL(I1)=-1.
           ISVRCL(I1)=ISVI
           ICORR(I1)=0
           VAL (I1) = 0.0
C***
C     IF OBJECTIVE FUNCTION VALUE + SHADOW PRICE EXCEEDS TOLERANCE,
C     INDICATE UPWARD DIRECTION INFEASIBLE
           IF(ATAB(1,1)+ATAB(1,ISVI)-TLRNCE)410,409,409
       409 TPVAL(I1)=1000.
           IF(I1-1)4101,4101,4095
      4095 ISVN(I1)=0
           GO TO 4132
       410 TPVAL(I1)=1.
C***
           IF(I1-1)4100,4101,4100
C     SAVE ENTIRE TABLEAU OR ONLY COLUMN CORRESPONDING TO CURRENT
C     NONBASIC VARIABLE, DEPENDING ON SIZE OF PROB AND 2ND DIM OF SAVTAB
      4100 IF(I1-I1ROW)4132,4101,4101
      4101 L=KSVN(I1)
           DO 412 J=1,M
           ISVROW(J,I1)=IROW(J)
           DO 411 K=1,N
           I=L+K-1
           IF(J-1)4105,4105,411
      4105 SAVTAB(M+1,I)=ICOL(K)
       411 SAVTAB(J,I)=ATAB(J,K)
       412 CONTINUE
           ISVN(I1)=N
```

```
              KSVN(I1+1)=L+N
     4132 ICOL(ISVI)=ICOL(N)
              DO 4135 J=1,M
     4135 ATAB(J,ISVI)=ATAB(J,N)
              N=N-1
              GO TO 5000
C         CHOOSE NEXT INTEGER VARIABLE TO BE CONSTRAINED FROM
C         AMONG BASIC VARIABLES IN CURRENT TABLEAU
      420 CONTINUE
              IF(I1-I1ROW)4204,600,4205
     4204 I1ROW=I1
     4205 INDCT7=1
      421 AMAX = -X1
              IF ( M .LT. 2) GO TO 457
              DO 425 I2=2,M
              IF(IROW(I2))425,425,422
      422 IF(IROW(I2)-1000)423,424,424
      423 IF(IROW(I2)-NZR1VR)4241,4241,425
      424 IF(IROW(I2)-1000-NZR1VR)4241,4241,425
     4241 AMAX2 = 1.0E35
              AMAX3 = -1.0E35
              AJO = -ATAB(I2,1) + ADELT
              ALW = AINT(AJO)
              AUP=ALW+1.
              IF(N-1)426,426,4240
     4240 DO 4246 I3=2,N
              IF(ATAB(I2,I3))4244,4246,4242
     4242 RTIO=ATAB(1,I3)/ATAB(I2,I3)
              IF(RTIO-AMAX2)4243,4246,4246
     4243 AMAX2=RTIO
              GO TO 4246
     4244 RTIO2=ATAB(1,I3)/ATAB(I2,I3)
              IF(RTIO2-AMAX3)4246,4246,4245
     4245 AMAX3=RTIO2
     4246 CONTINUE
              IF ( AMAX3 + 1E34) 430, 430, 4247
     4247 IF (AMAX2 - 1E34) 4248, 429, 429
     4248 DIFF1 =DABS (AMAX2 * (ATAB(I2,1) + ALW))
              DIFF2 =DABS (AMAX3 * (ATAB(I2,1) + AUP))
              DIFF =DABS (DIFF1 - DIFF2)
              IF(DIFF-AMAX)425,425,4249
     4249 AMAX=DIFF
              SVALW=ALW
              ISVI2=I2
              IF(DIFF1-DIFF2)4251,4251,4252
     4251 ANDCT4=0.
              GO TO 425
     4252 ANDCT4=1.
      425 CONTINUE
      457 CONTINUE
              ALW=SVALW
              I2=ISVI2
              VAL(I1)=ALW+ANDCT4
              BTMVL(I1)=VAL(I1)-1.
     4255 TPVAL(I1)=VAL(I1)+1.
              GO TO 432
C***
C         IF NO. OF COLS=1 AND RIGHT HAND SIDE=0, DONT GO TO LP
```

```
      426 IF (DABS( ATAB(I2,1) + ALW) - ADELT) 427, 427, 5100
      427 BTMVL(I1)=-1.
          TPVAL(I1)=1000.
          VAL(I1)=ALW
          IVAR(I1)=IROW(I2)
          IROW(I2)=0
          GO TO 5000
C***
C         CONSTRAINING VARIABLE IN LOWER DIRECTION INFEASIBLE
      429 BTMVL(I1)=-1.
          IF (DABS ( ATAB(I2,1) + ALW) - ADELT ) 4295, 4295, 4296
     4295 ANDCT4=0.
          VAL(I1)=ALW+ANDCT4
          GO TO 4255
     4296 TPVAL(I1)=ALW+2.
          ANDCT4=1.
          GO TO 431
C***
C         CONSTRAINING VARIABLE IN UPPER DIRECTION INFEASIBLE
      430 TPVAL(I1)=1000.
          BTMVL(I1)=ALW-1.
          ANDCT4=0.
      431 VAL(I1)=ALW+ANDCT4
C***
C     SAVE ENTIRE TABLEAU
      432 JSVN=N
          L=KSVN(I1)
      438 DO 439 I3=1,M
          ISVROW(I3,I1)=IROW(I3)
          DO 439 I4=1,N
          I6=L+I4-1
          IF(I3-1)4385,4385,439
     4385 SAVTAB(M+1,I6)=ICOL(I4)
      439 SAVTAB(I3,I6)=ATAB(I3,I4)
          ISVN(I1)=N
          KSVN(I1+1)=L+N
          ATAB(I2,1)=ATAB(I2,1)+VAL(I1)
          ISVRCL(I1)=I2
          IVAR(I1)=IROW(I2)
          ICORR(I1)=1
          IROW(I2)=0
          IF (DABS ( ATAB(I2,1)) - ADELT) 433, 433, 434
      433 ATAB (I2,1) = 0.0
      434 INDCTR=2
C***
C     RETURN TO CARRY OUT LP
          IF(IOUT1)240,254,240
C     INFINITE RETURN
      435 IF(ANDCT4)4355,4352,4355
     4352 BTMVL(I1)=-1.
          GO TO 5120
     4355 TPVAL(I1)=1000.
          GO TO 5120
C***
C     FINITE RETURN
      437 GO TO 5000
C     TEST FOR ANY INTEGER VARIABLES LEFT TO BE CONSTRAINED
     5000 IF(I1-NZR1VR)5050,550,550
```

```
C       INCREMENT POINTER AND RETURN TO CONSTRAIN NEXT INTEGER VARIABLE
 5050   I1=I1+1
        IF(IOUT1)5051,402,5051
 5051   WRITE(NO,70)I1
        GO TO 402
C***
C       DECREMENT POINTER AND CONSTRAIN CURRENT VARIABLE TO
C       CURRENT VALUE + OR - 1
 5100   I1=I1-1
        IF(IOUT1)5110,5115,5110
 5110   WRITE(NO,75)I1
 5115   IF(I1)995,995,5120
 5120   IF(IVAR(I1)-1000)5151,5151,5152
 5151   K=IVAR(I1)
        GO TO 5153
 5152   K=IVAR(I1)-1000
 5153   I2=ISVRCL(I1)
 5155   IF(BTMVL(I1))516,517,517
  516   IF(TPVAL(I1)-UPBND(K))518,518,5100
  517   IF(TPVAL(I1)-UPBND(K))530,530,525
C***
C       TOP END FEASIBLE
  518   INDCT5=1
 5181   IF(ICORR(I1))5198,5182,5198
 5182   IF(I1-I1ROW)5183,5198,5198
 5183   INDCT8=1
        IF(I1-1)5185,5198,5185
 5185   INDCT5=4
        ISVI1=I1-1
        I1=1
        GO TO 5198
 5190   DO 5194 I3=1,ISVI1
        I4=ISVRCL(I3)
        ICOL(I4)=ICOL(N)
        DO 5193 J=1,M
        IF(VAL(I3)-1.)5193,5191,5192
 5191   ATAB(J,1)=ATAB(J,1)+ATAB(J,I4)
        GO TO 5196
 5192   ATAB(J,1)=ATAB(J,1)+VAL(I3)*ATAB(J,I4)
 5196   INDCT8=2
 5193   ATAB(J,I4)=ATAB(J,N)
        N=N-1
 5194   CONTINUE
 5195   I1=ISVI1+1
        INDCT5=1
        GO TO 521
C***
C       RETRIEVE SAVED TABLEAU
 5198   N=ISVN(I1)
        L=KSVN(I1)
        DO 5199 I3=1,M
        IROW(I3)=ISVROW(I3,I1)
        DO 5199 I4=1,N
        I6=L+I4-1
        IF(I3-1)5197,5197,5199
 5197   ICOL(I4)=SAVTAB(M+1,I6)
 5199   ATAB(I3,I4)=SAVTAB(I3,I6)
 5205   GO TO (521,526,531,5190),INDCT5
```

```
    521 VAL(I1)=TPVAL(I1)
        TPVAL(I1)=TPVAL(I1)+1.
        IF(ICORR(I1))541,522,541
    522 DO 523 I3=1,M
        ATAB(I3,1)=ATAB(I3,1)+(VAL(I1)*ATAB(I3,I2))
        IF (DABS ( ATAB(I3,1)) - ADELT) 5225, 5225, 523
   5225 ATAB(I3,1)=0.
    523 ATAB(I3,I2)=ATAB(I3,N)
        ICOL(I2)=ICOL(N)
        N=N-1
        IF(ATAB(1,1)-TLRNCE)5235,5100,5100
   5235 IF(I1-I1ROW)650,5415,5415
C***
C       BOTTOM END FEASIBLE
    525 INDCT5=2
        GO TO 5198
    526 VAL(I1)=BTMVL(I1)
        BTMVL(I1)=BTMVL(I1)-1.
        GO TO 541
C***
C       BOTH ENDS FEASIBLE
    530 INDCT5=3
        GO TO 5198
    531 AMAX2 = 1.0E35
        AMAX3 = -1.0E35
        DO 536 I3=2,N
        IF(ATAB(I2,I3))534,536,532
    532 RTIO=ATAB(1,I3)/ATAB(I2,I3)
        IF(RTIO-AMAX2)533,536,536
    533 AMAX2=RTIO
        GO TO 536
    534 RTIO2=ATAB(1,I3)/ATAB(I2,I3)
        IF(RTIO2-AMAX3)536,536,535
    535 AMAX3=RTIO2
    536 CONTINUE
        IF(AMAX2-1.E35)538,537,537
C***
C       BOTTOM END INFEASIBLE
    537 BTMVL(I1)=-1.
        GO TO 521
    538 IF(AMAX3+1.E35)539,539,540
C***
C       TOP END INFEASIBLE
    539 TPVAL(I1)=1000.
        GO TO 526
    540 DIFF1 =DABS ( AMAX2 * (ATAB(I2,1) + BTMVL (I1)))
        DIFF2 =DABS ( AMAX3 * (ATAB(I2,1) + TPVAL (I1)))
        IF(DIFF1-DIFF2)526,526,521
    541 ATAB(I2,1)=ATAB(I2,1)+VAL(I1)
        IROW(I2)=0
        IF (DABS ( ATAB(I2,1)) - ADELT) 5412, 5412, 5415
   5412 ATAB(I2,1)=0.
   5415 INDCTR=3
        IF(IOUT1)240,2505,240
C***
C       INFINITE RETURN
    542 GO TO (544,547,543),INDCT5
    543 IF(TPVAL(I1)-VAL(I1)-1.)545,544,545
```

```
      544 TPVAL(I1)=1000.
          GO TO 5120
      545 IF(VAL(I1)-BTMVL(I1)-1.)546,547,546
C***
      546 CONTINUE
      547 BTMVL(I1)=-1.
          GO TO 5120
C***
C         FINITE RETURN
      548 GO TO 5000
C         FEASIBLE INTEGER SOLUTION OBTAINED
      550 TLRNCE=ATAB(1,1)
          SOLMIN=1.
C***
C         WRITE CURRENT BEST MIXED INTEGER SOLUTION
          IF ( IOUT3) 552, 553, 552
      552 ZOPT =DABS( ATAB( 1,1))
          WRITE (NO, 35) ZOPT, ICNTR
      553 DO 560 I = 1, NZR1VR
          IF(IVAR(I))554,560,554
      554 IF(IVAR(I)-1000)555,555,557
      555 J=IVAR(I)
          T(J)=VAL(I)
          GO TO 560
      557 J=IVAR(I)-1000
          T(J)=UPBND(J)-VAL(I)
      560 CONTINUE
          WRITE (NO, 19)
      565 WRITE (NO, 18) (I, I = 1, NM1)
          WRITE (NO, 10) (T(I), I = 1, NM1)
          GO TO 5115
      600 GO TO (605,4205),INDCT7
      605 INDCTR=4
          IF(IOUT1)240,254,240
C***
C         INFINITE RETURN
      610 GO TO 5100
C***
C         FINITE RETURN
      615 INDCT7=2
          GO TO 402
C***
C         IF USING SECOND SOLUTION METHOD, SAVE TABLEAU MODIFIED
C         FOR NONZERO VALUE OF NONBASIC VARIABLE IN TBSAV
      650 DO 655 I=1,M
          ITBROW(I)=IROW(I)
          DO 655 J=1,N
      655 TBSAV(I,J)=ATAB(I,J)
          DO 660 J=1,N
      660 ITBCOL(J)=ICOL(J)
          JSVN=N
          INDCTR=5
          IF(IOUT1)240,254,240
C***
C         INFINITE RETURN
      665 GO TO (544,5120),INDCT8
C         FINITE RETURN
C***
```

```
C         IF USING SECOND SOLUTION METHOD, RETRIEVE MODIFIED TABLEAU FROM
C         TBSAV, AS THIS CORRESPONDS TO SAVED COLUMNS FOR I1 LESS THAN I1ROW
  670 N=JSVN
      DO 675 I=1,M
      IROW(I)=ITBROW(I)
      DO 675 J=1,N
  675 ATAB(I,J)=TBSAV(I,J)
      DO 680 J=1,N
  680 ICOL(J)=ITBCOL(J)
      GO TO 5000
C***
C         OUTPUT FINAL SOLUTION.
  995 IF(ITOL)996,9976,996
  996 IF(SOLMIN-1.E35)9976,997,997
  997 ITOL=ITOL+1
      TLRNCE=FLOAT(ITOL)*PCTTOL*ATAB11+ATAB12
      N=ISVN(1)
      DO 9972 I=1,M
      IROW(I)=ISVROW(I,1)
      DO 9972 J=1,N
 9972 ATAB(I,J)=SAVTAB(I,J)
      DO 9973 K=1,N
 9973 ICOL(K)=SAVTAB(M+1,K)
      GO TO 400
  998 CONTINUE
 9976 WRITE (NO, 50)
 1001 WRITE (NO, 65) ICNTR
  999 NMRUNS=NMRUNS-1
      IF(NMRUNS)68,1000,68
 1000 CALL EXIT
      END
```

WAGNER EX. 2, X1 THRU X4 INTEGER, 10 UPPER BND

PRINT CONTROL PARAMETERS
 1   1

ROWS X COLUMNS AND NO. OF INTEGER VARIABLES
 2 X 8   4

UPPER BOUND ON VARIABLE 1 TO N
 0.1000 02 0.1000 02 0.1000 02 0.1000 02   0.0   0.0   0.0   0.0

CONSTRAINT TYPES IN ROW ORDER
 -1

MATRIX FORMAT CODE
 0

INPUT TABLEAU ECHO, CONSTRAINT VALUE LEFT, BY ROW.
   0.0   -0.6000 02 -0.6000 02 -0.4000 02 -0.1000 02 -0.2000 02 -0.1000 02 -0.3000 01
   0.1000 02   0.3000 01   0.5000 01   0.4000 01   0.1000 01   0.4000 01   0.3000 01   0.1000 01

INITIAL WORKING TABLEAU
        1001   1002   1003   1004   1005   1006   1007
 -0.54700D 05 0.60000D 02 0.60000D 02 0.40000D 02 0.10000D 02 0.20000D 02 0.10000D 02 0.30000D 01
  0.81200D 04 -0.30000D 01 -0.50000D 01 -0.40000D 01 -0.10000D 01 -0.40000D 01 -0.30000D 01 -0.10000D 01

CONTINUOUS SOLUTION COMPLETE

FINAL TABLEAU FOR CONTINUOUS SOLUTION
    0   2   5   3   6   7   -1
 -0.20000D 03 0.40000D 02 0.10000D 02 0.60000D 02 0.40000D 02 0.50000D 02 0.17000D 02 0.20000D 02
 -0.66670D 01 -0.16667D 01 -0.33330D 00 -0.33330D 01 -0.13330D 01 -0.13330D 01 -0.10000D 01 -0.33330D 00

OBJECTIVE FUNCTION =   200.0000000   AT ITERATION   7

STRUCTURAL VARIABLES: X(I)

I =	1	2	3	4	5	6	7
	0.333D 01	0.0	0.0	0.0	0.0	0.0	0.0

OBJECTIVE FUNCTION =   185.0000000   AT ITERATION   8

STRUCTURAL VARIABLES: X(I)

I =	1	2	3	4	5	6	7
	0.300D 01	0.0	0.0	0.0	0.250D 00	0.0	0.0

OBJECTIVE FUNCTION =   190.0000000   AT ITERATION   9

STRUCTURAL VARIABLES: X(I)

I =	1	2	3	4	5	6	7
	0.300D 01	0.0	0.0	0.100D 01	0.0	0.0	0.0

OPTIMALITY ESTABLISHED

END OF PROBLEM, ITERATION NO.   9

IV. ZERO-ONE PROGRAMMING (ZONE ALGORITHM)*

A. Purpose

This procedure finds the minimum (or maximum) of a multivariable, linear function subject to linear constraints; the values of the variables are restricted to zero and unity:

Minimize (or Maximize) $F = C_1X_1 + C_2X_2 + \ldots + C_NX_N$

Subject to $A_{i1}X_1 + A_{i2}X_2 + \ldots + A_{iN}X_N \leq, =, \geq B_i$ $i = 1, 2, \ldots, M$

and $X_j = 0$ or 1

where the A_{ij}, B_i, and C_j are given constants and the X_j are the decision variables.

B. Method

The procedure is a modification of the original Balas algorithm (1) utilizing simplex iteration techniques and surrogate constraints, as developed by Glover (22), Geoffrion (21), and Peterson (36). The algorithm proceeds as follows:

1) The constraints are examined to find variables that are restrained to have values between zero and one.

2) These variables are inserted in the tableau as 1, and their complement 0, and the tableau solved by the simplex method for F.

3) Those combinations yielding minimum (or maximum) F are retained and steps 1) and 2) repeated.

4) Variables found to have a lower limit greater than zero are set to 1, those with an upper limit less than one are set to 0.

5) When bounds do not exist, then the existing constraints are combined to form a new, surrogate constraint in an attempt to establish a bound; steps 1) through 5) are repeated.

6) If no bound exists, the program advises of an unbounded problem.

A flow diagram illustrating the above procedure is shown in Figure 2.IV.

---

*The program in this section is used by permission of Dr. Thomas L. Yates, Oregon State University. This program is a modification of an earlier code developed by Dr. C. C. Peterson, Purdue University.

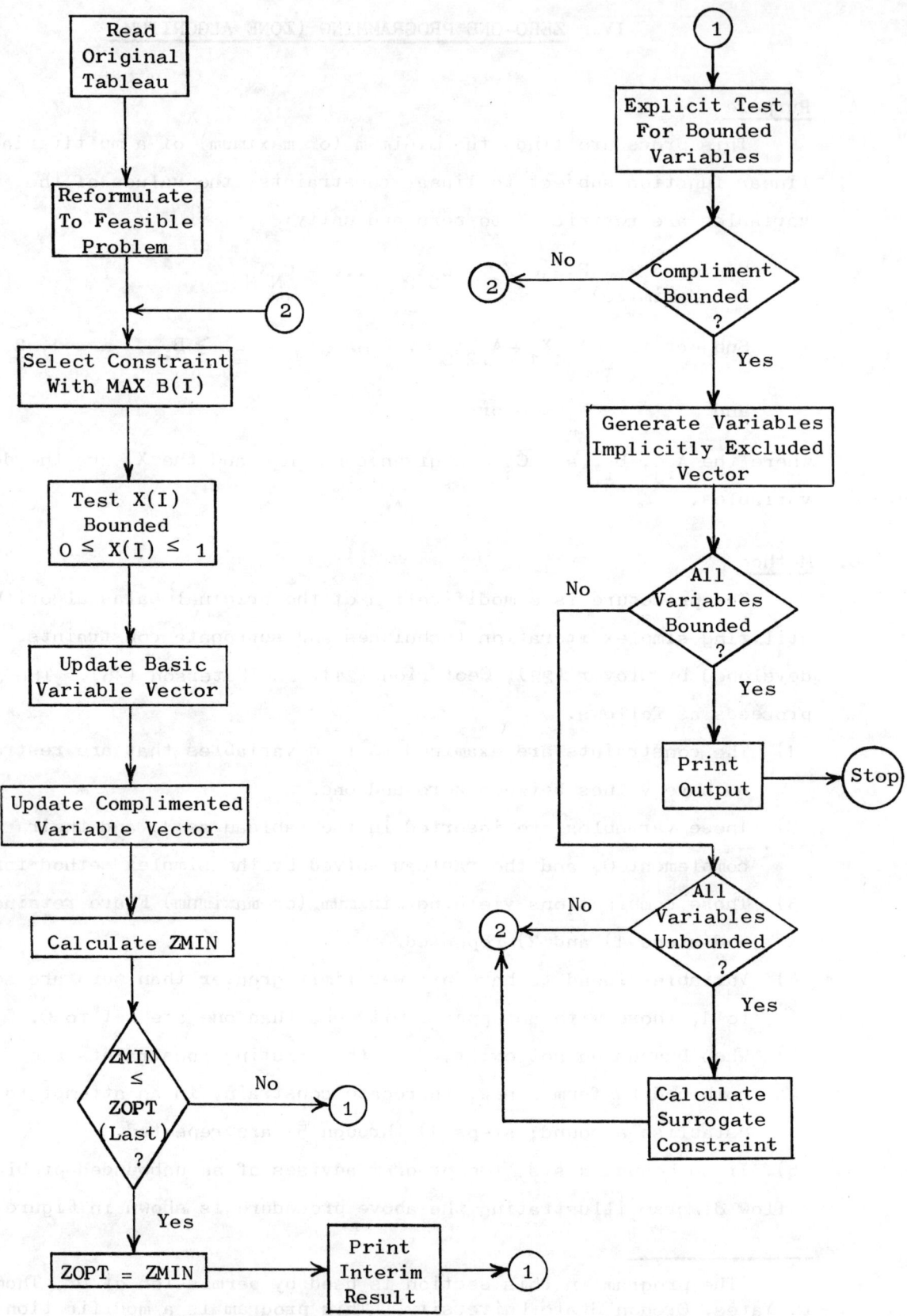

Figure 2.IV. Linear Integer (ZONE ALGORITHM) Logic Diagram

C. Program Description

1) Usage:

The program consists of a main program and two subroutines, INPUT and ZERO1. The main program establishes storage space for the original tableau and all working vectors, performs the algorithm, and outputs the results. Subroutine INPUT is used to read in the original problem. Subroutine ZERO1 is used to reformulate the problem into a feasible tableau for the algorithm.

2) Subroutines Required:

SUBROUTINE INPUT is called from the main program to read the problem size parameters, iteration limit, and type of problem. It also reads in all the coefficients for the objective function, the constraint rows, type constraints, and right hand sides for the constraints.

SUBROUTINE ZERO1 is called from the main program. The minimize (maximize) code is tested and if the problem is to be maximized the objective row is complemented since the basic algorithm is formulated for minimization only. Each constraint row specified as $\leq B_i$ is complemented. Each constraint row specified as $= B_i$ adds a complemented, identical row to the original tableau to force equality. The objective coefficients are tested and, if negative, are complemented as in the whole column of the array coefficients. The complemented array coefficients are then added to their respective constraint values. This completes reformulation of the tableau.

3) Description of Parameters:

NN	Number of variables
MM	Number of constraints
MSMAX	Maximum number of iterations, 100 suggested
MNMX	Type of optimization. Code is: 0 Minimize Objective Fct. 1 Maximize Objective Fct.
C	Objective coefficients vector, size = NN + 1 (Integer)
A	Matrix coefficients, size = MM X NN (Integer)
INEQ	Constraint type vector, size = MM. Code is: $i \geq B_i$; $2 \leq B_i$; $3 = B_i$
NI	Card reader unit number - Define in main program
NO	Printer unit number - Define in main program

B Constraint values vector, size = MM (Integer)

CV Complemented variables vector, size NN
(Used on error exit only)

4) **DIMENSION Requirements:**

DIMENSION statement in main program and COMMON statements in main and subroutines should be modified to meet at least minimum size requirements of the particular problem. The parameters included in the following statements conform to the definitions above:

DIMENSION JS(NN), Y(MM), JSUL(NN), JP(NN), NS(NN), ME(NN),
    V(NN), MF(NN), JR(MM,NN), NR(MM), SORT(NN), JOPT(NN)

COMMON NN, MM, MNMX, INEQ(MM), A(MM,NN), B(MM), C(NN + 1),
    INVC(NN), MSMAX

5) Input Formats:

CARD TYPE	FORMAT	CONTENTS
1	(4I5)	NN, MM, MSMAX, MNMX
2	(6X,I6,(10I6))	C(J), (A(I,J),I = 1, MM)

(If MM > 10 then additional cards per column required in FORMAT 10I6)
(NN*([MM/10] + 1) TYPE 2 cards required.)

3	(I1,1X, I6)	INEQ(I), B(I)

(MM TYPE 3 cards required.)

6) **Output:**

SUBROUTINE INPUT prints out the initial problem parameters, the tableau less constraint values in the order read in, and then the constraint types and values.

  Intermediate results are printed for each feasible, explicit set of variables tested.

  The final results show the number of iterations to complete the solution, basic variables set equal to 1, and the objective function value.

7) **Summary of User Requirements:**

a) Formulate the problem in a standard simplex (LP) format with constraint values on the right hand side of the tableau and without adding slack or artificial variables.

b) Determine values for NN, MM, MSMAX, MNMX, NI, and NO.

c) Define values for each INEQ per type of constraint (ordering not required).

d) Rewrite tableau in column order, objective coefficient first, less constraint column for easy entry to data cards.

e) Alter the FORMAT statements as required by the particular problem under study.

D. <u>Test Problem</u>

The following problem is from Wagner (46, p. 497). The calculations were performed on an IBM 360/65 computer.

Maximize: $60X_1 + 60X_2 + 40X_3 + 10X_4 + 20X_5 + 10X_6 + 3X_7$

Subject to: $3X_1 + 5X_2 + 4X_3 + 1X_4 + 4X_5 + 3X_6 + 1X_7 \leq 10$

and all $X_j = 0, 1$

Parameters: NN = 7; MM = 1; MSMAX = 20; MNMX = 1

Algorithm Answers: F = 133
$$X_1 = X_2 = X_4 = X_7 = 1$$
$$X_3 = X_5 = X_6 = 0$$

Number of Iterations: 7

Central Processor Time: 4 seconds

The listing and output for this problem are contained in the following section.

E. Program Listings and Example Output

```
          INTEGER BMOST
          INTEGER Y, A, B, C, V, SORT
          INTEGER TEMP,ZMIN,CJSUM,ZOPT,SUMCR,APOS,YASUM,CFSUM
          DIMENSION JS(78),Y(31),JSUL(78),JP(78),NS(78),ME(78),V(78),MF(78)
          DIMENSION JR(31,78),NR(31),SORT(78),JOPT(78)
          COMMON NN,MM,MNMX,INEQ(31),A(31,78),B(31),C(78),INVC(78),MSMAX
11        FORMAT (1H ,1X,2HJ=,39I3)
12        FORMAT (1H ,3HCV=,39I3)
13        FORMAT (1H1,31X,57HINTEGER PROGRAMMING PROBLEM, REFORMULATED BALAS
         1 ALGORITHM)
14        FORMAT (1H ,55X,4HN = , I10)
   20     FORMAT (1H0,6X,13HITERATION NO.,I7,10X,12HOBJ. VALUE =,I9)
   22     FORMAT (1H0,3X, 6HFINISH)
23        FORMAT (1H ,3X, 20HSTOPPED FOR MS LIMIT)
   25     FORMAT (1H0,5X,22HBASIC VARIABLES    X(J)//10X,6H    J =, (39I3))
   26     FORMAT (1H0,10X,5HZOPT=, I8)
          NO = 6
  993 DO 43 I = 1, 78
          JP(I) = 0
          JS(I)=0
          JSUL(I)=0
          SORT(I)=0.
          DO 42 J=1,31
42        JR(J,I)=0
43        INVC(I)=0
C***      READ PROBLEM, WRITE INITIAL TABLEAU
          CALL INPUT
          CALL ZERO1
          WRITE (NO,13)
          WRITE (NO,14) NN
          DO 49 I=1,MM
          NR(I)=0
          DO 45 J=1,NN
          IF ( A(I,J)) 45, 44, 44
44        NR(I)=NR(I)+1
          IABC=NR(I)
          IF ( A (I,J)) 4400, 4400, 4401
 4400 SORT (IABC) = 7777777
          GO TO 4402
4401  SORT(IABC)=C(J)/A(I,J)
4402  JR(I,IABC)=J
45        CONTINUE
          IF (NR(I)-1) 49,49,46
46        IAB=NR(I)-1
          DO 48 K=1,IAB
          KP1=K+1
          IABC=NR(I)
          DO 48 L=KP1,IABC
          IF (SORT(K)-SORT(L)) 48,48,47
47        TEMP=SORT(K)
          SORT(K)=SORT(L)
          SORT(L)=TEMP
          ITEMP=JR(I,K)
```

```
              JR(I,K)=JR(I,L)
              JR(I,L)=ITEMP
    48        CONTINUE
    49        CONTINUE
              ZMIN = 7777777
              MS=1
   300  BMOST = B(MM)
              IMAX=1
              DO 304 I=1,MM
              IF (BMOST-B(I)) 302,304,304
   302        BMOST=B(I)
              IMAX=I
   304        CONTINUE
   305        IABC=NR(IMAX)
              DO 325 I=1,IABC
              JS(I)=JR(IMAX,I)
              DO 319 K=1,MM
              Y(K)=-B(K)
              DO 318 L=1,I
              IAB=JS(L)
   318        Y(K)=Y(K)+A(K,IAB)
              IF (Y(K)) 325,319,319
   319        CONTINUE
              CJSUM=0.
              DO 323 J=1,I
              IAB=JS(J)
              CJSUM=CJSUM+C(IAB)
   323        CONTINUE
              JRME=I
              GO TO 72
   325        CONTINUE
              JRME=NR(IMAX)
              GO TO 100
    52        MS=MS+1
C***     ITERATION COUNTER, EXIT IF MSMAX EXCEEDED
              IF (MS-MSMAX) 54,54,53
    53   WRITE (NO,23)
              WRITE (NO,20) MS,ZMIN
              WRITE (NO,11) (JS(J),J=1,39)
              IF (JRME-39) 59,59,51
    51   WRITE (NO,11) (JS(J),J=40,JRME)
    59        K=1
              DO 354 J=1,NN
              IF (INVC(J)) 354,354,352
   352        JP(K)=J
              K=K+1
   354        CONTINUE
              WRITE (NO,12) (JP(K),K=1,39)
              IF (NN-39) 356,356,355
   355   WRITE (NO,12) (JP(K),K=40,NN)
   356   GO TO 993
    54        DO 57 IY=1,NN
              I=IY
              IF (JS(I)) 55,58,55
    55        IF (NN-1) 57,56,57
    56        I=I+1
              GO TO 58
    57        CONTINUE
    58        JRME=I-1
```

```
          CJSUM=0
          DO 64 I=1,JRME
          IF (JS(I)) 64,64,62
62        IAB=JS(I)
          CJSUM=CJSUM+C(IAB)
64        CONTINUE
          DO 70 I=1,MM
          Y(I)=-B(I)
          DO 68 J=1,JRME
          IF (JS(J)) 68,68,67
67        IAB=JS(J)
          Y(I)=Y(I)+A(I,IAB)
68        CONTINUE
70        CONTINUE
          DO 71 I=1,MM
          IF (Y(I)) 100,71,71
71        CONTINUE
          IF (ZMIN-CJSUM) 73,73,72
72        ZMIN=CJSUM
   73     WRITE (NO,20) MS,ZMIN
83        DO 76 J=1,JRME
76        JOPT(J)=JS(J)
          IAB=JRME+1
          DO 79 J=IAB,NN
79        JOPT(J)=0
74        IF (JSUL(JRME)) 75,92,92
75        DO 80 I=1,JRME
          J=JRME+1-I
          IF (JSUL(J)) 80,85,85
80        CONTINUE
C***      PROBLEM COMPLETE, WRITE OUTPUT
          WRITE (NO,22)
          K=1
          DO 84 J=1,NN
          IF (INVC(J)) 84,84,82
82        JP(K) =J
          K=K+1
84        CONTINUE
   88     WRITE (NO,20) MS, ZMIN
          DO 400 K=1,NN
400       JP(K)=0
          DO 405 J=1,NN
          IF (JOPT(J)) 401,405,402
401       IAB=-JOPT(J)
          JP(IAB)=JOPT(J)
          GO TO 405
402       IAB=JOPT(J)
          JP(IAB)=JOPT(J)
405       CONTINUE
          K=1
          DO 412 J=1,NN
          IF (JP(J)) 406,406,407
406       IF (INVC(J)) 412,412,409
407       IF (INVC(J)) 409,409,412
409       JOPT(K) =J
          K=K+1
412       CONTINUE
          IF (K-1-39) 413,413,414
```

```
413     IAD=K-1
        WRITE (NO,25) (JOPT(J),J=1,IAD)
        GO TO 415
414     WRITE (NO,25) (JOPT(J),J=1,39)
        IAB=K-1
        WRITE (NO,25) (JOPT(J),J=40,IAB)
415     ZOPT=0
        IAB=K-1
        DO 418 J=1,IAB
        IABC = IABS( JOPT(J))
        IF (INVC(IABC)) 417,417,416
416     ZOPT = ZOPT - C(IABC)
        GO TO 418
417     ZOPT = ZOPT + C(IABC)
418     CONTINUE
        IF (MNMX) 420,420,419
419     ZOPT=-ZOPT
420     WRITE (NO,26) ZOPT
        GO TO 993
85      JSUL(J)=-1
        JS(J)=-JS(J)
        IBC=J+1
        DO 90 K=IBC,JRME
        JS(K)=0
90      JSUL(K)=0
        GO TO 52
92      JS(JRME)=-JS(JRME)
        JSUL(JRME)=-1
        GO TO 52
100     DO 102 J=1,NN
102     NS(J)=1
        DO 106 J=1,JRME
        IF (JS(J)) 103,106,104
103     JNEG=-JS(J)
        NS(JNEG)=0
        GO TO 106
104     IABC=JS(J)
        NS(IABC)=0
106     CONTINUE
        DO 120 I=1,NN
        IF (NS(I)) 120,120,108
108     IF (C(I)-ZMIN+CJSUM) 120,110,110
110     NS(I)=0
120     CONTINUE
121     DO 122 I=1,NN
122     ME(I)=1
        DO 140 I=1,MM
        IF (Y(I)) 132,140,140
132     DO 135 J=1,NN
        IF (A(I,J)) 135,135,134
134     ME(J) =0
135     CONTINUE
140     CONTINUE
        DO 150 J=1,NN
        IF (NS(J)) 150,150,144
144     IF (ME(J)) 150,150,146
146     NS(J) =0
150     CONTINUE
```

```
          DO 160 J=1,NN
          IF (NS(J)) 160,160,167
160       CONTINUE
          GO TO 74
167       MARKF=0
          DO 183 I=1,MM
          IF (Y(I)) 169,183,183
169       SUMCR=0
          APOS=0
          IABC=NR(I)
          DO 180 N=1,IABC
          IAB=JR(I,N)
          IF (NS(IAB)) 180,180,171
171       IAB=JR(I,N)
          SUMCR=SUMCR+C(IAB)
          APOS=APOS+A(I,IAB)
          IF (SUMCR-ZMIN+CJSUM) 176,177,177
176       IF (APOS+Y(I)) 180,181,183
177       IF (APOS+Y(I)) 74,74,183
180       CONTINUE
          GO TO 74
181       IF (N-NR(I)) 173,182,182
173       IABC=NR(I)
          DO 174 K=N,IABC
          IAB=JR(I,K)
          IF (NS(IAB)) 174,174,183
174       CONTINUE
182       MARKF=1
183       CONTINUE
          IF (MARKF) 190,190,240
190       DO 210 J=1,NN
          IF (NS(J)) 210,210,192
192       YASUM=0
          DO 198 I=1,MM
          IF (Y(I)+A(I,J)) 193,193,198
193       YASUM=YASUM+Y(I)+A(I,J)
198       CONTINUE
          V(J)=YASUM
210       CONTINUE
          DO 230 IY=1,NN
          J=IY
          IF (NS(J)) 230,230,218
218       IF (NN-J) 235,235,219
219       L=J+1
          DO 226 K=L,NN
          IF (NS(K)) 226,226,220
220       IF (V(J)-V(K)) 230,226,226
226       CONTINUE
          DO 229 I=L,NN
          IF (NS(I)) 229,229,225
225       IF (V(J)-V(I)) 227,227,229
227       IF (C(J)-C(I)) 229,229,228
228       J=I
229       CONTINUE
          GO TO 235
230       CONTINUE
235       JS(JRME+1) = J
          GO TO 52
```

```
240         DO 241 J=1,NN
241         MF(J) =0
            DO 256 I=1,MM
            IF (Y(I)) 243,256,256
243         APOS=0
            DO 248 J=1,NN
            IF (NS(J)) 248,248,245
245         IF (A(I,J)) 248,248,246
246         APOS=APOS+A(I,J)
248         CONTINUE
            IF (APOS + Y(I)) 256,249,256
249         DO 255 K=1,NN
            IF (NS(K)) 255,255,250
250         IF (A(I,K)) 255,255,251
251         MF(K)=1
255         CONTINUE
256         CONTINUE
            CFSUM=0
            DO 262 I=1,NN
            IF (MF(I)) 262,262,260
260         CFSUM=CFSUM+C(I)
262         CONTINUE
            IF (CJSUM+CFSUM-ZMIN) 263,74,74
263         IF (MS-1) 264,264,265
264         J=1
            GO TO 267
265         J=JRME+1
267         DO 268 I=1,NN
            IF (MF(I)) 268,268,266
266         JS(J)=I
            J=J+1
268         CONTINUE
            GO TO 52
            END


            SUBROUTINE ZERO1
            COMMON NN,MM,MNMX,INEQ(31),A(31,78),B(31),C(78),INVC(78),MSMAX
            INTEGER A,B,C,SUMA
            MMM=MM
            IF (MNMX) 2,4,2
2           DO 3 J=1,NN
3           C(J) =-C(J)
4           DO 25 J=1,MM
            IF (INEQ(J)-1) 25,25,5
5           IF (INEQ(J) -2) 25,8,6
6           IF (INEQ(J) -3) 25,15,15
8           DO 10 K=1,NN
10          A(J,K) =-A(J,K)
            B(J)=-B(J)
            GO TO 25
15          MMM=MM+1
            DO 20 K=1,NN
20          A(MMM,K)=-A(J,K)
            B(MMM) =-B(J)
```

```
      25    CONTINUE
            MM=MMM
            DO 30 J=1,NN
            IF (C(J)) 27,30,30
      27    C(J)=-C(J)
            INVC(J) =1
            DO 28 I=1,MM
      28    A(I,J) =-A(I,J)
      30    CONTINUE
            DO 35 I=1,MM
            SUMA=0
            DO 34 J=1,NN
            IF (INVC(J)) 34,34,32
      32    SUMA=SUMA+A(I,J)
      34    CONTINUE
            B(I)=B(I)+SUMA
      35    CONTINUE
            RETURN
            END




            SUBROUTINE INPUT
            COMMON NN,MM,MNMX,INEQ(31),A(31,78),B(31),C(78),INVC(78),MSMAX
            INTEGER A,B,C
      10    FORMAT (1H1)
      15    FORMAT (4I5)
      16    FORMAT (6X,I6,(10I6))
      27    FORMAT (I1,1X,I6)
      33 FORMAT(7X,15HPROBLEM LIMITS:,5X,11HVARIABLES =,I5,
          1 13HCONSTRAINTS =,I5,12HITERATIONS =,I5,11HOPT. CODE =,I5)
      34 FORMAT (1H0,6X,10HCOLUMN NO.,6X,11HOBJ. COEFF.,18X,13HCONST. COEFF
          1.//)
      28    FORMAT (7X,I6,10X,I6,11X,(10I6))
      36    FORMAT (1H0,//7X,15HCONSTRAINT CODE,6X,5HVALUE)
      32    FORMAT (7X,I8,7X,I9)
            NI = 5
            NO = 6
C***    READ PROBLEM LIMITS:  NN = NO. OF VARIABLES, MM = NO. OF CONSTRAINTS,
C       MXMAX = ITERATION LIMIT, MNMX = TYPE OPTIMIZATION: CODE =
C                       0---MINIMIZE
C                       1---MAXIMIZE
            READ (NI,15) NN,MM,MSMAX,MNMX
            IF ( NN .EQ. 99) CALL EXIT
            DO 29 J=1,NN
C       READ COLUMNS OF TABLEAU , OBJECTIVE COEFFICIENT FIRST.
            READ(NI,16) C(J),(A(I,J),I=1,MM)
      29    CONTINUE
            DO 17 I=1,MM
C       READ CONSTRAINT TYPE: CODE =
C                       1   GREATER THAN OR =
C                       2   LESS THAN OR =
C                       3   = ONLY
C       THEN CONSTRAINT FOR ROW I
            READ (NI,27) INEQ(I),B(I)
```

```
17      CONTINUE
        WRITE (NO, 10)
        WRITE (NO,33) NN,MM,MSMAX,MNMX
        WRITE (NO, 34)
        DO 30 J=1,NN
30      WRITE (NO,28) J, C(J), (A(I,J), I = 1, MM)
        WRITE (NO, 36)
        DO 31 I=1,MM
31      WRITE (NO,32) INEQ(I),B(I)
        RETURN
        END
```

PROBLEM LIMITS: VARIABLES =   7 CONSTRAINTS = 1
ITERATIONS =   200 PT. CODE =   1

COLUMN NO.	OBJ. COEFF.	CONST. COEFF.
1	60	3
2	60	5
3	40	4
4	10	1
5	20	4
6	10	3
7	3	1

CONSTRAINT CODE	VALUE
2	10

INTEGER PROGRAMMING PROBLEM, REFORMULATED BALAS ALGORITHM
               N =   7

ITERATION NO.   1   OBJ. VALUE =   83

ITERATION NO.   4   OBJ. VALUE =   73

ITERATION NO.   11   OBJ. VALUE =   70

FINISH

ITERATION NO.   14   OBJ. VALUE =   70

BASIC VARIABLES   X(J)

    J =   1   2   4   7

ZOPT=   133

Chapter 3

QUADRATIC PROGRAMMING

Several variations of the linear programming problem were presented in Chapter 2. In all those variations, the objective function and all constraints were expressed as linear algebraic functions of the decision variables. This chapter is concerned with the problem of maximizing or minimizing a quadratic objective function subject to linear constraints.

The quadratic programming problem is different from the linear programming problem only in that the objective function may include quadratic and product expressions of the decision variables. That is to say, x_j^2 and $X_j X_k$ ($j \neq k$) terms are permitted in the objective function.

Several solution methods have been proposed for the special case of the quadratic programming problem in which the objective function is a concave function. Two such methods are presented in this chapter:

    I. Wolfe (WOLFE ALGORITHM)
    II. Quadratic Differential (QUADIF ALGORITHM)

I. WOLFE (WOLFE ALGORITHM)*

A. **Purpose**

This program finds the minimum of a multivariable quadratic objective function of the form:

$$\text{Minimize} \quad Z = \sum_{j=1}^{N} C_j X_j + \frac{1}{2} \sum_{j=1}^{N} \sum_{k=1}^{N} Q_{jk} X_j X_k$$

Subject to linear constraints:

$$\sum_{j=1}^{N} A_{ij} X_j \leq B_i \qquad i = 1, 2, \ldots, M$$

$$X_j \geq 0 \qquad j = 1, 2, \ldots, N$$

B. **Method**

The procedure, a modification of the simplex method, was developed by Wolfe (48) and programmed by Bates (2). The algorithm proceeds as follows:

1) A basic feasible solution to the constraint set is found such that the resulting values of the state variable are all non-negative.

2) The objective function is separated into its linear and quadratic terms:

$$Z = \sum_{j=1}^{N} C_j X_j + H[X_i, X_j]$$

3) The quadratic function, H, is decomposed into an N × N matrix by inspection or by partial derivatives:

$$H[X_i, X_j] = \underline{X}_j \, \underline{Q} \, \underline{X}_i$$

where $\underline{X}_j$ ans $\underline{X}_i$ are N element row and column vectors, respectively, and

$$Q_{ij} = \frac{1}{2} \frac{\partial H}{\partial (X_{ij})}$$

4) A simplex algorithm imbedded in the program then finds the minimum of the augmented tableau.

---

*The computer code included in this section was developed by H. T. Bates of Kansas State University. Used by permission.

C. **Program Description**
   1) <u>Usage</u>:

   The program consists only of a main program. All array and vector sizes are specified in DIMENSION statements, which may be altered for the particular problem. The program is designed for minimization. Maximization may be obtained simply by reversing the sign of each term in the objective function. The final value of the objective function printed by the program will have the wrong sign. The values of the decision variables, however, will have the correct sign. Several problems can be solved on one computer run.

   2) <u>Subroutines Required</u>:

   None

   3) <u>Description of Parameters</u>:

N	Number of structural variables
M	Number of constraints
ITMAX	Maximum number of iterations allowed
MTR	Code for type of trace desired: 0 - only final result printed 1 - prints each iteration tableau 2 - prints each iteration tableau and present pivot data
A	M X N array of constraint coefficients
B	M vector of constraint limits (must all be non-negative)
C	N X N array of quadratic coefficients from objective function
P	N vector of linear term coefficients from objective function
TITLE	80 character alphanumeric problem description
T	Simplex array of size (M+N) X (2M+3N+1)
NPRINT	Device code number for line printer
NREAD	Device code number for card reader

   4) <u>DIMENSION Requirements</u>:

   The DIMENSION statements in the main program must be large enough to meet the requirements of the largest problem set to be solved on a computer run. The parameters included in the following statements conform to the Input Parameter definitions, above:

   DIMENSION A(M, N), B(M), C(N, N), P(N), T(M+N, 2M+3N+1),
   COST(2M+3N+1), DIFF(2M+3N+1), TT(2M+3N+1), PRFIT(2M+3N+1),
   RATIO(M+N), IB(M+N), III(2M+3N), OPP(M+N), TITLE(20)

5) <u>Input Formats:</u>

CARD TYPE	FORMAT	CONTENT
1	(20A4)	Title of Problem
2	(4I10)	N,M,ITMAX,MTR
3	(8F 0.4)	((A(I,J), J=1, N), I=1, M)

(If N > 8, additional Type 3 Cards are required for each constraint. One set of Type 3 Cards required for each constraint.)

4	(8F10.4)	(B(J), J=1, M)

(If M > 8, additional Type 4 Cards are required.)

5	(8F10.4)	((C(I,J), J=1, N), I=1, N)

(If N > 8, additional Type 5 Cards are required in each set. One set of Type 5 Cards required for each variable in the problem.)

6	(8F10.4)	(P(I), I=1, N)

(If N > 8, additional Type 6 Cards are required.)

7		Blank card as program delimiter.

Problem Layout:

$$\text{Min } Z = P(J) * X(J) + X(I) * C(I,J) * X(J)$$

$$\text{Subject to: } A(I,J) * X(J) \leq B(I)$$

6) <u>Output:</u>

The program prints out the input data with appropriate labels, the value of the objective function, the non-zero values of real variables, Lagragians, slacks, and gradients plus the shadow prices (opportunity costs) for the real variables.

If MTR = 1 each tableau, T(I,J), is printed

If MTR = 2 each tableau, T(I,J), and the pivot trace information are printed.

7) <u>Summary of User Requirements:</u>

a) If problem is maximization, reverse all signs of terms in objective function (final objective function value will be printed with true sign reversed).

b) Convert problem to one having an initial basic feasible solution. This requires all constraints to be of the form

$$\sum_{i=1}^{N} A_{ij} X_j \leq B_i \qquad j = 1, \ldots, M$$

and all B_i non-negative.

c) Determine coefficients for C array from quadratic portion of objective function:

$$\begin{array}{c c} & \begin{array}{cccc} X_1 & X_2 & \cdots & X_N \end{array} \\ \begin{array}{c} X_1 \\ X_2 \\ \vdots \\ X_N \end{array} & \left[\begin{array}{cccc} C_{11} & \tfrac{1}{2}C_{12} & \cdots & \tfrac{1}{2}C_{1N} \\ \tfrac{1}{2}C_{21} & C_{22} & \cdots & \tfrac{1}{2}C_{2N} \\ \vdots & \vdots & & \vdots \\ \tfrac{1}{2}C_{N1} & \tfrac{1}{2}C_{N2} & \cdots & C_{NN} \end{array} \right] \end{array}$$

d) Specify values of N, M, ITMAX, and MTR.

e) Adjust DIMENSION statements to largest problem set to be solved on this computer run. Specify NREAD and NPRINT in main program.

f) Adjust FORMAT statements, if required.

D. Test Problem

The following problem was taken from Wilde and Beightler (47), p. 98, and was run on an IBM 360/65:

$$\text{Min } Z = X_1^2 - 2X_1X_2 - 2X_1X_3 + 5X_2^2 + 6X_2X_3 + 6X_3^2 - 6X_1 - 2X_2$$

Subject to:
$$16X_1 + 8X_2 + 4X_3 \leq 75$$
$$X_1 + X_2 + X_3 \leq 6$$

N = 3, M = 2, ITMAX = 10, MTR = 2

$$A = \begin{bmatrix} 16 & 8 & 4 \\ 1 & 1 & 1 \end{bmatrix}$$

$$B = \begin{bmatrix} 75 & 6 \end{bmatrix}$$

$$C = \begin{bmatrix} 1 & -1 & -1 \\ -1 & 5 & 3 \\ -1 & 3 & 6 \end{bmatrix}$$

$$P = \begin{bmatrix} -6 & -2 & 0 \end{bmatrix}$$

Solution: $Z_{min} = -13.25$
$X_1 = 4.125$
$X_2 = 0.875$
$X_3 = 0.25$

Number of iterations = 5

Central Processor Time = 5 seconds

The listing and output for this problem are presented in the following section.

E. Program Listings and Example Output

```
C       QUADRATIC PROGRAM BY THE WOLFE METHOD.
C       MINIMIZES OBJECTIVE FCT (Z)
C       Z = P(J) * X(J) + X(I) * C(I,J) * X(J)
C       THE CONSTRAINTS ARE
C       A(I,J) * X(J) .LE. B(I)
C       ALL X(J) .GT. 0
        DIMENSION A(20,20),B(20),C(20,20),P(20),T(40,101),COST(101)
        DIMENSION DIFF(101),TT(101),PRFIT(101),RATIO(101),IB(101),III(101)
        DIMENSION OPP(20),TITLE(20)
        DATA BLNK/4H    /
  100   FORMAT (1X,10F10.3)
  101   FORMAT (8F10.4)
  102   FORMAT (/18X,4HC(J),3X,9(1X,F8.3,1X)/(14X,10(F9.2,1X)))
  103   FORMAT (6X,4HC(I),3X,4HT(I),10(4X,I2,4X)/(17X,10(4X,I2,4X)))
  104   FORMAT(2X,F8.2,2X,F11.2,1X,9(1X,F8.2,1X)/(14X,10(1X,F8.2,1X)))
  105   FORMAT(//5X,4HZ(J),3X,F11.2,1X,9(F9.2,1X)/14X,10(F9.2,1X))
  106   FORMAT(//5X,3HC-Z,6X,10(F9.2,1X)/14X,10(F9.2,1X))
  107   FORMAT (//9X, 26HTHE MINIMUM VALUE OF Z IS=, E16.8)
  108   FORMAT (//9X, 37HTHE OPTIMUM POINTS ARE PRINTED BELOW  )
  109   FORMAT (//9X, 43HTHE REST OF THE VARIABLES ARE EQUAL TO ZERO)
  128   FORMAT(//,6X,4H  I ,8X,4HX(I)//)
  110   FORMAT (9X,I2,1X,4X,E16.8)
  111   FORMAT (1H1, 4X,5HTABLE,3X,I3)
  112   FORMAT (13X, 36HTHE OBJECTIVE FUNCTION IS UNBOUNDED.)
  113   FORMAT (9X, 4HC-Z(, I2, 1X, 2H)=, E16.8)
  114   FORMAT (//3X,5HIB(I))
  115   FORMAT (//9X,25HTHE OPPORTUNITY COSTS ARE/)
  116   FORMAT (//9X, 42HTHE REST OF THE OPPORTUNITY COSTS ARE ZERO)
  117   FORMAT (5X, 8HRATIO(I))
  118   FORMAT (//4X, 17HNUMBERS 1 THROUGH, I3, 32H ARE ORDINARY VARIABLES
       1, NUMBERS,I3,8H THROUGH,I3,17H ARE LAGRANGIANS.)
  119   FORMAT (8X,3HIPR,7X,3HIPC,7X,3HKCK)
  120   FORMAT (7X,4HTEST,7X,5HPIVOT,6X,9HDIFF(IPC))
  121   FORMAT(8I10)
  122   FORMAT (/4X,7HNUMBERS,I3,8H THROUGH,I3,20H ARE SLACKS, NUMBERS,
       1I3,8H THROUGH,I3,15H ARE GRADIENTS.)
  300   FORMAT (4I10)
  302   FORMAT (3X,I4,F20.6)
 1005   FORMAT(20A4)
 1006   FORMAT (1H1, 10X, 20A4)
        NREAD = 5
        NPRINT = 6
  999   READ ( NREAD,1005) (TITLE(L1), L1 = 1, 20)
        IF ( TITLE(1).EQ. BLNK) GO TO 30
        WRITE(NPRINT,1006)(TITLE(L1),L1=1,20)
        READ (NREAD,300) N,M,ITMAX,MTR
        DO 215 I = 1, M
        READ (NREAD, 101) (A(I,J), J = 1, N)
  215   CONTINUE
        READ (NREAD, 101) (B(J), J = 1,M)
        DO 220 I = 1, N
        READ (NREAD,101) (C(I,J), J = 1, N)
  220   CONTINUE
```

```
      READ(NREAD,101)(P(I),I=1,N)
      IF (MTR) 350, 350, 351
  351 WRITE (NPRINT,123)
  123 FORMAT (1H0,34HN,M,ITERATION LIMIT,TRACE IN ORDER)
      WRITE(NPRINT,300) N, M, ITMAX, MTR
      WRITE ( NPRINT,124)
  124 FORMAT (1H0,10H  A MATRIX)
      DO 225 I = 1, M
      WRITE (NPRINT, 100) (A(I,J), J = 1, N)
  225 CONTINUE
      WRITE (NPRINT, 125)
  125 FORMAT (1H0,24H  B VECTOR (CONSTRAINTS))
      WRITE(NPRINT,100)(B(J),J=1,M)
      WRITE (NPRINT, 126)
  126 FORMAT (1H0,22H  C MATRIX (OBJ. FCT.))
      DO 230 I = 1, N
      WRITE (NPRINT, 100) (C(I,J), J = 1, N)
  230 CONTINUE
      WRITE (NPRINT, 127)
  127 FORMAT (1H0,24H  P VECTOR (COST COEFF.))
      WRITE(NPRINT,100)(P(I),I=1,N)
  350 MP1=M+1
      MM1=M-1
      NP1=N+1
      NP2=N+2
      MN = M + N
      MNM1=MN-1
      MNP1=MN+1
      MNP2=MN+2
      NV=MN+N
      NVP1=NV+1
      NVP2=NV+2
      NY = NV + M
      NYP1=NY+1
      NYP2=NY+2
      NZ=NY+N
      NC=NZ+1
      NZP2=NZ+2
      DO 180 I=1,MN
      DO 180 J=1,NC
  180 T(I,J)=0.0
      DO 182 I=1,M
  182 T(I,1)=B(I)
      DO 183 I=MP1,MN
      J=I-M
  183 T(I,1)=-P(J)
      DO 184 I=1,M
      DO 184 J=1,N
      JP1=J+1
  184 T(I,JP1)=A(I,J)
      DO 185 I=1,N
      DO 185 J=1,N
      IPM=I+M
      JP1 = J + 1
  185 T(IPM,JP1)=2.*C(I,J)
      DO 186 I=MP1,MN
      IMM=I-M
      DO 186 J=NP2,MNP1
```

```
          JMN=J-N-1
  186 T(I,J) = A(JMN,IMM)
      DO 187 I=1,MN
      IJ = I + NVP1
      DO 187 J = NYP2, NC
      IF(J-IJ)187,179,187
  179 T(I,J)=1.
  187 CONTINUE
      DO 188 I = MP1, MN
      IJ = I - M + MNP1
      DO 188 J = MNP2, NC
      IF(J-IJ)188,178,188
  178 T(I,J)=-1.
  188 CONTINUE
      DO 208 I=1,MN
      OPP(I) = T(I,1)
  208 CONTINUE
      DO 340 J=1,NZ
  340 COST(J)=0.0
      DO 189 I=1,M
      J=NP1+I
  189 COST(J)=T(I,1)
      DO 190 J=NYP2,NC
  190 COST(J)=9999.
      NN=NZ-MN
      DO 25 KK=1,NZ
   25 III(KK)=KK
      DO 1 I = 1, MN
    1 IB(I)=NN+I
      K=0
C     ITERATION START
   19 K=K+1
      DO 2 J=1,NC
    2 PRFIT(J)=0.
      DO 3 J=1,NC
      SUM=0.
      DO 4 I = 1, MN
      JJ=IB(I)+1
    4 SUM=SUM+COST(JJ)*T(I,J)
      PRFIT(J)=SUM
    3 DIFF(J)=COST(J)-PRFIT(J)
      IF(MTR)555,666,555
  555 WRITE(NPRINT,111)K
C     PRINT TABLE IF DESIRED.
      WRITE(NPRINT,102)(COST(J),J=2,NC)
      WRITE(NPRINT,103)(III(KK),KK=1,NZ)
      DO 26 I=1,MN
      JJ=IB(I)+1
   26 WRITE(NPRINT,104)COST(JJ),(T(I,J),J=1,NC)
      WRITE(NPRINT,105)(PRFIT(J),J=1,NC)
      WRITE(NPRINT,106)(DIFF(J),J=2,NC)
C     FIND THE PIVOT ELEMENT --- T(IPR,IPC)
  666 IPC=0
      TEST=0.
C     FIND THE VARIABLE WITH THE LARGEST PROFIT
      DO 5 I =2,NC
  235 IF ( DIFF(I) - TEST) 6, 5, 5
    6 TEST=DIFF(I)
```

```
         IPC=I
    5 CONTINUE
      IF(IPC)99,99,7
    7 KCK=0
      DO 8 I=1,MN
      IF(T(I,IPC))32,32,20
   20 RATIO(I) = T(I,1) / T(I,IPC)
      GO TO 8
   32 KCK=KCK+1
      IF(KCK-MN)21,31,21
   21 RATIO(I)=1.E20
    8 CONTINUE
C     REMOVE LIMITING VARIABLE
      DO 9 I=1,MN
      IF(RATIO(I))9,10,10
   10 IF(RATIO(I).GT.10000.)RATIO(I)=10000.
      TEST=RATIO(I)
      IPR=I
      GO TO 11
    9 CONTINUE
   11 DO 12 I = 1, MN
      IF(TEST-RATIO(I))12,12,13
   13 TEST=RATIO(I)
      IPR=I
   12 CONTINUE
C     START PIVOTING AND INTRODUCING NEW VARIABLE INTO SOLUTION
      PIVOT=T(IPR,IPC)
      DO 15 J=1,NC
   15 T(IPR,J)=T(IPR,J)/PIVOT
      DO 171 I=1,MN
      IF(I-IPR)17,171,17
   17 DO 18 J=1,NC
   18 TT(J)=T(IPR,J)*T(I,IPC)/T(IPR,IPC)
      DO 172 J=1,NC
  172 T(I,J)=T(I,J)-TT(J)
  171 CONTINUE
      COST(IPR)=COST(IPC)
      IB(IPR)= IPC-1
C     TRACE OUTPUT IF DESIRED.
      IF(MTR-1)205,205,86
   86 WRITE(NPRINT,114)
      DO 87 I=1,MN
   87 WRITE (NPRINT,300) I, IB(I)
      WRITE(NPRINT,119)
      WRITE(NPRINT,121)IPR,IPC,KCK
      WRITE(NPRINT,120)
      WRITE(NPRINT,100)TEST,PIVOT,DIFF(IPC)
      WRITE(NPRINT,117)
      DO 88 I=1,MN
   88 WRITE (NPRINT,302) I, RATIO(I)
C     RECOMPUTE COSTS
  205 DO 176 J=1,NYP1
  176 COST(J)=0.
      DO 197 I=1,MN
      IF(IB(I)-MN)192,192,195
  192 JJ=IB(I)+MNP1
      GO TO 198
  195 IF(IB(I)-NY)196,196,197
```

```
  196 JJ=IB(I)-MNM1
      GO TO 198
  198 COST(JJ)=T(I,1)
  197 CONTINUE
      IF(K - ITMAX) 19, 830, 830
   99 SUM=0.
      DO 201 I=1,MN
      IN=IB(I)
      IF(IN.GT.N)GO TO 201
      SUM=SUM+P(IN)*T(I,1)
  201 CONTINUE
      FRST=SUM
      SUM=0.
      DO 202 I=1,MN
      DO 202 J=1,MN
      IN=IB(I)
      IF(IN.GT.N)GO TO 202
      JN=IB(J)
      IF(JN.GT.N)GO TO 202
      SUM = SUM + C(IN, JN) * T(I,1) * T(J,1)
  202 CONTINUE
      SCND=SUM
      OBJ=FRST+SCND
      WRITE(NPRINT,107)OBJ
      WRITE(NPRINT,118)N,NP1,MN
      WRITE(NPRINT,122)MNP1,NY,NYP1,NZ
      WRITE(NPRINT,108)
      WRITE (NPRINT,128)
      DO 28 I=1,MN
      IF ( T( I,1)) 27, 28, 27
   27 WRITE(NPRINT,110)IB(I),T(I,1)
   28 CONTINUE
      WRITE(NPRINT,109)
      WRITE(NPRINT,115)
      DO 53 I=1,MN
      IF (OPP(I)) 52, 53, 52
   52 WRITE (NPRINT,113) I, OPP(I)
   53 CONTINUE
      WRITE(NPRINT,116)
      GO TO 999
   31 WRITE(NPRINT,112)
      GO TO 999
  830 WRITE(NPRINT,831)
  831 FORMAT(1X,24HITERATION LIMIT EXCEEDED   )
   30 STOP
      END
```

EXAMPLE FROM BEIGHTLER, P. 98

N,M,ITERATION LIMIT,TRACE IN ORDER
     3   2   10   2

   A MATRIX
     16.000   8.000   4.000
      1.000   1.000   1.000

   B VECTOR (CONSTRAINTS)
     75.000   6.000

   C MATRIX (OBJ. FCT.)
      1.000   -1.000   -1.000
     -1.000   5.000   3.000
     -1.000   3.000   6.000

   P VECTOR (COST COEFF.)
     -6.000   -2.000   0.0

TABLE   1

C(I)	T(I)	C(J) 0.0	0.0	0.0	0.0	75.000	6.000	0.0	0.0	0.0	0.0
		1	2	3	4	5	6	7	8	9	10
		11	12	13							
0.0	75.00	0.0	16.00	8.00	4.00	0.0	0.0	0.0	0.0	0.0	0.0
0.0	0.0	0.0	0.0	0.0	0.0						
0.0	6.00	6.00	1.00	1.00	1.00	0.0	0.0	0.0	0.0	0.0	0.0
0.0	0.0	0.0	0.0								
9999.00	0.0	2.00	-2.00	-2.00	16.00	1.00	-1.00	0.0	0.0	0.0
0.0	0.0	0.0	0.0								
9999.00	0.0	1.00	10.00	6.00	8.00	1.00	0.0	-1.00	0.0	0.0
0.0	0.0	0.0	0.0								
9999.00	0.0	-2.00	6.00	12.00	4.00	1.00	0.0	0.0	-1.00	0.0
0.0	0.0	1.00	0.0								

Z(J) 79992.00 -19998.00 139986.00 159984.00 279972.00 29997.00 -9999.00 -9999.00 -9999.00 0.0
      0.0 9999.00 9999.00

C-Z   19998.00 ********* ********* ********* -29991.00 9999.00 9999.00 9999.00 9999.00 0.0
      0.0 ********* 0.0

IB(I)
1   9
2   10
3   11
4   12
5   4

IPR IPC   PIVOT   KCK   DIFF(IPC)
TEST 5   4.000 ***********   2
0.0

RATIO(I)
1   10000.000000
2 ****************
3   0.375000
4   0.250000
5   0.0

(Iteration 2 omitted.)

117

TABLE   3

C(I)	T(I)	C(J)									
		0.0	0.0	0.0	9999.00	0.0	65.400	5.400	0.600	0.0	0.0	0.500
		11	12	3	13	4	5	6	7	8	9	10
0.30	1	65.40	0.0	49.60	9999.00	84.00	0.0	4.80	1.60	0.0	-6.40	0.0
0.0		0.0	-1.60	0.0		6.40	0.0	0.30	0.10	0.0	-0.40	0.0
0.0		5.40	0.0	3.60		6.00	0.0	-0.30	-0.10	0.0	0.40	0.0
0.0		0.60	-0.10	0.0		-0.40	0.0	-0.40	0.20	0.0	1.20	0.0
0.5555		0.80	1.00	-2.60		-5.00	0.0	0.10	-0.05	-1.00	-0.05	0.0
		0.0	0.10	0.0		-0.40						
		0.0	0.0	3.20		-8.00						
65.40		0.30	-0.20	1.00		-1.20	1.00					
		0.0	0.0	0.20		0.50						
		0.0	0.05	0.0		0.05						
Z(J)		8038.43	0.0	32024.74	-79934.00	0.0	65.40	-3991.62	1997.01	-9999.00	11993.60	0.0
		0.0	-1997.01	9999.00	-11993.60							
C-Z		0.0	-32024.74	79934.00	0.0	0.0	3997.02	-1996.41	9999.00	-11993.60	0.30	
		11996.01	0.0	21992.60								

IB(I)	
1	9
2	10
3	1
4	2
5	4

IPR	IPC	PIVOT	KCK	DIFF(IPC)
4	3	3.200 -32024.74	1	-32024.742

TEST
0.250

RATIO(I)	
1	1.318548
2	1.500000
3	************
4	0.250000
5	1.499993

(Iteration 4 omitted.)

TABLE 5

C(J)	0.0	0.0	0.0	1.000	0.750	4.125	0.875	0.250	0.0
T(I)	9999.00	9999.00	9999.00	5	6	7	8	9	10
	11	12	13	4					
C(I) 0.0	1.00	0.0	0.0	-208.00	-15.00	11.50	2.50	1.00	0.0
0.0	0.0	-11.50	-2.50	-15.00	-1.12	0.81	0.19	0.13	0.0
0.0	0.75	0.0	0.0	11.50	0.81	-0.66	-0.09	-0.06	0.0
0.0	0.0	-0.81	-0.19	-1.00	0.09	1.00	0.16	0.06	0.0
0.0	4.12	1.00	0.0	2.50	0.19	-0.09	-0.16	-0.06	0.0
0.0	0.87	0.66	-0.09	1.00	0.12	-0.06	0.06	-0.13	0.0
0.0	0.0	0.09	1.00						
0.0	0.25	0.16	-0.06						
0.0	0.0	0.06	0.13						
Z(J) 0.0	0.0	0.0	0.0	0.0	0.0	0.0	0.0	0.0	0.0
C-Z 0.0	0.0	0.0	1.00	0.75	4.12	0.87	0.25	0.0	
9999.00	9999.00	9999.00							

THE MINIMUM VALUE OF Z IS= -0.1324999E 02

NUMBERS 1 THROUGH 3 ARE ORDINARY VARIABLES, NUMBERS 4 THROUGH 5 ARE LAGRANGIANS.
NUMBERS 6 THROUGH 10 ARE SLACKS, NUMBERS 11 THROUGH 13 ARE GRADIENTS.

THE OPTIMUM POINTS ARE PRINTED BELOW

I	X(I)
9	0.1000167sE 01
10	0.7500114sE 00
1	0.4124990sE 01
2	0.8749981sE 00
3	0.2499923E 00

THE REST OF THE VARIABLES ARE EQUAL TO ZERO

THE OPPORTUNITY COSTS ARE

C-Z(1)= 0.7500000E 02
C-Z(2)= 0.6000000E 01
C-Z(3)= 0.6000000E 01
C-Z(4)= 0.2000000E 01

THE REST OF THE OPPORTUNITY COSTS ARE ZERO

II. QUADRATIC DIFFERENTIAL (QUADIF ALGORITHM)*

A. Purpose

This program finds the minimum of a multivariable quadratic objective function of the form:

$$\text{Minimize} \quad Z = \sum_{j=1}^{N} C_j X_j + \frac{1}{2} \sum_{j=1}^{N} \sum_{k=1}^{N} X_j Q_{jk} X_k$$

Subject to linear constraints:

$$\sum_{j=1}^{N} A_{ij} X_j \leq B_i \qquad i = 1, 2, \ldots, M$$

$$X_j \geq 0 \qquad j = 1, 2, \ldots, N$$

B. Method

The procedure follows the development in Wilde and Beightler (47), Chapter 3. While that development is general and not restricted to quadratic objective functions, the particular method included here is restricted to quadratic objective functions and linear constraints. The algorithm proceeds as follows:

1) An initial basic feasible solution to the constraint set is found.

2) Calculate the partial derivatives, v_m, of the current set of state variables.

3) Denote v_i as the smallest decision derivative. Calculate and set v_h as the largest decision derivative having its decision variable, $d_h > 0$. If there are no negative v_m, set $v_i = 0$. If all positive v_m have $d_h = 0$, set $v_h = 0$.

4) Calculate $v = v_i + v_h$. If $v = 0$, the optimal solution has been found. If $v = -$, increase d_i, holding all other decision variables constant. If $v = +$, decrease d_h, holding all other decision variables constant. Continue this process until:

   a) Some state variable, S_p, becomes zero; then change state set (d_r replaces S_p) and return to Step 2), or

   b) v_r becomes zero; $r = h$ if v is positive, $r = i$ if v is negative; return to Step 2), or

---

*The computer code included in this section was developed by M. P. Terrell and V. Sumaria of Oklahoma State University. Used by permission.

c) d_h becomes zero; return to Step 2).

The above procedure is illustrated in Figure 3.II.

C. Program Description

1) Usage:

The program consists only of a main program. All array and vector sizes are specified in DIMENSION statements, which should be altered for each particular problem. The program is designed for minimization. Maximization may be obtained simply by reversing the sign of each term in the objective function. The final value of the objective function printed by the program will have the wrong sign. The values of the decision variables, however, will have the correct sign. Several problems can be solved on one computer run.

2) Subroutines Required:

None

3) Description of Parameters:

N	Number of structural variables
K	Number of constraints
ITMAX	Maximum number of iterations allowed
D	N element vector of linear objective function coefficients
C	N N array of quadratic coefficients from objective function
B	K element vector of constraint limits (must all be non-negative)
A	(K) (N+K) array of constraint coefficients
Z	80 character alphanumeric problem description
NI	Card reader unit number
NO	Printer unit number

4) DIMENSION Requirements:

The DIMENSION statements in the main program must be large enough to meet the requirements of the largest problem set to be solved on a computer run. The parameters included in the following statement conform to the Input Parameter definitions, above:

DIMENSION T(N,N), IS(K), A(K,N+K), B(K), C(N,N), ALPHA(K,N+K), S(K), V(N+K), B1(K), A1(K,N+K), Z(20), NBV(N), XBV(N), VNBV(N), Q(N+K, N+K), X(N+K), D(N)

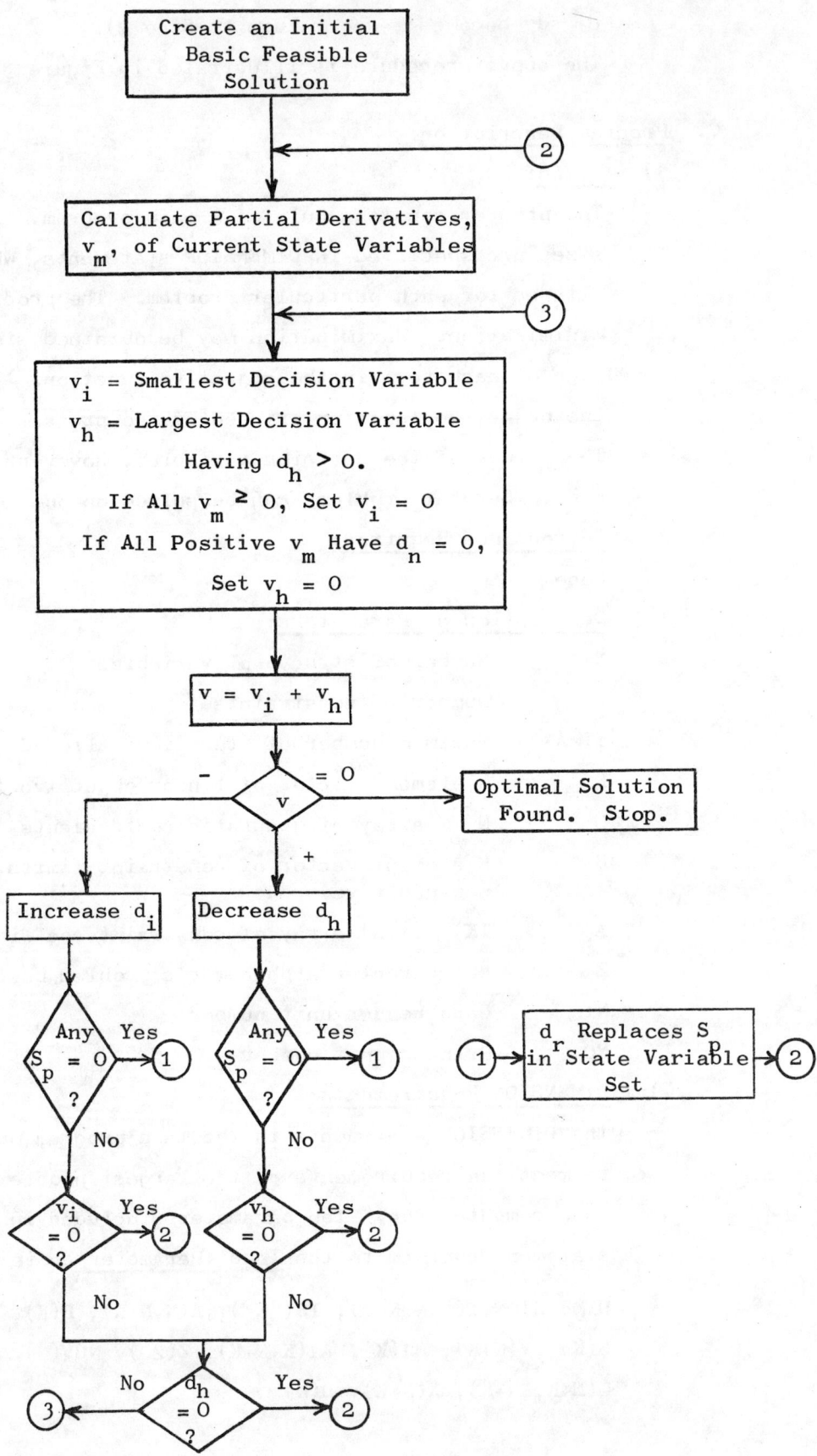

Figure 3.II. Quadratic Differential (QUADIF) Algorithm Logic Diagram

5) <u>Input Formats:</u>

CARD TYPE	FORMAT	CONTENTS	
1	(20A4)	Z alphanumeric problem description	
2	(3I5)	N, K, ITMAX	
3	(8F10.0)	D(I)	} I = 1,N
4	(8F10.0)	C(I,J), J=1, N	

(Card Types 3 and 4 entered as a set. One value on Card Type 3; N values on Card Type 4. If N > 8, additional Card Type 4 required per set. A total of N sets are required.)

5	(8F10.0)	B(J)	} J = 1,K
6	(8F10.0)	A(J,I), I=1, (N+K)	

(Card Types 5 and 6 entered as a set. One value on Card Type 5; (N+K) values on Card Type 6. If (N+K) > 8, additional Card Type 6 required per set. A total of K sets are required.)

Problem Layout:

$$\text{Min} \quad Z = D(J) * X(J) + X(I) * C(I,J) * X(J)$$

$$\text{Subject to:} \quad A(I,J) * X(J) \leq B(I)$$

Slack variables must be entered explicitly into matrix A.

6) <u>Output:</u>

The program prints out the input data with appropriate labels, and on each iteration prints out the current T matrix, the values of all variables, which are decision variables, which are state variables, and the complete tableau.

7) <u>Summary of User Requirements:</u>

a) If the problem is maximization, reverse all signs of terms in objective function (final objective function value will be printed with wrong sign).

b) Convert problem to one having an initial basic feasible solution. This requires all constraints to be of the form

$$\sum_{i=1}^{N} A_{ij} X_j \leq B_i \qquad j = 1, \ldots, M$$

and all B_i non-negative.

c) Determine coefficients for C array from quadratic portion of objective function:

$$\begin{array}{c} & \begin{array}{cccc} X_1 & X_2 & \cdots & X_N \end{array} \\ \begin{array}{c} X_1 \\ X_2 \\ \vdots \\ X_N \end{array} & \left[\begin{array}{cccc} C_{11} & \tfrac{1}{2}C_{12} & & \tfrac{1}{2}C_{1N} \\ \tfrac{1}{2}C_{21} & C_{22} & & \tfrac{1}{2}C_{2N} \\ \vdots & \vdots & & \vdots \\ \tfrac{1}{2}C_{N1} & \tfrac{1}{2}C_{N2} & \cdots & C_{NN} \end{array} \right] \end{array}$$

d) Specify values of N, K, and ITMAX.

e) Adjust DIMENSION statements to largest problem set to be solved on this computer run.

f) Specify NI and NO. Adjust FORMAT statements, if required.

D. <u>Test Problem</u>

The following problem is from Wilde and Beightler (47, p. 98). The calculations were performed on an IBM 360/65 computer.

Minimize: $Y = X_1^2 - 2X_1X_2 - 2X_1X_3 + 5X_2^2 + 6X_2X_3 + 6X_3^2 - 6X_1 - 2X_2$

Subject to: $16X_1 + 8X_2 + 4X_3 \leq 75$

$X_1 + X_2 + X_3 \leq 6$

all $X_j \geq 0$

Parameters: $N = 3$; $K = 2$; ITMAX $= 10$

Algorithm Answers: $Y = -13.250$

$X_1 = 4.117$

$X_2 = .873$

$X_3 = .250$

Number of Iterations: 9

Central Processor Time: 3 seconds

The listing and output for this problem are contained in the following section.

E. Program Listings and Example Output

```
C       ****************** DIFFERENTIAL      ALGORITHM    ********************
C       *                                                                    *
        DIMENSION T(10,10),IS(10)
        DIMENSION A(10,10),B(10),C(10,10),ALPHA(10,20),S(10),V(20)
        DIMENSION B1(10),A1(10,20),Z(20)
        DIMENSION NBV(10),XNBV(10),VNBV(10)
        DIMENSION Z2(115),Z3(10),SYMB(2),Z1(1)
        DIMENSION Q(20,20),X(20),D(10)
        DATA Z1,Z2,Z3,SYMB/1H|,115*1H= ,10*1HX,1H+,1HB/
        NI = 5
        NO = 6
C
C       ********************************************************
C
C          READ INPUT VALUES AND INITIALIZE VARIABLES
        IPROB=0
      1 READ(NI,75) Z
        READ(NI,10) N,K,ITMAX
        NK=N+K
        ITER=-1
        IPROB=IPROB+1
        NPT=12*(NK+1)
        Z3(10)=SYMB(2)
        Y=0.0
        DO 120 I=1,N
        READ(NI,20) D(I)
        READ(NI,20) (C(I,J),J=1,N)
        X(I)=0.0
        NBV(I)=I
        XNBV(I)=X(I)
    120 CONTINUE
        DO 130 J=1,K
        READ(NI,20) B(J)
        II=J+N
        X(II)=B(J)
        IS(J)=II
        S(J)=X(II)
        READ(NI,20) (A(J,I),I=1,NK)
    130 CONTINUE
        DO 110 I=1,NK
        DO 110 J=1,NK
        IF(I.GT.N.OR.J.GT.N)GO TO 111
        Q(I,J)=2.*C(I,J)
        GO TO 110
    111 Q(I,J)=0.0
    110 CONTINUE
        DO 140 I=1,K
        S(I)=B(I)
        DO 140 J=1,NK
        ALPHA(I,J)=A(I,J)
        V(J)=0.0
    140 CONTINUE
C          CALCULATE DERIVATIVES
```

```
    2 DO 150 I=1,N
      V(I)=D(I)
      DO 150 J=1,N
      V(I)=V(I)+2.*C(I,J)*X(J)
      T(I,J)=Q(I,J)
      VNBV(I)=V(I)
  150 CONTINUE
C
C            ******************************************************
C
C            WRITE INPUT VALUES
      WRITE(NO,44) IPROB,Z
      WRITE(NO,50) N,K
      WRITE(NO,55)
      WRITE(NO,76)
      DO 115 I=1,N
      WRITE(NO,60) (C(I,J),J=1,N)
  115 CONTINUE
      WRITE(NO,77)
      WRITE(NO,90) (D(I),I=1,N)
      WRITE(NO,65)
      WRITE(NO,78)
      WRITE(NO,97) (Z3(I),I,I=1,NK),Z3(10)
      DO 116 J=1,K
      WRITE(NO,60) (ALPHA(J,I),I=1,NK),B(J)
  116 CONTINUE
      WRITE(NO,96) (Z2(I2),I2=1,NPT)
      WRITE(NO,60) (V(I),I=1,NK),Y
      WRITE(NO,61)
      WRITE(NO,71)
      DO 112 I=1,NK
      WRITE(NO,60) (Q(I,J),J=1,NK)
  112 CONTINUE
C
C            ******************************************************
C
C            WRITE OUTPUT OF EVERY ITERATION
      WRITE(NO,45)
    3 ITER=ITER+1
      IF(ITER.GE.ITMAX) GO TO 900
      WRITE(NO,46) ITER
      WRITE(NO,1110)
      DO 1200 I=1,N
 1200 WRITE(NO,60) (T(I,J),J=1,N)
      Y=0.0
      DO 149 I=1,N
      Y=Y+D(I)*X(I)
      DO 149 J=1,N
      Y=Y+X(J)*C(I,J)*X(I)
  149 CONTINUE
      WRITE(NO,61)
      WRITE(NO,70) (I,X(I),I=1,NK)
      WRITE(NO,61)
      WRITE(NO,47) (I,NBV(I),I=1,N)
      WRITE(NO,61)
      WRITE(NO,48) (J,IS(J),J=1,K)
      WRITE(NO,61)
      WRITE(NO,91) ITER
```

```
      WRITE(NO,97) (Z3(I),I,I=1,NK),Z3(10)
      DO 155 J=1,K
      WRITE(NO,60) (ALPHA(J,I),I=1,NK),B(J)
  155 CONTINUE
      WRITE(NO,96) (Z2(I2),I2=1,NPT)
      WRITE(NO,60) (V(I),I=1,NK),Y
      WRITE(NO,61)
      WRITE(NO,61)
C
C     ********************************************************
C
C
C        SELECT MINIMUM AND MAXIMUM DERIVATIVES
      VI=0.0
      VH=0.0
      DO 160 I=1,N
      IF(VNBV(I).LT.VI)GO TO 151
      IF(VNBV(I).GT.VH.AND.XNBV(I).GT.0.0)GO TO 152
      GO TO 160
  151 VI=V(I)
      ILESS=I
      GO TO 160
  152 VH=V(I)
      IHIGH=I
  160 CONTINUE
  161 VT=VI+VH
C
C
C     ********************************************************
C
C        DETERMINE MINIMUM ALLOWABLE INCREMENT IN DECISION VARIABLE
      IF(VT)180,170,190
  170 IF(VI.EQ.0.0.AND.VH.EQ.0.0)GO TO 900
  180 DX1=.100E10
      IVAR=NBV(ILESS)
      DO 200 J=1,K
      IF(ALPHA(J,IVAR).EQ.0.0)GO TO 200
      DX2=B(J)/ALPHA(J,IVAR)
      IF(DX2.LT.DX1.AND.DX2.GT.0.0)GO TO 181
      GO TO 200
  181 DX1=DX2
      IK=J
      IKT=IS(J)
  200 CONTINUE
      DX2=-VNBV(ILESS)/T(ILESS,ILESS)
      IST=ILESS
      IF(DX2.LE.DX1.AND.DX2.GT.0.0)GO TO 300
      GO TO 350
  190 DX1=.100E10
      IVAR=NBV(IHIGH)
      DO 191 J=1,K
      IF(ALPHA(J,IVAR).EQ.0.0)GO TO 191
      DX2=B(J)/ALPHA(J,IVAR)
      IF(DX2.LT.0.0.AND.ABS(DX2).LT.ABS(DX1))GO TO 192
      GO TO 191
  192 DX1=DX2
      IK=J
      IKT=IS(J)
  191 CONTINUE
```

```
              DX2=-VNBV(IHIGH)/T(IHIGH,IHIGH)
              IST=IHIGH
              IF(DX2.LT.0.0.AND.ABS(DX2).LE.ABS(DX1))GO TO 320
              GO TO 350
C
C
C             ***********************************************************
C
C             INCREMENT DECISION VARIABLE WHEN IST TH DERIVATIVE
C             IS DRIVEN TO ZERO
          320 DX1=DX2
              VNBV(IST)=0.0
              XNBV(IHIGH)=XNBV(IHIGH)+DX1
              IV=IHIGH
              GO TO 330
          300 DX1=DX2
              VNBV(IST)=0.0
              XNBV(ILESS)=XNBV(ILESS)+DX1
              IV=ILESS
C             CALCULATE NEW TABLEAU
          330 DO 340 J=1,K
              II=IS(J)
              B(J)=B(J)-ALPHA(J,IVAR)*DX1
              IF(ALPHA(J,II).EQ.0.0)GO TO 340
              S(J)=B(J)/ALPHA(J,II)
          340 CONTINUE
              DO 341 J=1,N
              ID=NBV(J)
              IF(J.EQ.IST)GO TO 342
              VNBV(J)=VNBV(J)+T(J,IV)*DX1
          342 X(ID)=XNBV(J)
              V(ID)=VNBV(J)
          341 CONTINUE
              DO 343 I=1,K
              ID=IS(I)
              X(ID)=S(I)
              V(ID)=0.0
          343 CONTINUE
              GO TO 3
C
C             ***********************************************************
C
C
C             CHANGE IN STATE SET AND CALCULATE NEW T MATRIX
C             IK TH CONSTRAINT IS ZERO DUE TO IST TH VARIABLE
          350 CONTINUE
              WRITE(NO,62) IK,IST
C             CALCULATE NEW TABLEAU
              B1(IK)=B(IK)/ALPHA(IK,IST)
              DO 351 J=1,NK
              A1(IK,J)=ALPHA(IK,J)/ALPHA(IK,IST)
          351 CONTINUE
              DO 352 J=1,K
              IF(J.EQ.IK)GO TO 352
              DO 353 I=1,NK
              FACTOR=-ALPHA(J,IST)
              A1(J,I)=ALPHA(J,I)+A1(IK,I)*FACTOR
              B1(J)=B(J)+B1(IK)*FACTOR
          353 CONTINUE
```

```
  352 CONTINUE
      DO 355 J=1,K
      B(J)=B1(J)
      S(J)=B(J)
      DO 355 I=1,NK
      ALPHA(J,I)=A1(J,I)
  355 CONTINUE
C     ***********************************************
C         CALCULATE NEW DERIVATIVES
      DO 550 I=1,N
      VNBV(I)=VNBV(I)+T(I,IST)*(S(IK)/ALPHA(IK,IVAR))
  550 CONTINUE
      VOLD=VNBV(IST)
      NBV(IST)=IKT
      IS(IK)=IVAR
      VNBV(IST)=0.0
      DO 560 J=1,N
      ID=NBV(J)
      VFAC=-ALPHA(IK ,ID)/ALPHA(IK,IVAR)
      VNBV(J)=VNBV(J)+VOLD*VFAC
  560 CONTINUE
C     ***********************************************
C         CALCULATE NEW T MATRIX
      DO 520 I=1,N
      ID=NBV(I)
      DO 520 J=1,N
      JD=NBV(J)
       T(I,J)=Q(ID,JD)
      DO 520 M=1,K
      MK=IS(M)
      T(I,J)=T(I,J)-Q(ID,MK)*ALPHA(M,JD)-Q(JD,MK)*ALPHA(M,ID)
      DO 520 MM=1,K
      JK=IS(MM)
      T(I,J)=T(I,J)+(ALPHA(MM,ID)*Q(MK,JK)*ALPHA(MM,JD))
  520 CONTINUE
      XNBV(IST)=0.0
      S(IK)=X(IVAR)+B(IK)
      B(IK)=S(IK)
      DO 570 J=1,N
      ID=NBV(J)
      X(ID)=XNBV(J)
      V(ID)=VNBV(J)
  570 CONTINUE
      DO 580 I=1,K
      ID=IS(I)
      X(ID)=S(I)
      V(ID)=0.0
  580 CONTINUE
      GO TO 3
C     ***********************************************
C         WRITE OUTPUT OF FINAL ITERATON
  900 WRITE(NO,61)
      WRITE(NO,81)
      WRITE(NO,70) (I,X(I),I=1,NK)
      WRITE(NO,80)
      WRITE(NO,97) (Z3(I),I,I=1,NK),Z3(10)
      DO 400 J=1,K
      WRITE(NO,60) (ALPHA(J,I),I=1,NK),B(J)
```

```fortran
  400 CONTINUE
      WRITE(NO,96) (Z2(I2),I2=1,NPT)
      WRITE(NO,60) (V(I),I=1,NK),Y
      WRITE(NO,45)
      GO TO 1
C
C
C           ***********************************************************
   10 FORMAT(3I5)
   20 FORMAT(8F10.0)
   44 FORMAT('1'//,15X,'INPUT',5X,'PROBLEM NO.',I3,5X,20A4)
   45 FORMAT('1')
   46 FORMAT(//10X,'OUTPUT OF ITERATION',I3//)
   47 FORMAT(5X,'DECISION VARIABLE(',I2,')= X(',I2,')')
   48 FORMAT(5X,'STATE VARIABLE(',I2,')= X(',I2,')')
   50 FORMAT(//5X,'NO. OF VARIABLES=',I2,5X,'NO. OF CONSTRAINTS=',I2/)
   55 FORMAT(5X,'THE OBJECTIVE FUNCTION IS F=XTCX+DTX'/10X,'WHERE XT IS
     1X TRANSPOSE'/16X,'DT IS D TRANSPOSE'/16X,' C IS COEFFICIENT MATRIX
     2'/16X,' D IS COEFFICIENT VECTOR'/)
   60 FORMAT(6X,10G12.5)
   61 FORMAT(/)
   62 FORMAT(//5X,I2,'TH CONSTRAINT IS ZERO DUE TO ',I2,'TH VARIABLE'/)
   65 FORMAT(//5X,' THE CONSTRAINT EQUATIONS ARE AX=B'/)
   70 FORMAT(5X,'X(',I2,')=',F10.3)
   71 FORMAT(/25X,'MATRIX Q'/)
   75 FORMAT(20A4)
   76 FORMAT(15X,'MATRIX C'/)
   77 FORMAT(//10X,'VECTOR D'/)
   78 FORMAT(30X,'INITIAL TABLEAU'/)
   80 FORMAT(30X,'FINAL TABLEAU'/)
   81 FORMAT(//15X,'FINAL OUTPUT'/)
   90 FORMAT(10X,G10.3)
   91 FORMAT(30X,'TABLEAU NO.',I3/)
   96 FORMAT(6X,115A1)
   97 FORMAT(    10(10X,A1,I1),14X,A1)
 1110 FORMAT(15X,'MATRIX T'/)
  999 STOP
      END
```

INPUT   PROBLEM NO. 1   BEIGHTLER EXAMPLE

NO. OF VARIABLES= 3   NO. OF CONSTRAINTS= 2

THE OBJECTIVE FUNCTION IS F=XTCX+DTX
   WHERE XT IS X TRANSPOSE
         DT IS D TRANSPOSE
          C IS COEFFICIENT MATRIX
          D IS COEFFICIENT VECTOR

MATRIX C

1.0000	-1.0000	-1.0000
-1.0000	5.0000	3.0000
-1.0000	3.0000	6.0000

VECTOR D

-6.00
-2.00
 0.0

THE CONSTRAINT EQUATIONS ARE AX=B

INITIAL TABLEAU

X1	X2	X3	X4	X5	B
16.000	8.0000	4.0000	1.0000	0.0	75.000
1.0000	1.0000	1.0000	0.0	1.0000	6.0000
=========	=========	========	========	========	========
-6.0000	-2.0000	0.0	0.0	0.0	0.0

MATRIX Q

2.0000	-2.0000	-2.0000	0.0	0.0
-2.0000	10.000	6.0000	0.0	0.0
-2.0000	6.0000	12.000	0.0	0.0
0.0	0.0	0.0	0.0	0.0
0.0	0.0	0.0	0.0	0.0

OUTPUT OF ITERATION 0

MATRIX T

```
  2.0000       -2.0000       -2.0000
 -2.0000       10.000         6.0000
 -2.0000        6.0000       12.000
```

```
X( 1)=      0.0
X( 2)=      0.0
X( 3)=      0.0
X( 4)=     75.000
X( 5)=      6.000
```

DECISION VARIABLE(1)= X(1)
DECISION VARIABLE(2)= X(2)
DECISION VARIABLE(3)= X(3)

STATE VARIABLE(1)= X(4)
STATE VARIABLE(2)= X(5)

TABLEAU NO. 0

X1	X2	X3	X4	X5	B
16.000	8.0000	4.0000	1.0000	0.0	75.000
1.0000	1.0000	1.0000	0.0	1.0000	6.0000
===
-6.0000	-2.0000	0.0	0.0	0.0	0.0

(Iterations 1 through 4 omitted.)

OUTPUT OF ITERATION 5

MATRIX Γ

```
  2.0000      -2.0000     -2.0000
 -2.0000      10.000       6.0000
 -2.0000       6.0000     12.000
```

X(1) =   4.033
X(2) =   0.800
X(3) =   0.233
X(4) =   3.133
X(5) =   0.933

DECISION VARIABLE(1) = X(1)
DECISION VARIABLE(2) = X(2)
DECISION VARIABLE(3) = X(3)

STATE VARIABLE(1) = X(4)
STATE VARIABLE(2) = X(5)

TABLEAU NO. 5

X1	X2	X3	X4	X5	B
16.000	8.0000	4.0000	1.0000	0.0	3.1333
1.0000	1.0000	1.0000	0.0	1.0000	0.93333
=========	=========	=========	=========	=========	=========
0.0	-0.66667	-0.46667	0.0	0.0	-13.221

(Iterations 6 through 8 omitted.)

OUTPUT OF ITERATION 9

MATRIX T

```
   2.0000      -2.0000      -2.0000
  -2.0000      10.000        6.0000
  -2.0000       6.000       12.000
```

X(1)=   4.117
X(2)=   0.867
X(3)=   0.250
X(4)=   1.200
X(5)=   0.767

DECISION VARIABLE(1)= X(1)
DECISION VARIABLE(2)= X(2)
DECISION VARIABLE(3)= X(3)

STATE VARIABLE(1)= X(4)
STATE VARIABLE(2)= X(5)

TABLEAU NO. 9

X1	X2	X3	X4	X5	B
16.000	8.0000	4.0000	1.0000	0.0	1.2000
1.0000	1.0000	1.0000	0.0	1.0000	0.76667
===========	===========	===========	===========	===========	===========
0.0	-0.66666E-01	-0.33333E-01	0.0	0.0	-13.250

FINAL OUTPUT

X(1)=   4.117
X(2)=   0.873
X(3)=   0.250
X(4)=   1.147
X(5)=   0.760

FINAL TABLEAU

X1	X2	X3	X4	X5	B
16.000	8.0000	4.0000	1.0000	0.0	1.1467
1.0000	1.0000	1.0000	0.0	1.0000	0.76000
===========	===========	===========	===========	===========	===========
-0.13333E-01	0.0	0.66665E-02	0.0	0.0	-13.250

Chapter 4

GEOMETRIC PROGRAMMING

Geometric programming is one of the more recent developments in optimization theory. It is based upon the mathematical properties of inequalities and is capable of solving certain problems involving nonlinear terms in both the objective function and constraints. It is sometimes possible to locate the optimum solution by simple inspection of the exponents in the objective function.

Wilde and Beightler (47, p. 28) describe the method as follows:

> Instead of seeking the optimum values of the independent variables first, geometric programming finds the optimal way to distribute the total cost among the various terms of the objective function. Once these optimal allocations are obtained, often by inspection of simple linear equations, the optimal cost can be found by routine calculation. If the cost is suitably attractive, one can then find the policy to attain it.

I. BLAU (GOMTRY ALGORITHM)*

A. Purpose

This program finds the minimum of a multivariable, nonlinear function of geometric form:

$$\text{Minimize} \quad y_o(\underline{x}) = \sum_{t=1}^{T_o} \sigma_{ot} C_{ot} \prod_{n=1}^{N} x_n^{a_{otn}}$$

Subject to constraints of geometric form:

$$\sum_{t=1}^{T_m} \sigma_{mt} C_{mt} \prod_{n=1}^{N} X_n^{a_{mtn}} \leq \sigma_m$$

$$\text{for } m = 1, 2, \ldots, M$$

σ_{ot} and $\sigma_{mt} = \pm 1$ (the sign of each term in the objective function and m^{th} constraint, respectively)

C_{ot} and $C_{mt} > 0$ (the coefficients of each term in the objective function and m^{th} constraint, respectively)

$x_n > 0$ (the independent variables)

$\sigma_m = \pm 1$ (the constant bound of the mth constraint)

a_{otn} and a_{mtn} are the exponents of the nth independent variable of the tth term of the objective function and m^{th} constraint, respectively

M is the number of constraints

T_o is the number of terms in the objective function

$T_1, T_2, \ldots, T_M$ are the number of terms in each constraint, 1 to M, respectively

$\sigma = \pm 1$ assumed sign of the objective function

B. Method

The basic procedure originates from the geometric programming technique of Zener (51). The particular algorithm included in this section was developed by Blau (6) and programmed by Garcia and Hogg (20).

---

* The computer code included in this section was developed by Alberto Garcia and Dr. Gary Hogg, University of Illinois at Urbana-Champaign. Used by permission of the authors.

The method proceeds as follows:

1) Enter problem as specified in Input Format section.

2) Determine initial weights:

$$Z = \sum_{t=1}^{T_o} \sigma_{ot} C_{ot} \prod_{n=1}^{N} x_n^{a_{otn}}$$

$$V = |Z|$$

$$\beta_{mt} = C_{mt} \prod_{n=1}^{N} x_n^{a_{mtn}} \qquad m = 0, 1, \ldots, M$$

$$\beta_{ot} = \beta_{ot}/V$$

3) Calculate the vector of orthogonal conditions for the objective, and the matrix of these conditions for the constraints:

$$K = \left[\sum_{t=1}^{T_m} \sigma_{mt} a_{mtn} \beta_{mt} \right] \qquad \begin{array}{l} m = 1, \ldots, M \\ n = 1, \ldots, N \end{array}$$

$$H_n = \sum_{t=1}^{T_o} \sigma_{ot} a_{otn} \beta_{ot} \qquad n = 1, \ldots, N$$

4) Evaluate initial multipliers

$$\underline{Y} = (K^T K)^{-1} K^T H \qquad (K^T \text{ denotes K transpose})$$

5) If this is the first iteration, go to Step 6); otherwise, determine new weights as in Step 2), new orthogonal conditions as in Step 3), and new multipliers, as follows:

$$\underline{Y}_{new} = \underline{Y}_{old} + \Delta \underline{Y}$$

Now proceed to Step 6).

6) Calculate the matrix T

$$T = \begin{bmatrix} \sum_{m=1}^{M} \left[\sum_{t=1}^{T_m} \sigma_{mt} a_{mti} a_{mtj} \beta_{mt} \right] Y_m & i = 1, \ldots, N \\ \\ -\sum_{t=1}^{T_o} \sigma_{ot} a_{mti} a_{mtj} \beta_{ot} & j = 1, \ldots, N \end{bmatrix}$$

7) Evaluate error

$$e_i = \begin{cases} \sum_{t=1}^{T_o} \sigma_{ot} a_{oti} \beta_{ot} \\ \quad - \sum_{m=1}^{M} \left[\sum_{t=1}^{T_m} \sigma_{mt} a_{mti} \beta_{mt} \right] \gamma_m & i = 1, \ldots, N \\ \sigma_o - \sum_{t=1}^{T_o} \sigma_{ot} \beta_{ot} & i = N+1 \\ 1 - \sum_{t=1}^{T_m} \sigma_{mt} \beta_{mt} & i = N+1+m; \\ & m = 1, \ldots, M \end{cases}$$

8) Formulate Newton-Raphson Matrix, R:

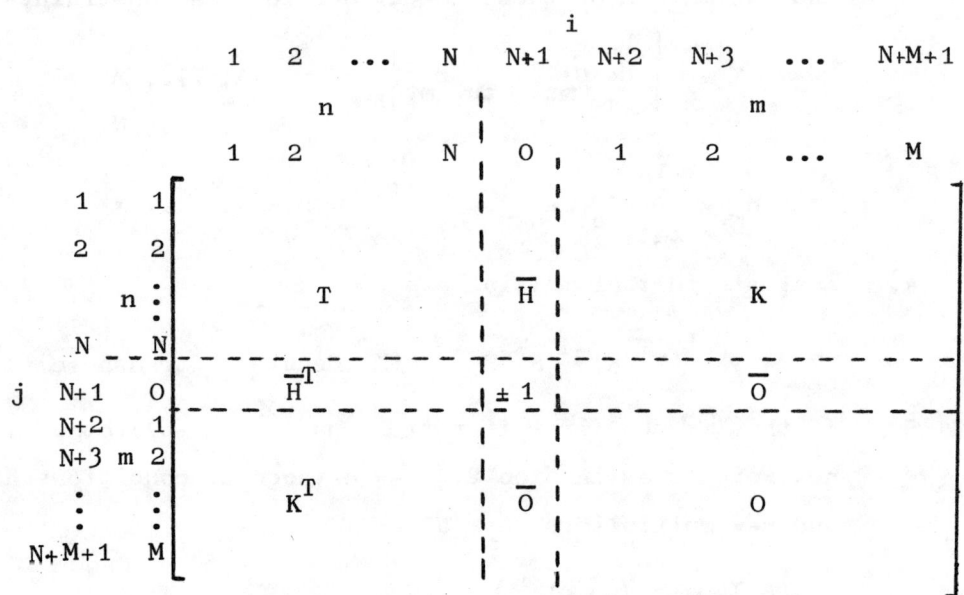

9) Invert the matrix R.

10) Find vector of adjustments:

$$R^{-1}\bar{e} = \begin{bmatrix} \Delta \ln \bar{x} \\ \Delta \ln V \\ \Delta \bar{\gamma} \end{bmatrix}$$

11) Calculate new values of independent variables:

$$\bar{x} = \bar{x} \exp(\Delta \ln \bar{x})$$
$$V = V \exp(\Delta \ln V)$$

12) Has solution converged to acceptable limit?

>Yes; print results and stop.
>
>No; go to Step 13.

13) Has maximum allowable iterations been reached?

>Yes; stop, print message.
>
>No; return to Step 5).

A flow diagram illustrating the above procedure is shown in Figure 4.I.

C. Program Description

1) Usage:

The program consists of a main program and three subroutines, GP3, GP10, and GP22. The main program reads the data, prints the results, and performs a major part of the algorithm.

2) Subroutines Required:

Subroutine GP3 calculates the absolute value of all terms in the problem. Subroutine GP10 inverts to Newton-Raphson matrix, or one of its submatrices. Subroutine GP22 calculates orthogonal conditions for the objective function and the constraints. All linkage between main program and the three subroutines is through COMMON statements.

3) Description of Parameters:

N	Total number of variables
M	Number of constraints
SIG	Assumed sign of the optimal value of the objective function
CONVRG	Convergence criterion
ITMAX	Maximum number of iterations permitted
KT	Vector giving number of terms per polynomial
C	Vector of coefficients
A	Array of exponents
X	Vector of initial estimates of optimal solution
W	Vector of weights (dual variables)
E	Error vector
AMD	Vector of LaGrange multipliers
PIVOT	Vector used in matrix inversion
R	Newton-Raphson matrix
NI	Card reader unit number
NO	Printer unit number

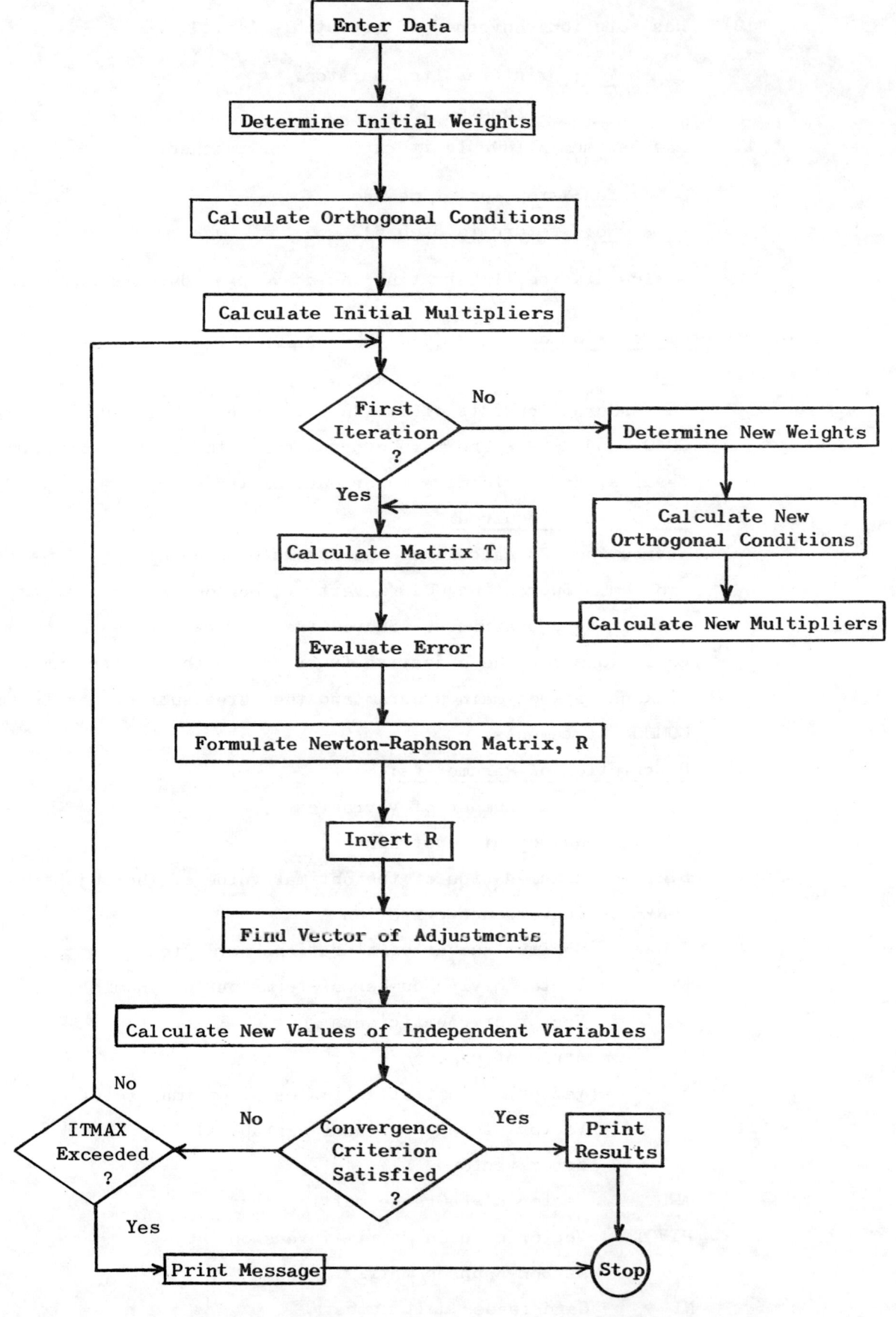

Figure 4.I. Geometric Programming (GOMTRY ALGORITHM) Logic Diagram

SUM   Objective function

V   Dual function

KTT   Total number of terms in problem

MR   Dimension of the matrix R

M1   Total number of polynomials

4) DIMENSION Requirements:

The DIMENSION and COMMON statements should be modified according to the requirements of each particular problem. The parameters included in the following statements conform to the Input Parameter definitions, above:

COMMON C(KTT)

COMMON X(N), KT(M1), W(KTT), A(KTT,N), R(KTT,KTT), PIVOT(KTT), AMD(M), E(KTT)

DIMENSION IPVOT(KTT), INDEX(KTT,2)

Also, FORMAT statements 102 and 107 and their associated READ and WRITE statements should be modified to conform to the requirements of each problem. (They are currently set for $N = 9$ variables.)

5) Input Formats:

CARD TYPE	FORMAT	CONTENTS
1	(19X,51H 51 blanks)	Problem name, ≤ 51 characters
2	(I4,I4,F5.1,F10.8,I5)	N, M, SIG, CONVRG, ITMAX
3	(I4)	KT(L), L = 1, M + 1

(One Type 3 card for each polynomial, including objective function.)

4	(E15.8)	C(I) (Coefficient for I^{th} Term)
5	(7F10.3)	A(I,J) (Exponent for J^{th} Variable in I^{th} Term)

(If J > 7, additional Type 5 cards required for each Term. A total of M + 1 sets of Type 4 and Type 5 cards required.)

6	(6X,F10.5)	X(J), J = 1, N

(One Type 6 card required for each variable.)

6) Output:

The problem name and certain statistical information (number of variables, number of restrictions, etc.) are printed. Also, the complete set of coefficients and exponents is printed. Thereafter, on each iteration, the current solution is printed. This is continued until the convergence criterion has been satisfied, after which the message "OPTIMAL RESULTS" is printed.

7) Summary of User Requirements:

a) Formulate problem according to requirements explained in Wilde and Beightler ( ).

b) Specify values for N, M, SIG, CONVRG, ITMAX, NI, and NO.

c) Specify initial solution, X(J).

d) Adjust DIMENSION and COMMON statements in main program and subroutines.

e) Modify FORMAT statements 102 and 107 and their associated READ and WRITE statements for the particular problem being solved.

D. Test Problem

The following program was run on an IBM 360/65:

Maximize $\quad y_0 = 0.3X_8 + X_9$

Subject to
$$y_1 = X_4^{-1} - 2X_1X_4 = 1$$
$$y_2 = 2X_1X_4^2X_5^{-1} - 1.055X_1^{1.667} = 1$$
$$y_3 = X_4X_6^{-1} - 2X_2X_6 = 1$$
$$y_4 = X_5X_7^{-1} + 2X_2X_6^2X_7^{-1} - 1.055X_2^{1.667} = 1$$
$$y_5 = X_6X_8^{-1} - 2X_3X_8 = 1$$
$$y_6 = X_7X_9^{-1} + 2X_3X_8^2X_9^{-1} - 1.-55X_3^{1.667} = 1$$
$$X_n > 0 \quad n = 1, \ldots, 9$$

N = 9, M = 6, SIG = +1, CONVRG = 0.000003, ITMAX = 20

VARIABLES	STARTING SOL.	OPTIMAL SOL.
X_1	0.371	0.232736
X_2	0.371	0.205088
X_3	0.371	0.191412
X_4	0.800	0.743021
X_5	0.200	0.235143
X_6	0.600	0.596885
X_7	0.300	0.354608
X_8	0.400	0.500852
X_9	0.400	0.422330
y_0	0.520	0.572586

Number of iterations: 8

Central processor time: 4 seconds

The listing and output for this problem are contained in the following section.

E. Program Listings and Example Output

```
C      SOLUTION TO A GENERALIZED GEOMETRIC PROGRAMMING PROBLEM
C
       COMMON N,M,SIG,M1,KTT,KTO,MR,N1,IT,MR1,V,SUM,NR
       COMMON C(16)
       COMMON X(9),KT(7),W(16),A(16,9),R(16,16),PIVOT(16),AMD(6),E(16)
       COMMON MM
       DIMENSION IPVOT(16),INDEX(16,2)
C
       NI = 5
       NO = 6
       WRITE(NO,399)
C
C      ************ INPUT TO THE PROBLEM *********************
C
       WRITE(NO,315)
       READ(NI,100) N,M,SIG
       KTT=0
       M1=M+1
C
C      NUMBER OF TERMS IN EACH POLYNOMIAL
C
       DO 200 L=1,M1
       READ(NI,101) KT(L)
  200  KTT = KTT + KT(L)
       KSIG=SIG
       IF(M-1)263,263,264
  263  WRITE(NO,197) N,M,KTT,KSIG
       GO TO 500
  264  WRITE(NO,109) N,M,KTT,KSIG
  500  CONTINUE
       WRITE(NO,301)
       WRITE(NO,305)
C
C      COEFFICIENTS AND EXPONENTS
C
       I2=0
       DO202L=1,M1
       I1=I2+1
       I2=I1+KT(L)-1
       DO202I=I1,I2
       READ(NI,199) C(I)
       READ(NI,102) A(I,1),A(I,2),A(I,3),A(I,4),A(I,5),A(I,6)
       READ(NI,102) A(I,7),A(I,8),A(I,9)
  202  WRITE(NO,107) L,I,C(I),A(I,1),A(I,2),A(I,3),A(I,4),A(I,5),A(I,6),
      1A(I,7),A(I,8),A(I,9)
C
C      INITIAL SOLUTION
C
       DO 203 J=1,N
  203  READ(NI,103) X(J)
C
C      **********************************************************
C
```

```
C       *************************************************************
        MR = M + N + 1
        N1=N + 1
        NR = N + 2
        MR1= MR+1
        KTO = KT(1)
        IT = 1
        MRX=16
        DO 213 IR=1,MRX
        DO 213 JR=1,MRX
    213 R(IR,JR) = 0.
C       *************************************************************
C
C       ****** EVALUATION OF THE INITIAL WEIGHTS ****************
C
C       THESE WEIGHTS WILL CHANGE AT EACH ITERATION
C
        CALL GP3
C
C       EVALUATION OF THE OBJECTIVE FUNCTION
C
        SUM =0.
        DO 206 I=1,KTO
    206 SUM=SUM+W(I)*(C(I)/ABS(C(I)))
C
C       INITIAL VALUE OF V
C
        V = ABS (SUM)
C
        DO 207 I = 1,KTO
    207 W(I)= W(I)/V
C
C       *************************************************************
    100 FORMAT(I4,I4,F5.1)
    109 FORMAT(1H0,9X,I3,10H VARIABLES,7X,I3,14H RESTRICTIONS ,7X,I3,
       19H TERMS   ,7X,14HOBJECTIVE SIGN,I3)
    197 FORMAT(1H0,9X,I3,10H VARIABLES,7X,I3,14H RESTRICTION  ,7X,I3,
       19H TERMS   ,7X,14HOBJECTIVE SIGN,I3)
    301 FORMAT(1H0,37X,26HCOEFFICIENTS AND EXPONENTS)
    305 FORMAT(1H0,1X,4HTERM,2X,11HCOEFFICIENT,3X,4HX(1),5X,4HX(2),5X,
       14HX(3),5X,4HX(4),5X,4HX(5),5X,4HX(6),5X,4HX(7),5X,4HX(8),5X,
       24HX(9))
    102 FORMAT(7F10.3)
    107 FORMAT(1H0,I2,1H-,I3,E13.6,9F9.4)
    199 FORMAT(E15.8)
    101 FORMAT(I4)
    103 FORMAT(6X,F10.5)
    315 FORMAT(10X,'**   GEOMETRIC PROGRAMMING PROBLEM  **')
    399 FORMAT(1H1)
C       *************************************************************
        CALL GP22
C
C       PRODUCT OF THE TRANSPOSE MATRIX OF ORTHOGONAL CONDITIONS FOR THE
C       CONSTRAINTS BY THE SAME MATRIX WITHOUT TRANSPOSING
C
        DO 215 JR=NR,MR
        DO 215 IR=NR,MR
        DO 215 KR= 1,N
```

```
  215 R(IR,JR) = R(IR,JR) + R(KR,IR) * R(KR,JR)
      MM=NR
C
C     INVERSION OF THE MATRIX FOUND ABOVE
C
      CALL GP10
C
C     PRODUCT OF THE TRANSPOSE MATRIX OF ORTHOGONAL CONDITIONS FOR
C     THE CONSTRAINTS BY THE VECTOR OF THE SAME CONDITIONS FOR THE
C     OBJECTIVE FUNCTION
C
      DO 224 JR=NR,MR
      DO 224 IR= 1,N
  224 R(N1,JR) = R(N1,JR) + R(IR,JR) * R(IR,N1)
C
C     INITIAL MULTIPLIERS
C
      DO 225 IR=NR,MR
      L = IR - N1
      AMD(L) = 0.
      DO 225 JR=NR,MR
  225 AMD(L) = AMD(L) + R(N1,JR) * R(IR,JR)
C
      DO 273 KKK=1,20
      IF(IT - 1)232,226,232
C
C     THE FIRST ITERATION IS NOT OVER
C
  226 CONTINUE
C
C     MATRIX T
C
      DO 231 J=1,N
      DO 231 JJ=1,N
      SU=0.
      DO 228 I=1,KT0
  228 SU = SU + C(I) / ABS(C(I)) * A(I,J) * A(I,JJ) * W(I)
      I2 = KT0
      SUD=0.
      DO 230 L=2,M1
      I1 = I2 + 1
      I2 = I1 + KT(L) - 1
      SUS = 0.
      DO 229 I=I1,I2
  229 SUS = SUS + C(I) / ABS(C(I)) * A(I,J) * A(I,JJ) * W(I)
      SUL = SUS * AMD(L - 1)
  230 SUD = SUD + SUL
  231 R(J,JJ) = SUD - SU
      GO TO 236
C
C     THE FIRST ITERATION IS OVER
C
  232 CONTINUE
C
C     NEW WEIGHTS GENERATED BY THE ALGORITHM
C
      CALL GP3
C
```

```
      SUM=0.
      DO 243 I=1,KTO
  243 SUM=SUM+W(I)*(C(I)/ABS(C(I)))
      DO 233 I=1,KTO
  233 W(I)=W(I)/V
C
C     NEW VECTOR AND MATRIX OF ORTHOGONAL CONDITIONS
C
      CALL GP22
C
      DO 235 L=1,M
      IR = L + N1
  235 AMD(L) = AMD(L) + PIVOT(IR)
      GO TO 226
  236 CONTINUE
C     ***************************************************************
C
C
C     EVALUATION OF THE ERROR
C
      SUMV=0.
      DO 286 I=1,KTO
  286 SUMV=SUMV+W(I)*(C(I)/ABS(C(I)))
      DO 254 IR=1,N
      E(IR)= 0.
      DO 253 JR=NR,MR
      J = JR - N1
  253 E(IR)=E(IR)+R(IR,JR)*AMD(J)
  254 E(IR)=R(IR,N1)-E(IR)
      E(N1)=SIG-SUMV
      I2 = KTO
      DO 256 L=2,M1
      I1 = I2 + 1
      I2 = I1 + KT(L) - 1
      SG = 0.
      DO 255 I=I1,I2
  255 SG=SG+W(I)*(C(I)/ABS(C(I)))
      IR=L+N1-1
  256 E(IR)= 1. - SG
C
C     ***************************************************************
C
      WRITE(NO,314) IT
      DO 396 J=1,N
  396 WRITE(NO,108) J,X(J)
      WRITE(NO,308)
      VV=V*SIG
      WRITE(NO,104) SUM
      WRITE(NO,105) VV
C
C     ***************************************************************
  104 FORMAT(34X,18HOBJECTIVE FUNCTION,1H=,F15.8)
  105 FORMAT(34X,18HDUAL FUNCTION     ,1H=,F15.8)
  108 FORMAT(1H0,42X,2HX(,I2,5H ) = ,F15.8)
  314 FORMAT(1H1,//,1X,42H******************************************
     11X,9HITERATION,I3,1X,41H*****************************
  308 FORMAT(1X)
C     ***************************************************************
C
```

```
C       FINAL EXPRESSION OF THE NEWTON-RAPHSON MATRIX
C
        DO 250 IR=N1,MR
        DO 250 JR=1,N
    250 R(IR,JR) = R(JR,IR)
        DO 251 IR= N1,MR
        DO 251 JR= N1,MR
    251 R(IR,JR)= 0.
        R(N1,N1) =-SIG
C       ****************************************************
C
C       INVERSION OF THE MATRIX R
C
        MM = 1
C
        CALL GP10
C
C       THE INVERSE OF THE MATRIX R IS MULTIPLIED BY THE VECTOR E IN ORDER
C       TO FIND THE VECTOR OF ADJUSTMENTS
C
        DO 257 IR=1,MR
        PIVOT(IR)=0.
        DO 257 JR=1,MR
    257 PIVOT(IR) = PIVOT(IR) + R(IR,JR) * E(JR)
C
C       THE NEW ITERATION STARTS HERE
C
        IT = IT +1
C
C       THE PRIMITIVE SOLUTION IS MODIFIED AND A NEW VALUE OF V IS FOUND
C
        DO 258 J=1,N
    258 X(J) = X(J) * EXP(PIVOT(J))
        V = V * EXP(PIVOT(N1))
        DO 298 IR=1,N
        DO 298 JR=N1,MR
    298 R(IR,JR)=0.
C
C       TEST OF THE SOLUTION
C
        IF(ABS(V-SIG*SUM)/V-0.000003)274,274,273
C
C       THE SOLUTION IS SATISFACTORY******************************
C
    274 GO TO 299
C
C       THE SOLUTION IS NOT SATISFACTORY**************************
C
    273 CONTINUE
        WRITE(NO,778)
        GO TO 897
    299 WRITE(NO,388)
    897 CONTINUE
        DO 395 J=1,N
    395 WRITE(NO,108) J,X(J)
        WRITE(NO,308)
        WRITE(NO,104) SUM
        VV=V*SIG
```

```
      WRITE(NO,105) VV
C
C     ****************************************************************
  388 FORMAT(1H0,45X,15HOPTIMAL RESULTS)
  778 FORMAT(30X,39HNOT ENOUGH CONVERGENCE IN 20 ITERATIONS)
C     ****************************************************************
C
      STOP
      END



      SUBROUTINE GP10
C
C     THE SUBROUTINE GP10 IS USED TO INVERT THE NEWTON AND RAPHSON
C     MATRIX,OR ONE SUBMATRIX OF IT
C
      COMMON N,M,SIG,M1,KTT,KTO,MR,N1,IT,MR1,V,SUM,NR
      COMMON C(16)
      COMMON X(9),KT(7),W(16),A(16,9),R(16,16),PIVOT(16),AMD(6),E(16)
      COMMON MM
      DIMENSION IPVOT(16),INDEX(16,2)
C
      DET = 1.
      DO 216 JR=MM,MR
  216 IPVOT(JR) = 0
      DO 241 IR=MM,MR
C
C     INVESTIGATION OF THE PIVOT
C
      T = 0.
      DO 220 JR=MM,MR
      IF(IPVOT(JR)-1)217,220,217
  217 DO 219 K=MM,MR
      IF(IPVOT(K)-1)218,219,239
  218 IF(ABS(T) - ABS(R(JR,K)))242,219,219
  242 IROW = JR
      ICOL = K
      T = R(JR,K)
  219 CONTINUE
  220 CONTINUE
      IPVOT(ICOL) = IPVOT(ICOL) + 1
C
C     THE PIVOT IS LOCATED ON THE DIAGONAL
C
      IF(IROW - ICOL)221,223,221
  221 DET = -DET
      DO 222 LL=MM,MR
      T = R(IROW,LL)
      R(IROW,LL) = R(ICOL,LL)
  222 R(ICOL,LL) = T
  223 INDEX(IR,1) = IROW
      INDEX(IR,2)=ICOL
      PIVOT(IR)=R(ICOL,ICOL)
      DET = DET * PIVOT(IR)
C
```

```
C       THE PIVOT ROW IS DIVIDED BY THE PIVOT
C
        R(ICOL,ICOL) = 1.
        DO 284 LL=MM,MR
  284   R(ICOL,LL) = R(ICOL,LL) / PIVOT(IR)
C
C       ROWS WITHOUT PIVOT ARE REDUCED
C
        DO 241 KL=MM,MR
        IF(KL-ICOL)245,241,245
  245   T = R(KL,ICOL)
        R(KL,ICOL) = 0.
        DO 240 LL = MM,MR
  240   R(KL,LL) = R(KL,LL) - R(ICOL,LL) * T
  241   CONTINUE
C
C       COLUMNS ARE EXCHANGED
C
        DO 238 IR=MM,MR
        LL = MR - IR + MM
        IF(INDEX(LL,1) - INDEX(LL,2))227,238,227
  227   JROW = INDEX(LL,1)
        JCOL = INDEX(LL,2)
        DO 237 K=MM,MR
        T = R(K,JROW)
        R(K,JROW) = R(K,JCOL)
        R(K,JCOL) = T
  237   CONTINUE
  238   CONTINUE
  239   RETURN
        END




        SUBROUTINE GP3
C
C       THE SUBROUTINE GP3 CALCULATES THE ABSOLUTE VALUE OF ALL THE TERMS
C       OF THE MODEL
C
        COMMON N,M,SIG,M1,KTT,KTO,MR,N1,IT,MR1,V,SUM,NR
        COMMON C(16)
        COMMON X(9),KT(7),W(16),A(16,9),R(16,16),PIVOT(16),AMD(6),E(16)
        COMMON MM
C
        I2 = 0
        DO 205 L=1,M1
        I1 = I2 + 1
        I2 = I1 + KT(L) - 1
        DO 205 I = I1,I2
        W(I) = 1.
        DO 204 J = 1,N
  204   W(I) = W(I) * X(J) ** A(I,J)
  205   W(I)=W(I)*ABS(C(I))
        RETURN
        END
```

```
      SUBROUTINE GP22
C
C     THE SUBROUTINE GP22 IS USED TO FIND THE VECTOR OF ORTHOGONAL
C     CONDITIONS FOR THE OBJECTIVE,AND THE MATRIX OF THESE CONDITIONS
C     FOR THE CONSTRAINTS
C
      COMMON N,M,SIG,M1,KTT,KTO,MR,N1,IT,MR1,V,SUM,NR
      COMMON C(16)
      COMMON X(9),KT(7),W(16),A(16,9),R(16,16),PIVOT(16),AMD(6),E(16)
      COMMON MM
C
      DO 212 IR=1,N
      I2=0
      DO 212  JR=N1,MR
      L= JR - N1 + 1
      I1 = I2+1
      I2= I1 + KT(L) -1
      J=IR
      DO 212 I=I1,I2
  212 R(IR,JR) = R(IR,JR) + C(I) / ABS(C(I)) * A (I,J) * W(I)
      RETURN
      END
```

******************* ITERATION   1 ************************

$$X(\ 1\ ) = \quad 0.37099999$$

$$X(\ 2\ ) = \quad 0.37099999$$

$$X(\ 3\ ) = \quad 0.37099999$$

$$X(\ 4\ ) = \quad 0.79999995$$

$$X(\ 5\ ) = \quad 0.19999999$$

$$X(\ 6\ ) = \quad 0.59999996$$

$$X(\ 7\ ) = \quad 0.29999995$$

$$X(\ 8\ ) = \quad 0.39999998$$

$$X(\ 9\ ) = \quad 0.39999998$$

OBJECTIVE FUNCTION=   0.51999986
DUAL FUNCTION   =   0.51999986

(Iterations 2 through 4 omitted.)

******************* ITERATION   5 ************************

$$X(\ 1\ ) = \quad 0.25755590$$

$$X(\ 2\ ) = \quad 0.20133948$$

$$X(\ 3\ ) = \quad 0.18803436$$

$$X(\ 4\ ) = \quad 0.73037565$$

$$X(\ 5\ ) = \quad 0.24962854$$

$$X(\ 6\ ) = \quad 0.59035367$$

$$X(\ 7\ ) = \quad 0.35967892$$

$$X(\ 8\ ) = \quad 0.49742591$$

$$X(\ 9\ ) = \quad 0.42460495$$

OBJECTIVE FUNCTION=   0.57383269
DUAL FUNCTION   =   0.57369047

(Iterations 6 and 7 omitted.)

******************* ITERATION   8 ***********************

$$X(1) = 0.23273540$$
$$X(2) = 0.20508897$$
$$X(3) = 0.19141293$$
$$X(4) = 0.74302167$$
$$X(5) = 0.23514313$$
$$X(6) = 0.59688616$$
$$X(7) = 0.35460883$$
$$X(8) = 0.50085258$$
$$X(9) = 0.42233050$$

OBJECTIVE FUNCTION= 0.57258624
DUAL FUNCTION   = 0.57258612

OPTIMAL RESULTS

$$X(1) = 0.23273557$$
$$X(2) = 0.20508915$$
$$X(3) = 0.19141281$$
$$X(4) = 0.74302143$$
$$X(5) = 0.23514301$$
$$X(6) = 0.59688568$$
$$X(7) = 0.35460865$$
$$X(8) = 0.50085229$$
$$X(9) = 0.42233086$$

OBJECTIVE FUNCTION= 0.57258624
DUAL FUNCTION   = 0.57258666

** GEOMETRIC PROGRAMMING PROBLEM **

9 VARIABLES   6 RESTRICTIONS   16 TERMS   OBJECTIVE SIGN 1

COEFFICIENTS AND EXPONENTS

TERM	COEFFICIENT	X(1)	X(2)	X(3)	X(4)	X(5)	X(6)	X(7)	X(8)	X(9)
1- 1	0.300000E 00	0.0	0.0	0.0	0.0	0.0	0.0	0.0	1.0000	0.0
1- 2	0.100000E 01	0.0	0.0	0.0	0.0	0.0	0.0	0.0	0.0	1.0000
2- 3	0.100000E 01	0.0	0.0	0.0	-1.0000	0.0	0.0	0.0	0.0	0.0
2- 4	-0.200000E 01	1.0000	0.0	0.0	1.0000	0.0	0.0	0.0	0.0	0.0
3- 5	0.200000E 01	1.0000	0.0	0.0	2.0000	-1.0000	0.0	0.0	0.0	0.0
3- 6	-0.105500E 01	1.6670	0.0	0.0	0.0	0.0	0.0	0.0	0.0	0.0
4- 7	0.100000E 01	0.0	0.0	0.0	1.0000	0.0	-1.0000	0.0	0.0	0.0
4- 8	-0.200000E 01	0.0	1.0000	0.0	0.0	0.0	1.0000	0.0	0.0	0.0
5- 9	0.100000E 01	0.0	0.0	0.0	0.0	1.0000	0.0	-1.0000	0.0	0.0
5- 10	0.200000E 01	0.0	1.0000	0.0	0.0	0.0	2.0000	-1.0000	0.0	0.0
5- 11	-0.105500E 01	0.0	1.6670	0.0	0.0	0.0	0.0	0.0	0.0	0.0
6- 12	0.100000E 01	0.0	0.0	0.0	0.0	0.0	1.0000	0.0	-1.0000	0.0
6- 13	-0.200000E 01	0.0	0.0	1.0000	0.0	0.0	0.0	0.0	1.0000	0.0
7- 14	0.100000E 01	0.0	0.0	1.0000	0.0	0.0	0.0	1.0000	0.0	-1.0000
7- 15	0.200000E 01	0.0	0.0	1.0000	0.0	0.0	0.0	0.0	2.0000	-1.0000
7- 16	-0.105500E 01	0.0	0.0	1.6670	0.0	0.0	0.0	0.0	0.0	0.0

Chapter 5

DYNAMIC PROGRAMMING

Dynamic Programming is a mathematical optimization technique used for making a series of interrelated decisions. Usually, a multi-stage decision process is transformed into a series of single-stage decision processes:

> Dynamic programming starts with a small portion of the problem and finds the optimal solution for this smaller problem. It then gradually enlarges the problem, finding the current optimal solution from the previous one, until the original problem is solved in its entirety.
>
> (28, p. 241)

In contrast to other mathematical programming techniques (such as linear programming), there does not exist a standard mathematical formulation of "the" dynamic programming problem. Dynamic programming is a general strategy for optimization rather than a specific set of rules. Consequently, the particular equations used must be developed to fit each problem.

A major distinction among dynamic programming problems is the nature of the decision variable. If the decision variable can take on any real value, the problem is said to be continuous. If the decision variable is restricted to integer values, the problem is said to be discrete. The two sections of this chapter reflect this distinction:

    I. Discrete (DP ALGORITHM)

    II. Continuous (DYNAM ALGORITHM)

I. DISCRETE (DP ALGORITHM)*

A. <u>Purpose</u>

This procedure determines the optimal sequence of decisions in a multi-stage decision process. The procedure is limited to a single decision variable. The solutions are restricted to integers and the number of sequential decisions must be finite. Constraints are permitted.

Dynamic programming is essentially recursive optimization. The typical dynamic programming recursive formulation is:

$$f_n(n,s,x) = g[R(n,s,x), f^*_{n-1}(s')]$$

where
- n = the stage of the problem
- s = the state of the system at stage n
- x = the decision (policy) being evaluated at stage n
- $R(n,s,x)$ = the immediate return associated with making decision x at stage n when the state of the system is s
- s' = the state of the system at stage n-1 resulting from decision x
- $f^*_{n-1}(s')$ = the return associated with the optimal sequence of decisions at stage n-1 when the state is s'

In most cases a function of $f^*_{n-1}(s')$ will be added to or multiplied by $R(n,s,x)$. On the first stage (n=1), this term is not included in the formulation.

At each stage, the results of the recursive formulation are calculated for all feasible values of x, and the optimal decisions are retained for subsequent use. The optimal value will be denoted $f^*_n(s)$ and the optimal decision as $x^*_n(s)$.

As noted in the introduction to this chapter, there does not exist a standard mathematical formulation for "the" dynamic programming problem. Each problem requires a particular recursive relationship. Nine different types of dynamic programming problems have been identified and pre-programmed recursive functions are included in the

---

*The algorithm described in this section was developed by E. G. Rider, a former graduate student at Arizona State University. The complete algorithm is described in (39).

computer packages. A large number of problems can be fitted to one of these recursive relationships. However, special instructions are provided for writing specific recursive relationships.

B. Method

The procedure is based upon the original work by Bellman (3) and the computational method of Rider (39). In contrast to the normal formulation of an optimization problem, where we try to solve the entire problem as a total entity, we develop a formulation that expresses the contribution to the total objective function of the results of a policy at one stage of the problem. This formulation is used to determine the optimal policy for all possible states of the system for the first stage of the problem.

Typically, the first stage in a dynamic programming formulation refers to the last decision which must be made in a series of sequential decisions. The results of this stage are then included in the next stage of the dynamic programming formulation. This recursive procedure is applied at each stage until the last stage is reached, at which point we are able to determine the optimal policy and value for the initial decision of our problem.

We can now determine the optimal policy for each sequential decision by retracing our steps through the set of optimal policies that were found for each state of the system at previous stages.

The method proceeds as follows:

1) Formulate the recursive relationship and all constraints. Constraints are typically stated by specifying upper and lower bounds of the state of the system and the decision values.

2) Compute $f_1^*(s)$ and $x_1^*(s)$ for n=1 (i.e., the final decision point). The values of $f_1^*(s)$ will be retained for use during the second stage computation. The values of $x_1^*(s)$ will be retained but not used until the final calculations are completed.

3) The solution procedure then moves backward stage by stage--each time finding the optimal policy for each state of that stage--until the optimal value $f_n^*(s)$ and decision $x_n^*(s)$ are found for the stage n=n (i.e., the first decision point). Keep in mind that s now refers to the initial state of the system since we performed a backwards analysis.

4) The optimal value of the other decisions is specified by the values of $x_n^*(s)$ for the previous n-1 stages according to the state of the system at these stages.

A flow diagram illustrating the above procedure is shown in Figure 5.I.

C. **Program Description**

1) <u>Usage:</u>

The program consists of a main program, one subroutine, and 19 function subprograms. If one of the pre-programmed recursive functions is suited to the problem, the user only needs to prepare data cards. If, however, a special recursive function must be defined, the user writes one function subprogram in FORTRAN. The name of this function is RECUR. The nine pre-programmed recursive functions are explained below:

Type 1: FUNCTION RECUR1(KN,KS,KX)

$$f_n(n,s,x) = C(s,x) + f_{n-1}^*(x)$$

C(s,x) is the penalty if minimizing, or the payoff if maximizing, associated with making decision x, given that one is in state s. A typical application would be in finding the shortest route through a network, and C represents distances.

Type 2: FUNCTION RECUR2(KN,KS,KX)

$$f_n(n,s,x) = P(n,x) + f_{n-1}^*(s-x)$$

P(n,x) is the payoff associated with allocating x units to stage n. A typical application would be in determining the optimal allocation of limited resources to several possible sources (stages).

Type 3: FUNCTION RECUR3(KN,KS,KX)

$$f(n,s,x) = P(n,x) \cdot f_{n-1}^*(s-x)$$

P(n,x) is the probability of mission failure if x units are assigned to component (stage) n. A typical application would be in determining the optimal assignment of researchers to research teams so as to minimize the probability of mission failure.

Type 4: FUNCTION RECUR4(KN,KS,KX)

$$f_n(n,s,x) = C(1,1) \cdot f_{n-1}^*(s+x) + (1-C(1,1)) \cdot f_{n-1}^*(s-x)$$

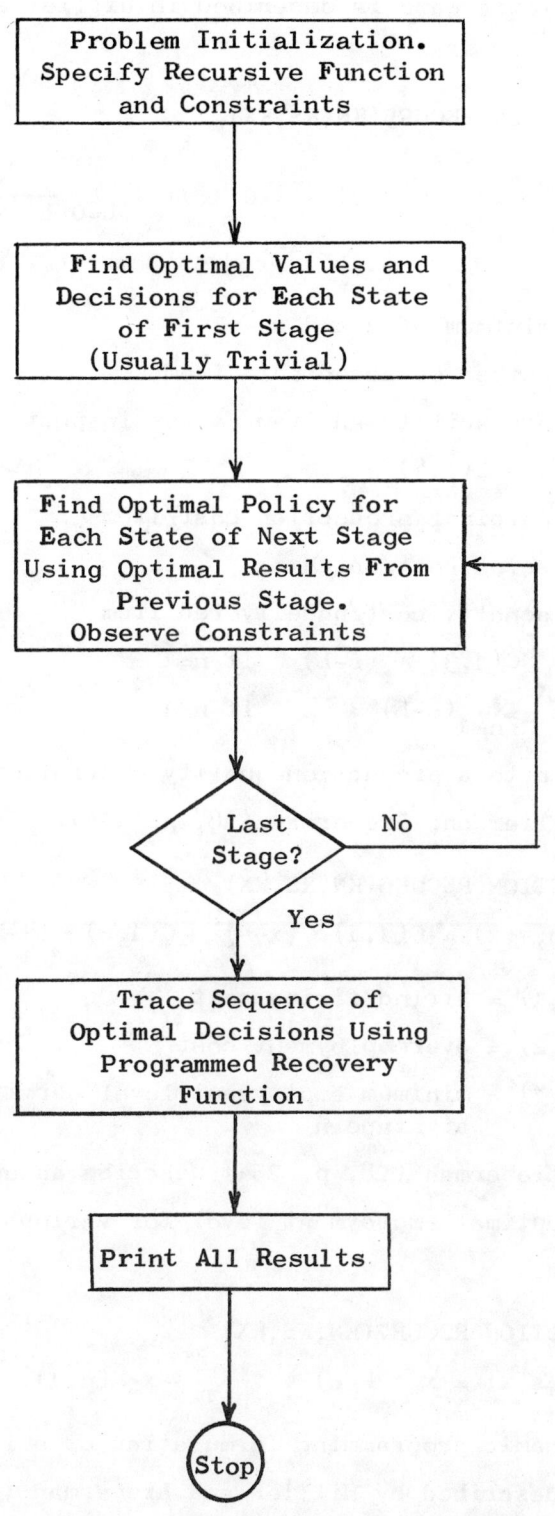

Figure 5.I. Discrete Dynamic Programming (DP ALGORITHM) Logic Diagram

$C(1,1)$ is the probability of success on each trial. An application to a probabilistic game is described in Hillier and Lieberman (28, p. 249).

Type 5: FUNCTION RECUR5(KN,KS,KX)

$$f_n(n,s,x) = C(1,2) + x \cdot C(1,1) + \sum_{L=0}^{U} \frac{x!}{L!(x-L)!} \cdot C(1,4)^L \cdot C(1,5)^{x-L} \cdot y$$

where U = minimum of x and s

L = number of acceptable items

$C(1,4)$ = probability an item passes inspection

$C(1,5) = 1 - C(1,4)$

$C(1,1)$ = marginal production cost/item

$C(1,2)$ = setup cost/batch

$C(1,3)$ = penalty cost/undelivered item

$$y = \begin{cases} C(1,3) \cdot (s-L) & \text{if } n=1 \\ f^*_{n-1}(s-L) & \text{if } n \neq 1 \end{cases}$$

An application to a production-quality control problem is described in Hillier and Lieberman (28, p. 251).

Type 6: FUNCTION RECUR6(KN,KS,KX)

$$f_n(n,s,x) = C(1,1) \cdot (x-s)^2 + C(1,2) \cdot (x-M(n,1)) + f^*_{n-1}(x)$$

where $C(1,1)$ = hiring/firing cost

$C(1,2)$ = overemployment cost

$M(n,1)$ = minimum employment level permitted at stage n

Hillier and Lieberman (28, p. 254) describe an application to finding the optimal employment level for various seasons assuming uneven demand.

Type 7: FUNCTION RECUR7(KN,KS,KX)

$$f(n,s,x) = x \cdot P(n) + f^*_{n-1}(s - x \cdot M(n,1))$$

This is a dynamic programming formulation of a linear programming problem, as described by Hillier and Lieberman (28, p. 259). P contains the coefficients of the objective function and M contains the coefficients of the constraint.

Type 8: FUNCTION RECUR8(KN,KS,KX)

$$f_n(n,s,x) = P(x) + S \cdot H + C(x) + f^*_{n-1}(s+x-D_n)$$

where
- P = production costs (regular and overtime as a function of x)
- H = inventory costs/unit/month
- C = setup cost
- D = demand in time period n

An application of this formulation is to the determination of the optimal number of units to produce in each time period when demand is known but variable.

Type 9: FUNCTION RECUR9(KN,KS,KX)

$$f_n(n,s,x) = P(n,x) \cdot f^*_{n-1}(s-M(n,x))$$

P(n,x) is the probability of component n functioning if x units of the component are placed in parallel. M(n,x) is the cost of installing x units of component n. The probability that the system will function is the product of the probabilities that the respective components will function. The objective is to determine the number of parallel units of each of the four components required to maximize the probability that the system will function, while not exceeding a specified maximum investment.

All of the above examples are described in Hillier and Lieberman (28, pp. 239-65) and are formulated for computer solution in Rider (39, Chapter IV).

Only one state variable and decision variable may be used. Other limitations are:

$$-999 \leq s \leq 9999$$
$$0 \leq x \leq 9999$$

At least one tape drive is required for program execution.

2) <u>Subprograms Required</u>:

In addition to the nine recursive function subprograms described above, there are 10 additional function subprograms and one subroutine.

SUBROUTINE (IE)

    Prints out certain error messages encountered during execution.

FUNCTION LOCATE (J)

Used for the storage and location of $f_n^*(\cdot)$, $f_{n-1}^*(\cdot)$, and $x_n^*(\cdot)$ by the main program.

FUNCTION LOCATD (MN,MD)

Allows user to input/locate only the data which will be used for his problem.

FUNCTION FSTAR (NEW,VL,VH,VI)

Returns the value of $f_{n-1}^*(s)$ and tests for feasibility.

FUNCTION NEWS1 (KN,KS,KX)

FUNCTION NEWS2 (KN,KS,KX)

FUNCTION NEWS3 (KN,KS,KX)

FUNCTION NEWS4 (KN,KS,KX)

FUNCTION NEWS5 (KN,KS,KX)

These five functions recover the sequence of optimal decisions, depending on the type of recursive relationship.

FUNCTION NEWS (KN,KS,KX)

This recovery function is written by the user to accompany his special recursive function RECUR. Can be ignored if RECUR is not written by user.

FUNCTION RECUR (KN,KS,KX)

This recursive relation function is written by the user if his problem cannot be formulated in one of the nine pre-programmed recursive forms (RECUR1 through RECUR9).

3) Description of Parameters:

NRUM	Number of problem solutions on this computer run
IDAY	Day
IMONTH	Month
IYEAR	Year
IUSER	User identification
IREM	Descriptive remarks
IFUNCT	Problem type (0-9)
MIN	= 0, maximization problem = 1, minimization problem
NTP	I/O Device code number for scratch tape drive. Define in main program

IOPT3	= 0, no intermediate results printed = 1, intermediate results printed
IOPT2	= 0, ties broken in favor of lower-valued decision = 1, ties broken in favor of higher-valued decision
IOPT1	= 0, normal calculation for stage 1 = 1, input optimal values and decisions for stage 1
NSTAGE	Number of stages
LOWD	Lowest permissible decision value
MAXD	Highest permissible decision value
LOWEST	Lowest state value
MAXS	Highest state value
INCRMT	Incremental step size
NCOLSP	Number of columns in array P
NCOLSM	Number of columns in array M
NPD	Number of elements in vector MPD
NPS	Number of elements in vector MPS
NROWSC	Number of rows in array C
NCOLSC	Number of columns in array C
P	Data array
C	Data array
M	Data array
MPD	Data vector-decision constraints for each stage
MPS	Data vector-state constraints for each stage
MTB	Data vector - MTB(I) = 1 if tie-breaker option (IOPT2) is desired at stage I
FSOPT	Data vector - if IOPT1 = 1, entries are values for $f_1^*(s)$
IDBEST	Data vector - if IOPT1 = 1, entries are values for $x_1^*(s)$
MAXSZE	Number of states
NCONST	Number of constraints

4) <u>DIMENSION Requirements</u>:

DIMENSION statement in main program may be modified according to the requirements of each particular problem. The parameters included in the following DIMENSION and COMMON statements conform to the Input Parameter definitions, above:

DIMENSION FSBEST(MAXSZE), IDBEST(MAXSZE), KTIES(NSTAGE), MTB(NSTAGE), KDEC(NSTAGE), IREM(12), LPS(NSTAGE), MPS(NDS), IUSER(5)

COMMON FSOPT(MAXSZE), LPD(NSTAGE), MPD(NPD), P(NSTAGE,NCOLSP), C(NROWSC,NCOLSC), M(NSTAGE,NCOLSM)

5) Input Formats:

CARD TYPE	FORMAT	CONTENTS
1	(2I2,A3,I4,5A4)	NRUN,IDAY,IMONTH,IYEAR,IUSER
2	(12A4)	IREM
3	(5I2)	IFUNCT,MIN,IOPT3,IOPT2,IOPT1
4	(I2,5I4)	NSTAGE,LOWD,MAXD,LOWEST,MAXS,INCRMT
5	(6I2)	NCOLSP,NCOLSM,NPD,NPS,NROWSC,NCOLSC
6	(20F4.0)	((P(I,J),J=1,NCOLSP),I=1,NSTAGE)

(If NCOLSP>20, more than one card required per row. Total of NSTAGE sets of cards required.)

7	(20F4.0)	((C(I,J),J=1,NCOLSC),I=1,NROWSC)

(If NCOLSP>20, more than one card required per row. Total of NROWSC sets of cards required.)

8	(20I4)	((M(I,J),J=1,NCOLSM)I=1,NSTAGE)

(If NCOLSM>20, more than one card required per row. Total of NSTAGE sets of cards required.)

9	(20I4)	(MPD(I),I=1,NPD)

(If NPD>20, additional Type 9 cards required.)

10	(20I4)	(MPS(I),I=1,NPS)

(If NPS>20, additional Type 10 cards required.)

11	(20I4)	(MTB(I),I=1,NSTAGE)

(If NSTAGE>20, additional Type 11 cards required.)

12	(20F4.0)	(FSOPT(I),I=1,NNS)

(If NNS>20, additional Type 12 cards required.)

13	(20I4)	(IDBEST(I),I=1,NNS)

(If NNS>20, additional Type 13 cards required.)

14	(12A4)	IREM
15	(12A4)	Blank card signifying end of data

6) Output:

The main program furnishes all printouts. Intermediate results are obtained if IOPT3=1 is specified in Card Type 3. All option selections are printed, along with constraints, original data and optimum solution.

7) Summary of User Requirements:

   a) Determine form of recursive relation. If it fits one of the pre-programmed functions, specify the problem type on

Card Type 3. If it does not, write RECUR and NEWS functions.

b) Specify decision and state constraints. Range of decisions are entered as LOWD and MAXD and range of states are entered as LOWEST and MAXS (Card Type 4). Enter all decision constraints into the vector MPD (Card Type 9) and all state constraints into the vector MPS (Card Type 10).

c) Organize problem data into the arrays P, C and/or M. Arrays P and C are used for floating point data and array M for integer-valued data. Arrays P and M are used when parameters differ for each stage of the problem.

d) Specify all program options, as explained in "Description of Parameters." Also, specify all parameters.

e) Adjust DIMENSION and COMMON statements in main program, if required.

f) Specify NTP (tape drive), NI (card reader) and NO (line printer) in main program.

D. Test Problem

The following test problem was taken from Hillier and Lieberman (28, p. 263) and was run on an IBM 360/65:

An electronic system consists of four components, each of which must function in order for the system to function. The reliability of the system can be improved by installing several parallel units in one or more of the components. The following table shows the probability that the components will function if they consist of 1, 2 or 3 units in parallel:

	No. of Units		
	1	2	3
Component 1	.60	.75	.85
Component 2	.40	.65	.80
Component 3	.70	.90	.95
Component 4	.50	.60	.80

The cost of installing 1, 2 or 3 parallel units in the respective components is given in the following table:

	No. of Units		
	1	2	3
Component 1	$6	$11	$15
Component 2	10	16	22
Component 3	5	10	14
Component 4	8	13	17

A maximum of $45 can be spent on components. The probability that the system will function is the product of the probabilities that the respective components will function. Determine the number of parallel units that should be installed in each of the four components in order to maximize the probability that the system will function.

Formulation:

This problem can be formulated as Problem Type 9.

$$f_n(n,s,x) = P(n,x) \cdot f^*_{n-1}(s - M(n,x))$$

$P(n,x)$ is the probability of component n functioning if x units of the component are placed in parallel. $M(n,x)$ is the cost of installing x units of component n.

Parameter and data values for this problem are given below:

IFUNCT = 9	problem type
MIN = 0	maximization
IOPT3 = 0	no intermediate results
IOPT2 = 0	normal tie-breaking
IOPT1 = 0	normal stage 1 calculation
NSTAGE = 4	number of stages
LOWD = 1	lowest number of parallel units
MAXD = 3	maximum number of parallel units
LOWEST = 0	lowest possible cost
MAXS = 45	highest permissible cost
INCRMT = 1	increment size
NCOLSP = 3	
NCOLSM = 3	
NPD = 0	no entries in MPD
NPS = 0	no entries in MPS

NROWSC = 0	C not used
NCOLSC = 0	C not used
P	enter table probability values
C	not used
M	enter table cost values
MPD	not used
MPS	not used
MTB	not used
FSOPT	not used
IDBEST	not used

A <u>stage</u> is a component of the system.

A <u>state</u> is the number of unallocated dollars at a particular decision point (a stage).

The <u>value</u> is the probability that the configuration of components functions.

The <u>decision variable</u> is the number of parallel units for each component (each stage).

<u>Solution</u>:

    2 units of component 1

    2 units of component 2

    2 units of component 3

    1 unit of component 4

Total cost: $45.00

Probability that system will function: 0.22

Central Processor Time: 2 seconds

The listing and output for this problem are contained in the following section.

E. Program Listings and Example Output

```
C       * GENERALIZED DYNAMIC PROGRAMMING PACKAGE. *
C       * POSITIVE VALUED INTEGER SOLUTIONS ONLY.
        DIMENSION FSBEST(2000),IDBEST(2000),KTIES(20),MTB(20),
       1 KDEC(20), IREM(12), LPS(20), MPS( 99),IUSER(5)
        COMMON FSOPT(2000),LPD(20), MPD( 99),
       1 P(20,   20), C(20,   20), M(20,   20),
       2 LOWD,MAXD,LOWEST,MAXS,INCRMT,N,NSTAGE,NCOLSC,
       3  NPD,LDHIGH,BIGM,IOPT3,IOPT2,IOPT1,NCOLSP,NCOLSM
        INTEGER HS,HD
        DATA IBF,MAXSZE,IDBEST/25,2000,2000*8888/
        DATA IBLANK,ITIED1,ITIED2/4H    ,1H ,1H*/
C
C       ALL FORMAT STATEMENT ARE GROUPED HERE.
C
530     FORMAT(1X,25I5)
C       ***START OF INPUT FORMAT STATEMENTS
900     FORMAT(2I2,A3,I4,5A4)
12      FORMAT(12A4)
930     FORMAT(6I2)
920     FORMAT(I2,5I4)
940     FORMAT(20F4.0)
950     FORMAT(20I4)
C       ***START OUTPUT FORMAT STATEMENTS.
905     FORMAT(1H1,///35X,'DYNAMIC PROGRAMMING',//15X,
       1 'RUN',I3,5X,'DATE',I3,1X,A3,I5,5X,'USER ',5A4,//)
15      FORMAT(15X,12A4)
915     FORMAT (/15X,'PROBLEM TYPE',I2,
       1/15X,'MINIMIZATION OPTION',I2,
       2  5X,'OUTPUT OPTION',I2,
       3/15X,'TIE-BREAKER OPTION',I2,
       4  6X,'STAGE 1 INPUT OPTION',I2,/)
925     FORMAT (  15X,'NUMBER OF STAGES =',I3,
       1/15X,'RANGE OF DECISIONS IS',I5,' TO',I5,
       2/15X,'RANGE OF STATES IS',I8,' TO',I5,
       3/15X,'INCREMENT =',I5,/)
945     FORMAT( 15X,'P(',I2,',J)',5F8.2,/,(22X,5F8.2))
980     FORMAT( 15X,'C(',I2,',J)',5F8.2,/,(22X,5F8.2))
955     FORMAT( 15X,'M(',I2,',J)',5I8,/,(22X,5I8))
960     FORMAT( 15X,'MPD (I)',5I8,/,(22X,5I8))
975     FORMAT( 15X,'MPS (I)',5I8,/,(22X,5I8))
985     FORMAT( 15X,'MTB (I)',5I8,/,(22X,5I8))
80      FORMAT(15X,'STAGE',3X,'STATE',3X,'OPTIMUM VALUE',3X,
       1 'OPTIMUM DECISION',//)
89      FORMAT(78X,1H.)
506     FORMAT( 16X,I2,5X,I4,5X,F12.2,8X,I4,A1 )
860     FORMAT(5(/16X,5(I4,'=X',I2,1X)))
875     FORMAT( 15X,'ALTERNATE OPTIMAL DECISIONS AT STAGE/S',
       1 3I3,/15X,17I3)
990     FORMAT( 15X,'* AFTER THE OPTIMUM DECISION INDICATES',
       1 /15X,'THAT ALTERNATE OPTIMAL DECISIONS EXIST.',
       2 /15X,'THE TIE-BREAKER OPTION CONTROLS SELECTION ',
       3 /15X,'OF THE LOWEST OR HIGHEST DECISION.')
```

```
C     *
C     * INPUT SECTION. *
C     *
      NI = 5
      NO = 6
      NTP = 4
      READ(NI,900) NRUNS,IDAY,IMONTH,IYEAR,(IUSER(I),I=1,5)
      NRUN = 1
      IBF1 = IBF - 1
C
C     PROGRAM STARTS HERE ON ALL RUNS EXCEPT THE FIRST.
C
10    WRITE(NO,905) NRUN,IDAY,IMONTH,IYEAR,(IUSER(I),I=1,5)
      READ(NI,12) (IREM(J),J=1,12)
      WRITE(NO,15)(IREM(J),J=1,12)
      BIGM= -9.0E50
      READ(NI,930) IFUNCT,MIN,IOPT3,IOPT2,IOPT1
      IF(MIN.GT.0) BIGM = -BIGM
      IF(IFUNCT.EQ.4.OR.IFUNCT.EQ.5) IOPT3 =1
      WRITE(NO,915)IFUNCT,MIN,IOPT3,IOPT2,IOPT1
      INCRMT = 1
      READ(NI,920) NSTAGE,LOWD,MAXD,LOWEST,MAXS,INC
      IF(INC.GT.0) INCRMT = INC
      WRITE(NO,925)NSTAGE,LOWD,MAXD,LOWEST,MAXS,INCRMT
      READ(NI,930) NCOLSP,NCOLSM,NPD,NPS,NROWSC,NCOLSC
      IF(NCOLSP.LE.0) GO TO 40
      DO 35  I = 1,NSTAGE
      READ(NI,940)(P(I,J), J = 1,NCOLSP)
35    WRITE(NO,945) I,(P(I,J),J=1,NCOLSP)
      WRITE(NO,89)
40    IF(NROWSC.LE.0.OR.NCOLSC.LE.0) GO TO 45
      DO 43  I = 1,NROWSC
      READ(NI,940)(C(I,J), J = 1,NCOLSC)
43    WRITE(NO,980) I,(C(I,J),J=1,NCOLSC)
      WRITE(NO,89)
45    IF(NCOLSM.LE.0) GO TO 50
      DO 48  I = 1,NSTAGE
      READ(NI,950)  (M(I,J), J = 1,NCOLSM)
48    WRITE(NO,955) I,(M(I,J),J=1,NCOLSM)
      WRITE(NO,89)
50    IF(NPD.LE.0) GO TO 65
      READ(NI,950) (MPD(I),I=1,NPD)
      WRITE(NO,960)(MPD(I),I=1,NPD)
      WRITE(NO,89)
      N = 1
      KMM = 1
      DO 60  I =1,NPD
      IF(MPD(KMM).LT.0) GO TO 57
      LPD(N) = KMM
      KMM = KMM + 1
      GO TO 59
57    LPD(N) = -(KMM + 1)
      KMM = KMM - MPD(KMM)
59    IF(KMM.GT.NPD) CALL ERROR(1)
      N = N + 1
      IF(N.GT.NSTAGE) GO TO 65
      KMM = KMM + 1
60    CONTINUE
```

```
     65     IF(NPS.LE.0) GO TO 70
            READ(NI,950) (MPS(I),I=1,NPS)
            WRITE(NO,975)(MPS(I),I=1,NPS)
            WRITE(NO,89)
            N = 1
            KMM = 1
            DO 69 I = 1, NPS
            IF(MPS(KMM).LT.0) GO TO 67
            LPS(N) = KMM
            KMM = KMM + 1
            GO TO 68
     67     LPS(N) = - (KMM + 1)
            KMM = KMM - MPS(KMM)
     68     IF(KMM.GT.NPS) CALL ERROR(2)
            N = N + 1
            IF(N.GT.NSTAGE) GO TO 70
            KMM = KMM + 1
     69     CONTINUE
     70     IF(IOPT2.GT.1) READ(NI,950) (MTB(I),I=1,NSTAGE)
            IF(IOPT2.GT.1)WRITE(NO,985) (MTB(I),I=1,NSTAGE)
            IF(IOPT2.GT.1) GO TO 55
            DO 52 I =1,NSTAGE
     52     MTB(I) = IOPT2
     55     NNS = LOCATE(MAXS)
            IF(NNS.GT.MAXSZE) CALL ERROR(3)
            IF(IOPT1.LE.0) GO TO 75
            READ(NI,940) (FSOPT(I),I=1,NNS)
            READ(NI,950) (IDBEST(I),I=1,NNS)
     75     WRITE(NO,905) NRUN,IDAY,IMONTH,IYEAR,(IUSER(I),I=1,5)
            WRITE(NO,15) (IREM(J),J=1,12)
            WRITE(NO,89)
     77     READ(NI,12) (IREM(J),J=1,12)
            WRITE(NO,15)(IREM(J),J=1,12)
            IF(IREM(1).NE.IBLANK) GO TO 77
C          *
C          * END OF INPUT SECTION. *
C          *
            REWIND NTP
            MODETP = -1
            MFLAG = 0
            WRITE(NO,80)
            N1 = 1
            IF(IOPT1.LE.0) GO TO 90
            DO 85 K = 1,NNS
            NS = K*INCRMT + LOWEST - INCRMT
     85     WRITE(NO,506) N1,NS,FSOPT(K),IDBEST(K)
            DO 83 KTAPE = 1,MAXSZE
            J = (KTAPE-1)*IBF+ 1
            J34 = J + IBF1
            WRITE(NTP,530) (IDBEST(L),L=J,J34)
            IF(J34.GE.NNS) GO TO 88
     83     CONTINUE
     88     WRITE(NO,89)
            N1 = 2
C
C          START MAIN LOOP OF PROGRAM.
C
     90     DO 1000 N = N1,NSTAGE
```

```
C         ****INITIALIZE FSBEST(I) AND IDBEST(I)
          DO 100 K = 1,NNS
          FSBEST(K) = BIGM
100       IDBEST(K) = 8888
C
C         INITIALIZE THE STATE VARIABLE NS.*
C
          IF(NPS.GT.0) GO TO 150
          NS = LOWEST
          HS = MAXS
          IF(N.EQ.NSTAGE.AND.NPS.EQ.-1) NS = MAXS
          IF(N.EQ.NSTAGE.AND.NPS.EQ.-2) HS = LOWEST
          GO TO 200
150       LOCS = LPS(N)
          IF(LOCS.LT.0) GO TO 175
          NS = MPS(LOCS)
          IF(NS.GT.MAXS.OR.NS.LT.LOWEST) CALL ERROR(5)
          HS = MPS(LOCS+1)
          IF(HS.GT.MAXS.OR.HS.LT.NS) CALL ERROR(6)
          GO TO 200
175       LOCS = -LOCS
          NS = MPS(LOCS)
          IF(NS.GT.MAXS.OR.NS.LT.LOWEST) CALL ERROR(7)
          K = LOCS -1
          LSHIGH = K-MPS(K)
C
C         NOW INITIALIZE THE DECISION VARIABLE ND.
C
200       I = LOCATE(NS)
          IDBEST(I) = 8888
          IF(NPD.GT.0) GO TO 250
          ND = LOWD
          HD = MAXD
          GO TO 300
250       LOCD = LPD(N)
          IF(LOCD.LT.0) GO TO 275
          ND = MPD(LOCD)
          IF(ND.GT.MAXD.OR.ND.LT.LOWD) CALL ERROR(9)
          HD = MPD(LOCD+1)
          IF(HD.GT.MAXD.OR.HD.LT.ND) CALL ERROR(10)
          GO TO 300
275       LOCD = -LOCD
          ND = MPD(LOCD)
          IF(ND.GT.MAXD.OR.ND.LT.LOWD) CALL ERROR(11)
          K = LOCD -1
          LDHIGH = K-MPD(K)
C
C         NOW USE THE RECURSIVE FUNCTION.
C
300       IF (IFUNCT.NE.0) GO TO 400
C
C         THE FUNCTION RECUR MAY BE WRITTEN BY THE USER.
C
          FNSXN = RECUR (N,NS,ND)
310       ITIED = 0
          IF(FNSXN.EQ.FSBEST(I) ) ITIED = 1
          IF(MIN.GT.0) GO TO 380
C
```

```
C       MAXIMIZING LOGIC.
C
        IF(MTB(N).LE.0) GO TO 325
        IF(FNSXN.GE.FSBEST(I)) GO TO 340
        GO TO 350
325     IF(FNSXN.LE.FSBEST(I)) GO TO 350
C
C       REPLACE WITH VALUE AND DECISION
C       FOR BEST DECISION TO DATE.
C
340     FSBEST(I) = FNSXN
        IDBEST(I)= ND
C
C       NOW INCREMENT THE DECISION VARIABLE.
C
350     IF(ITIED.EQ.0) GO TO 355
        IX = IDBEST(I)
        IDBEST(I) = -IABS(IX)
355     IF(NPD.LE.0.OR.LPD(N).GE.1) GO TO 375
        LOCD = LOCD + 1
        IF(LOCD.GT.LDHIGH) GO TO 500
        ND = MPD(LOCD)
        IF(ND.GT.MAXD.OR.ND.LT.LOWD) CALL ERROR(12)
        GO TO 300
375     ND = ND + INCRMT
        IF(ND.LE.HD) GO TO 300
        GO TO 500
C
C       MINIMIZING LOGIC.
C
380     IF(MTB(N).LE.0) GO TO 390
        IF(FNSXN.LE.FSBEST(I)) GO TO 340
        GO TO 350
390     IF(FNSXN.LT.FSBEST(I)) GO TO 340
        GO TO 350
C
C       LOOKUP OF APPROPRIATE RECURSIVE FUNCTION.
C
400     GO TO (410,420,430,440,450,460,470,480,490), IFUNCT
        CALL ERROR(14)
410     FNSXN = RECUR1 (N,NS,ND)
        GO TO 310
420     FNSXN = RECUR2 (N,NS,ND)
        GO TO 310
430     FNSXN = RECUR3 (N,NS,ND)
        GO TO 310
440     FNSXN = RECUR4 (N,NS,ND)
        GO TO 310
450     FNSXN = RECUR5 (N,NS,ND)
        GO TO 310
460     FNSXN = RECUR6 (N,NS,ND)
        GO TO 310
470     FNSXN = RECUR7 (N,NS,ND)
        GO TO 310
480     FNSXN = RECUR8 (N,NS,ND)
        GO TO 310
490     FNSXN = RECUR9 (N,NS,ND)
        GO TO 310
```

```
      500     IF(IOPT3.LE.0.AND.N.NE.NSTAGE) GO TO 510
      C
      C       OUTPUT OPTIONAL EXCEPT FOR LAST STAGE.
      C
              IX = IDBEST(I)
              ITIED = ITIED1
              IF(IX.GE.0) GO TO 505
              IX = -IX
              ITIED = ITIED2
              MFLAG = 1
      505     WRITE(NO,506) N,NS,FSBEST(I),IX,ITIED
              IF(N.EQ.NSTAGE) GO TO 600
      C
      C       NOW INCREMENT THE STATE VARIABLE.
      C
      510     IF(NPS.LE.0.OR.LPS(N).GE.1) GO TO 515
              LOCS = LOCS + 1
              IF(LOCS.GT.LSHIGH) GO TO 525
              NS = MPS(LOCS)
              IF(NS.GT.MAXS.OR.NS.LT.LOWEST) CALL ERROR(8)
              GO TO 200
      515     NS = NS + INCRMT
              IF(NS.LE.HS) GO TO 200
      525     IF(IOPT3.GT.0) WRITE(NO,89)
              IF(N.EQ.NSTAGE) GO TO 1000
      C
      C       SAVE THE OPTIMUM DECISIONS FOR THIS STAGE ON TAPE.
      C
              DO 535 KTAPE = 1,MAXSZE
              J = (KTAPE-1)*IBF + 1
              J34 = J + IBF1
              WRITE(NTP,530) (IDBEST(L),L=J,J34)
              IF(J34.GE.NNS) GO TO 540
      535     CONTINUE
      C
      C       UPDATE FSOPT(I) FOR USE AT NEXT STAGE.
      C
      540     DO 550 K = 1,NNS
      550     FSOPT(K) = FSBEST(K)
              GO TO 1000
      C
      C       SECTION FOR RECOVERY AND OUTPUT OF OPTIMAL DECISIONS.
      C
      600     N2 = NSTAGE
              JS = NS
              M2 = 1
              KDEC(M2) = IX
              NTY = 0
              IF(IDBEST(I).GE.0) GO TO 625
              NTY = NTY + 1
              KTIES(NTY) = N2
              GO TO 625
      C
      C       BACKSPACE TAPE TO POSITION IN FRONT OF NEXT STAGE.
      C
      610     DO 615 K = 1,KTAPE
      615     BACKSPACE NTP
      620     DO 622 K = 1,KTAPE
```

```
622       BACKSPACE NTP
C
C         NOW READ THE OPTIMUM DECISIONS FOR THE PREVIOUS STAGE.
C
623       DO 624 K = 1,KTAPE
          KK = (K-1)*IBF + 1
          K34 = KK + IBF1
          READ(NTP,530) (IDBEST(J),J=KK,K34)
624       CONTINUE
          KSS = LOCATE (JS)
          N2 = N2 -1
          M2 = M2 +1
          KDEC(M2)= IDBEST(KSS)
          IF(KDEC(M2).GE.0) GO TO 625
          KDEC(M2) = -KDEC(M2)
          NTY = NTY + 1
          KTIES(NTY) = N2
625       IF(N2.EQ.1) GO TO 850
          JX = KDEC(M2)
          IF(JX.EQ.8888) GO TO 510
          IF(IFUNCT.EQ.0) GO TO 700
          GO TO (720,710,710,510,510,720,750,740,730), IFUNCT
C
C         THE FUNCTION NEWS MAY BE WRITTEN BY THE USER.
C
700       JS = NEWS (N2,JS,JX)
          IF(JS.EQ.8888) GO TO 510
705       IF(N2.NE.NSTAGE)GO TO 610
          IF(MODETP.GT.0) GO TO 620
C
C         TAPE MUST BE PUT IN THE READING MODE.
C
          REWIND NTP
          MODETP = -MODETP
          IF(NSTAGE.LT.3) GO TO 623
          KB = (NSTAGE-2)*KTAPE
          DO 707 K = 1,KB
707       READ(NTP,530) (IDBEST(J),J=1,IBF)
          GO TO 623
C
C         LOOK UP THE RECOVERY FUNCTION.
C
710       JS = NEWS1 (N2,JS,JX)
          GO TO 705
720       JS = NEWS2 (N2,JS,JX)
          GO TO 705
730       JS = NEWS3 (N2,JS,JX)
          GO TO 705
740       JS = NEWS4 (N2,JS,JX)
          GO TO 705
750       JS = NEWS5 (N2,JS,JX)
          GO TO 705
C
C         NOW PRINT THE OPTIMAL DECISION FOR EACH STAGE.
C
850       N3 = NSTAGE + 1
          WRITE(NO,860) (KDEC(K),N3-K,K=1,NSTAGE)
```

```
      WRITE(NO,89)
      IF(NTY.EQ.0) GO TO 870
      WRITE(NO,875) (KTIES(K),K=1,NTY)
      WRITE(NO,89)
C
C     NOW REPOSITION TAPE.
C
870   IF(NSTAGE.LT.3) GO TO 510
      KB = (NSTAGE-2)*KTAPE
      DO 880 K = 1,KB
880   READ(NTP,530) (IDBEST(J),J=1,IBF)
      GO TO 510
C
C     END OF THE MAIN LOOP FOR FOR EACH STAGE.
C
1000  CONTINUE
      IF(MFLAG.GT.0) WRITE(NO,990)
      NRUNS = NRUNS - 1
      NRUN = NRUN + 1
      IF(NRUNS.GT.0) GO TO 10
      REWIND NTP
      CALL EXIT
      END




      SUBROUTINE ERROR(IE)
      NO = 6
      WRITE(NO,1) IT
1     FORMAT(///20X,'ERROR CONDITION',I3,
     1 //20X,'JOB TERMINATED.',/)
      CALL EXIT
      RETURN
      END

      FUNCTION LOCATE(IJ)
      COMMON FSOPT(2000),LPD(20), MPD( 99),
     1 P(20,  20), C(20,  20), M(20,  20),
     2 LOWD,MAXD,LOWEST,MAXS,INCRMT,N,NSTAGE,NCOLSC,
     3 NPD,LDHIGH,BIGM,IOPT3,IOPT2,IOPT1,NCOLSP,NCOLSM
      LOCATE = (IJ- LOWEST + INCRMT)/INCRMT
      RETURN
      END

      FUNCTION LOCATD(MN,MD)
      COMMON FSOPT(2000),LPD(20), MPD( 99),
     1 P(20,  20), C(20,  20), M(20,  20),
     2 LOWD,MAXD,LOWEST,MAXS,INCRMT,N,NSTAGE,NCOLSC,
     3 NPD,LDHIGH,BIGM,IOPT3,IOPT2,IOPT1,NCOLSP,NCOLSM
      IF(NPD.GT.0) GO TO 10
      L = LOWD
      GO TO 15
10    IF(LPD(MN).LE.0) GO TO 20
      K = LPD(MN)
      L = MPD(K)
```

```
15      LOCATD = (MD - L + INCRMT)/ INCRMT
19      RETURN
20      K = -LPD(MN)
        LOCATD = 0
25      JDEC = MPD(K)
        LOCATD = LOCATD + 1
        IF(JDEC.EQ.MD) GO TO 19
        K = K + 1
        IF(K.GT.LDHIGH) CALL ERROR(13)
        GO TO 25
        END


        FUNCTION FSTAR(NEW,VL,VH,V1)
        COMMON FSOPT(2000),LPD(20), MPD( 99),
       1 P(20,  20), C(20,  20), M(20,  20),
       2 LOWD,MAXD,LOWEST,MAXS,INCRMT,N,NSTAGE,NCOLSC,
       3  NPD,LDHIGH,BIGM,IOPT3,IOPT2,IOPT1,NCOLSP,NCOLSM
        IF(NEW.LT.LOWEST) GO TO 20
        IF(NEW.GT.MAXS  ) GO TO 25
        IF(N.EQ.1) GO TO 30
        K = (NEW - LOWEST + INCRMT)/INCRMT
        FSTAR = FSOPT(K)
        RETURN
20      FSTAR = VL
        RETURN
25      FSTAR = VH
        RETURN
30      FSTAR = V1
        RETURN
        END


        FUNCTION RECUR1(KN,KS,KX)
C       ******STAGECOACH PROBLEM.*********
        COMMON FSOPT(2000),LPD(20), MPD( 99),
       1 P(20,  20), C(20,  20), M(20,  20),
       2 LOWD,MAXD,LOWEST,MAXS,INCRMT,N,NSTAGE,NCOLSC,
       3  NPD,LDHIGH,BIGM,IOPT3,IOPT2,IOPT1,NCOLSP,NCOLSM
        LOCAD = LOCATD(KN,KX)
        LOCAS = LOCATE(KS)
        RECUR1 = C(LOCAS,LOCAD)
C       ****** TO ALLOW CHOICE OF MIN. OR MAX.****
        IF(RECUR1.EQ.0.) GO TO 99
        RECUR1 = RECUR1 + FSTAR(KX,BIGM,BIGM,0.)
        RETURN
99      RECUR1 = BIGM
        RETURN
        END


        FUNCTION RECUR2(KN,KS,KX)
C       **** ALLOCATION OF RESOURCES, RETURNS TABULATED.
        COMMON FSOPT(2000),LPD(20), MPD( 99),
       1 P(20,  20), C(20,  20), M(20,  20),
       2 LOWD,MAXD,LOWEST,MAXS,INCRMT,N,NSTAGE,NCOLSC,
       3  NPD,LDHIGH,BIGM,IOPT3,IOPT2,IOPT1,NCOLSP,NCOLSM
        KD = LOCATD(KN,KX)
        RECUR2 = P(KN,KD) + FSTAR(KS-KX,BIGM,BIGM,0.)
        RETURN
        END
```

```
      FUNCTION RECUR3 (KN,KS,KX)
C     *** ASSIGNMENT OF PERSONNEL TO MINIMIZE
      COMMON FSOPT(2000),LPD(20), MPD( 99),
     1 P(20,  20), C(20,  20), M(20,  20),
     2 LOWD,MAXD,LOWEST,MAXS,INCRMT,N,NSTAGE,NCOLSC,
     3 NPD,LDHIGH,BIGM,IOPT3,IOPT2,IOPT1,NCOLSP,NCOLSM
C         PROBABILITY OF FAILURE .
      KD = LOCATD(KN,KX)
      RECUR3 = P(KN,KD)*FSTAR(KS-KX,BIGM,BIGM,1.)
      RETURN
      END

      FUNCTION RECUR4(KN,KS,KX)
C     **** MAXIMIZING PROBABILITY OF WINNING
      COMMON FSOPT(2000),LPD(20), MPD( 99),
     1 P(20,  20), C(20,  20), M(20,  20),
     2 LOWD,MAXD,LOWEST,MAXS,INCRMT,N,NSTAGE,NCOLSC,
     3 NPD,LDHIGH,BIGM,IOPT3,IOPT2,IOPT1,NCOLSP,NCOLSM
C         MORE THAN MAXS IN N PLAYS.
      RECUR4 = C(1,1)*FSTAR(KS+KX,BIGM,1.,0.) +
     1 (1.-C(1,1))*FSTAR(KS-KX,BIGM,1.,0.)
      RETURN
      END


      FUNCTION RECUR5(KN,KS,KX)
C     **** PRODUCING BATCHES WITH POSSIBLE DEFECTIVES.
      COMMON FSOPT(2000),LPD(20), MPD( 99),
     1 P(20,  20), C(20,  20), M(20,  20),
     2 LOWD,MAXD,LOWEST,MAXS,INCRMT,N,NSTAGE,NCOLSC,
     3 NPD,LDHIGH,BIGM,IOPT3,IOPT2,IOPT1,NCOLSP,NCOLSM
      IF(KX.NE.0) GO TO 10
      IF(KN.EQ.1) V1 = KS*C(1,3)
      RECUR5 = FSTAR(KS,BIGM,BIGM,V1)
      RETURN
10    LM = KS
      IF(KX.LT.KS) LM = KX
C     INCLUDE SETUP COST AND PRODUCTION COST.
      RECUR5 = C(1,2) + KX*C(1,1)
      IF(KN.EQ.1) V1 = KS*C(1,3)
      RECUR5 = RECUR5+ C(1,5)**KX * FSTAR(KS,BIGM,BIGM,V1)
      IF(LM.EQ.0) GO TO 26
      FACTX = 1.
      DO 13 J = 1,KX
13    FACTX = FACTX*J
      FACTM = 1.
      DO 25  L = 1,LM
      IXM = KX - L
      FACTXM = 1.
      DO 15 J = 1,IXM
15    FACTXM = FACTXM*J
      FACTM = FACTM*L
      TEMP1 = FACTX/(FACTM*FACTXM)
      TEMP2 = ( C(1,4)**L ) * ( C(1,5)**IXM )
      IF(KN.EQ.1) V1 = (KS-L)*C(1,3)
      TEMP3 = FSTAR(KS-L,BIGM,BIGM,V1)
      RECUR5 = RECUR5 + TEMP1*TEMP2*TEMP3
25    CONTINUE
```

```
26      RETURN
        END

        FUNCTION RECUR6(KN,KS,KX)
C       **** CHANGING EMPLOYMENT LEVELS.
        COMMON FSOPT(2000),LPD(20), MPD( 99),
       1 P(20,   20), C(20,   20), M(20,   20),
       2 LOWD,MAXD,LOWEST,MAXS,INCRMT,N,NSTAGE,NCOLSC,
       3  NPD,LDHIGH,BIGM,IOPT3,IOPT2,IOPT1,NCOLSP,NCOLSM
        IF(KX.LT.M(KN,1) ) GO TO 99
        RECUR6=C(1,1)*(KS-KX)**2 + C(1,2)*(KX-M(KN,1)) +
       1       FSTAR(KX,BIGM,BIGM,0.)
        RETURN
99      RECUR6 = BIGM
        RETURN
        END

        FUNCTION NEWS4 (KN,KS,KX)
        COMMON FSOPT(2000),LPD(20), MPD( 99),
       1 P(20,   20), C(20,   20), M(20,   20),
       2 LOWD,MAXD,LOWEST,MAXS,INCRMT,N,NSTAGE,NCOLSC,
       3  NPD,LDHIGH,BIGM,IOPT3,IOPT2,IOPT1,NCOLSP,NCOLSM
        NEWS4  = KS + KX - M(KN,1)
        RETURN
        END

        FUNCTION NEWS5(KN,KS,KX)
        COMMON FSOPT(2000),LPD(20), MPD( 99),
       1 P(20,   20), C(20,   20), M(20,   20),
       2 LOWD,MAXD,LOWEST,MAXS,INCRMT,N,NSTAGE,NCOLSC,
       3  NPD,LDHIGH,BIGM,IOPT3,IOPT2,IOPT1,NCOLSP,NCOLSM
        NEWS5 = KS-KX*M(KN,1)
        RETURN
        END

        FUNCTION NEWS  (KN,KS,KX)
C       THIS IS A DUMMY FUNCTION TO BE REPLACED BY USER
C       IF NECESSARY TO WRITE FUNCTION RECUR.
        NEWS = 8888
C       * IF 8888 IS DETECTED BY MAIN PROGRAM,
C          NO RECOVERY WILL BE ATTEMPTED.
        RETURN
        END

        FUNCTION RECUR (KN,KS,KX)
C       * THIS IS A DUMMY FUNCTION WHICH IS TO BE REPLACED
C       BY THE USER IF NECESSARY.
        COMMON FSOPT(2000),LPD(20), MPD( 99),
       1 P(20,   20), C(20,   20), M(20,   20),
       2 LOWD,MAXD,LOWEST,MAXS,INCRMT,N,NSTAGE,NCOLSC,
       3  NPD,LDHIGH,BIGM,IOPT3,IOPT2,IOPT1,NCOLSP,NCOLSM
        RECUR = BIGM
        CALL ERROR(15)
        RETURN
        END
```

```
      FUNCTION NEWS3 (KN,KS,KX)
      COMMON FSOPT(2000),LPD(20), MPD( 99),
     1 P(20,   20), C(20,   20), M(20,   20),
     2 LOWD,MAXD,LOWEST,MAXS,INCRMT,N,NSTAGE,NCOLSC,
     3  NPD,LDHIGH,BIGM,IOPT3,IOPT2,IOPT1,NCOLSP,NCOLSM
      IF(NCOLSM.GT.1) GO TO 5
      NEWS3 = KS-KX*M(KN,1)
      RETURN
5     K  = LOCATD(KN,KX)
      NEWS3  = KS-M(KN,K)
      RETURN
      END


      FUNCTION RECUR8 (KN,KS,KX)
C     ****GENERAL SOLUTION FOR PRODUCTION VS. INVENTORY.
      COMMON FSOPT(2000),LPD(20), MPD( 99),
     1 P(20,   20), C(20,   20), M(20,   20),
     2 LOWD,MAXD,LOWEST,MAXS,INCRMT,N,NSTAGE,NCOLSC,
     3  NPD,LDHIGH,BIGM,IOPT3,IOPT2,IOPT1,NCOLSP,NCOLSM
      RECUR8 = FSTAR(KS+KX-M(KN,1),BIGM,BIGM,0.)
      IF(RECUR8.EQ.BIGM) GO TO 20
      IF(KX.GT.M(KN,2)) GO TO 5
      CP = KX*P(KN,1)
      GO TO 10
5     CP     = M(KN,2)*P(KN,1)+(KX-M(KN,2))*P(KN,2)
10    RECUR8 = RECUR8 + KS*P(KN,3) + CP
      IF(KX.GT.0) RECUR8 = RECUR8 + P(KN,4)
      IF(KN.EQ.1) GO TO 25
20    RETURN
25    IF(KS+KX.EQ.M(KN,1) ) GO TO 20
      RECUR8 = BIGM
      RETURN
      END


      FUNCTION RECUR9(KN,KS,KX)
C     ****GENERAL SOLUTION FOR SYSTEM RELIABILITY.
      COMMON FSOPT(2000),LPD(20), MPD( 99),
     1 P(20,   20), C(20,   20), M(20,   20),
     2 LOWD,MAXD,LOWEST,MAXS,INCRMT,N,NSTAGE,NCOLSC,
     3  NPD,LDHIGH,BIGM,IOPT3,IOPT2,IOPT1,NCOLSP,NCOLSM
      NEW  = NEWS3(KN,KS,KX)
      IF(NCOLSP.EQ.1) GO TO 5
      K  = LOCATD(KN,KX)
      RECUR9 =      P(KN,K )
      GO TO 10
5     RECUR9 = 1. - (1.-P(KN,1))**KX
10    RECUR9 = RECUR9 * FSTAR(NEW,BIGM,BIGM,1.)
      RETURN
      END
```

```
      FUNCTION RECUR7(KN,KS,KX)
C     **** MAXIMIZING A POLYNOMIAL OBJECTIVE FUNCTION.
      COMMON FSOPT(2000),LPD(20), MPD( 99),
     1 P(20,  20), C(20,  20), M(20,  20),
     2 LOWD,MAXD,LOWEST,MAXS,INCRMT,N,NSTAGE,NCOLSC,
     3  NPD,LDHIGH,BIGM,IOPT3,IOPT2,IOPT1,NCOLSP,NCOLSM
      RECUR7 = FSTAR(KS-KX*M(KN,1),BIGM,BIGM,0.)
      IF(KX.EQ.0.OR.RECUR7.EQ.BIGM) GO TO 15
      DO 10 K = 1,NCOLSP
10    RECUR7 = RECUR7 +      P(KN,K)*KX**(K-1)
15    RETURN
      END



      FUNCTION NEWS1 (KN,KS,KX)
      NEWS1   = KS-KX
      RETURN
      END

      FUNCTION NEWS2 (KN,KS,KX)
      NEWS2  = KX
      RETURN
      END
```

DYNAMIC PROGRAMMING

RUN 1   DATE 10 JAN 1973   USER   JOE H MIZE

SHORTEST ROUTE PROBLEM

PROBLEM TYPE 1
MINIMIZATION OPTION 1   OUTPUT OPTION 1
TIE-BREAKER OPTION 0   STAGE 1 INPUT OPTION 0

NUMBER OF STAGES = 4
RANGE OF DECISIONS IS   2 TO   10
RANGE OF STATES IS   1 TO   10
INCREMENT =   1

C(1,J)	2.00	4.00	3.00	
C(2,J)	7.00	4.00	6.00	
C(3,J)	3.00	2.00	4.00	
C(4,J)	4.00	1.00	5.00	
C(5,J)	1.00	4.00	0.00	
C(6,J)	6.00	3.00	0.00	
C(7,J)	3.00	3.00	0.00	
C(8,J)	3.00	0.00	0.00	
C(9,J)	4.00	0.00	0.00	
MPD (I)	-1	10	8	9	5
	7	2	4		
MPS (I)	8	9	5	7	2
	4	-1	1		

DYNAMIC PROGRAMMING

RUN 1   DATE 10 JAN 1973   USER JOE H MIZE

SHORTEST ROUTE PROBLEM

END OF DATA

STAGE	STATE	OPTIMUM VALUE	OPTIMUM DECISION
1	8	3.00	10
1	9	4.00	10
2	5	4.00	8
2	6	7.00	9
2	7	6.00	8
3	2	11.00	5*
3	3	7.00	5
3	4	8.00	5*
4	1	11.00	3*

    3=X 4   5=X 3   8=X 2   10=X 1

ALTERNATE OPTIMAL DECISIONS AT STAGE/S 4

* AFTER THE OPTIMUM DECISION INDICATES
THAT ALTERNATE OPTIMAL DECISIONS EXIST.
THE TIE-BREAKER OPTION CONTROLS SELECTION
OF THE LOWEST OR HIGHEST DECISION.

II. CONTINUOUS DYNAMIC PROGRAMMING (DYNAM ALGORITHM)*

A. Purpose

This program finds the optimum of a continuous, multivariable, nonlinear function subject to constraints:

Optimize $\quad F(r_1, r_2, \ldots, r_N)$

Subject to $\quad G_{j,k} \leq X_{j,k} \leq H_{j,k} \quad j = 1, 2, \ldots, N$
$\qquad\qquad\qquad\qquad\qquad\qquad\quad k = 1, 2, \ldots, M$

The problem must be structured in a serial stage format, where N is the number of stages. The variables r in F are functions of the stage decisions, $X_{j,k}$, and the stage states, S_j. The upper and lower constraints $H_{j,k}$ and $G_{j,k}$ for $X_{j,k}$ are either constants or functions of the state variables S_j. The state variables are in turn constrained, either between constants or functions of the input and output states. The algorithm will handle initial value, final value, and initial-final value problems.

B. Method

The procedure is based on the "principle of optimality" stated by Richard Bellman (3). This principle relates the properties of decisions made in sequence. Whatever the initial decisions are, the remaining decisions will result in an optimum with regard to the first decision. This idea of sequencing decisions enables a complex problem of many decisions, or variables, to be broken down into a sequence of smaller problems with only a few variables. Each smaller problem is termed a stage with its optimal decisions resulting in a "return" to the overall system objective function. This method of structuring sequential stages has proven effective in solving large problems of many variables. If only a few variables are involved or the problem is small, other, simultaneous, methods should be more efficient. The algorithm is written to solve "serial" type problems, i.e., all stages are in direct sequence.

---

*Computer code developed by Paul S. Inglish, Arizona State University, Tempe, Arizona.

A diagram is shown below.

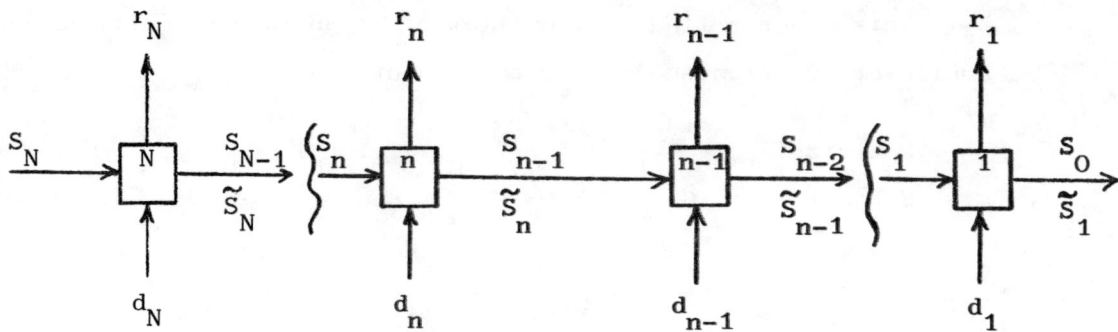

The algorithm proceeds as follows:

1) At stage 1 the upper and lower limits on the input state are evaluated. The interval from the high limit to the low limit is divided into a series of equidistant finite steps.

2) Beginning at the lower limit on the state, S_1, the optimum return is found over the range of decisions d_1:

$$f_1(S_1) = \underset{d_1}{\text{OPT}} \{r_1(S_1, d_1)\}$$

The complex method of Box (7) is used to find each stage optimum (see section I in Chapter 10).

3) The state value is then incremented and the optimum found for this value of S_1.

4) The process is repeated until the upper bound on S_1 is reached.

5) Starting with n=2 the process moves backwards through the stage sequence. The limits on S_n are evaluated and the state interval divided into equidistant steps.

6) The optimum return is found for each value of S_n by the formula

$$f_n(S) = \underset{d_n}{\text{OPT}} \{r_n(S_n, d_n) \; \theta \; f_{n-1}(S_n, d_n)\}$$

where θ is the operator of separability, either addition or multiplication.

7) Recursion is followed until stage N is reached and the optimum found for all values of S_N.

8) The algorithm then follows the optimum path forward to recover the optimum decisions as follows:

    a) For initial value and two-point problems the value of S_N is found in the table with the resulting f_N and optimum decisions. The transformation between S_N and S_{N-1} is used to find the optimum S_{N-1} and its corresponding values. This forward path is followed through all stages until S_1 and the corresponding r_1 and optimum decisions are found.

    b) For the final value problem, at stage N the optimum S_N is found in the table and the process followed as outlined above.

9) Once the optimum path and all values have been determined, the state values, returns, and decisions are printed for each stage.

A flow sheet illustrating the above procedure is given in Figure 5.II.

C. Program Description

1) Usage:

The program consists of a main program (DYNAM), three general subroutines (COMPLX, CHECK, CENTR), and five user supplied subroutines, (FUNC, CONST, SCONST, RETURS, TRANS). Initial guesses for the decisions at each stage, number of steps for the state variables, solution parameters, dimension limits, and printer code designation are passed to the subroutines from DYNAM. The main program coordinates the stage optimizations and performs the recursion to find the optimum path. Final returns, decision values, and state values are printed from the main line. Intermediate printouts are provided in subroutine COMPLX, if the user desires.

2) Subroutines Required:

SUBROUTINE COMPLX (N,M,K, ITMAX, ALPHA, BETA, GAMMA, DELTA, X, F, IT, IEV2, NO, IPRINT, R, G, H, XC) called from main program; coordinates special purpose subroutines (CHECK, CENTR, FUNC, CONST) and performs Complex Box search for single stage optimums.

SUBROUTINE CHECK (N,M,K,X,G,H,I,KODE,XC, DELTA, K1) checks all points against explicit and implicit constraints and applies correction if violations are found.

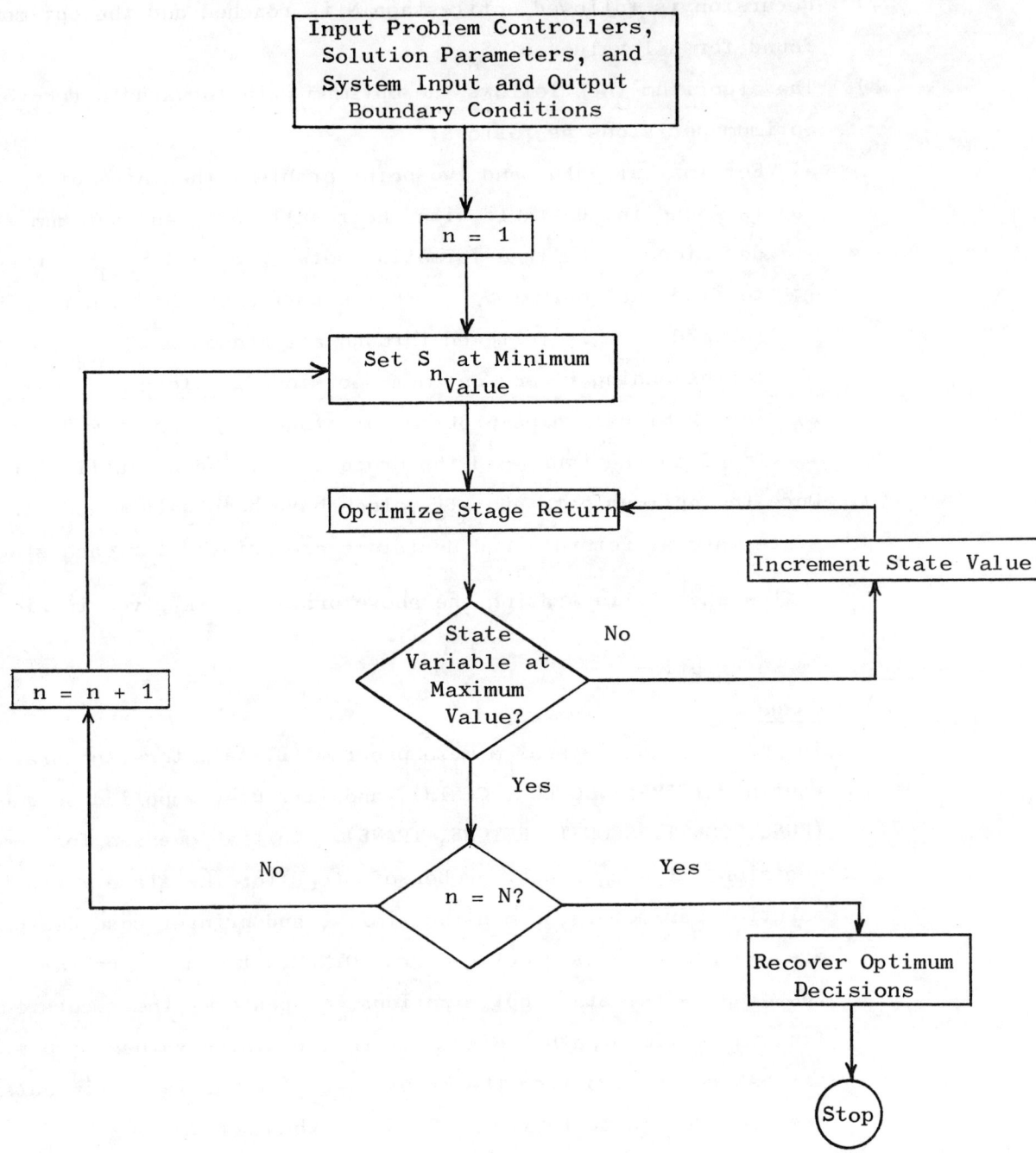

Figure 5.II. Continuous Dynamic Programming (DYNAM ALGORITHM) Logic Diagram

SUBROUTINE CENTR (N,M,K,IEV1, I, XC, X, K1) calculates the centroid of points in complex.

SUBROUTINE FUNC (N,M,K,X,F,I) specifies the function to be optimized at each stage, normally either the sum or product of the returns to that point (user supplied).

SUBROUTINE CONST (N,M,K,X,G,H,I) specifies explicit and implicit constraint limits on the decision values (user supplied), order explicit constraints first.

SUBROUTINE SCONST (N,SCON,KODE) specifies the upper and lower bound on the state variables (user supplied).

SUBROUTINE RETURS (X,I) specifies the return at each stage as a function of the state and decision variables (user supplied).

SUBROUTINE TRANS (IBAC, SMAX, DECMAX, SO, KODE) specifies the stage transformations (user supplied).

3) Description of Parameters:

NSTAGE	Number of stages in system - Define in main program
ITMAX	Maximum number of iterations for stage optimizations - Define in main program
IPRINT	Code to control printing of intermediate values. IPRINT = 2 causes the tables of values to print after each stage is optimized. IPRINT = 1 causes all search values and tables to be printed. IPRINT = 0 suppresses printing until final solution is obtained - Define in main program
IPROB	Code for defining type of problem. IPROB = 1 for initial value problem. IPROB = 2 for final value problem. IPROB = 3 for initial - final (or two-point) problem - Define in main program
IOPT	Code for sign of function at optimum. IOPT =-1 for minimum. IOPT =+1 for maximum - Define in main program
SN	Value of system input state - Define in main program
NDECIS	Number of decision variables at each stage - Define in main program
M	Number of constraints on decision variables at each stage - Define in main program
NSTEP	Number of intervals that range of state variables is to be divided into - Define in main program
DECIS	Decision variables - Define initial guesses in main program
K	Number of points in complex - Defined in main program

ALPHA	Reflection factor - Defined in main program	
BETA	Convergence parameter - Defined in main program	
GAMMA	Convergence parameter - Defined in main program	
DELTA	Explicit constraint violation correction - Defined in main program	
F	Objective function - Defined in subroutine FUNC	
IT	Iteration number - Defined in subroutine COMPLX	
IEV2	Index of point with highest function value - Defined in subroutine COMPLX	
IEV1	Index of point with lowest function value - Defined in subroutine COMPLX	
NI	Card reader unit number - Define in main program	
NO	Printer unit number - Define in main program	
R	Random numbers between 0 and 1 - Defined in subroutine COMPLX	
G	Lower constraints - Define in subroutine CONST	
H	Upper constraints - Define in subroutine CONST	
XC	Centroid - Defined in subroutine CENTR	
I	Point index - Defined in subroutine COMPLX	
KODE	Key used to determine if implicit constraints are provided - Defined in subroutine COMPLX. Key used to determine if high or low constraint is sought - Defined in main program. Key used to determine if system output is sought - Defined in main program	
K1	Do loop limit - Defined in subroutine COMPLX	
SCON	Value of state variable limit - Defined in subroutine SCONST	
IBAC	Index of stage - Defined in main program	
SMAX	Optimum state variable values - Defined in main program and subroutine TRANS	

4) **DIMENSION and COMMON Requirements:**

The DIMENSION statements in the main program and subroutines should be modified according to the requirements of each particular system problem. The parameters included in the following DIMENSION statements conform to the Input Parameter definitions above and the following:

NDM	Maximum number of decisions at any stage
NSM	Maximum number of intervals for state variables
MAIN	DIMENSION NDECIS (NSTAGE), X (NSTAGE, NDM), F(NSTAGE), DECMAX (NSTAGE,NDM), NTOT(NSTAGE), SMAX(NSTAGE), R(K,NDM), G(NDM), H(NDM), XC(NDM), RETMAX(NSTAGE)

FUNC   DIMENSION SFUNC(NSTAGE), XFUNC(NSTAGE,NDM)
RETURS   DIMENSION X(NSTAGE,NDM)
TRANS   DIMENSION SMAX(NSTAGE), DECMAX(NSTAGE,NDM)

Common statements in DYNAM, FUNC, CONST, SCONST, and RETURS should be modified according to the requirements of the system. The parameters are the same as those listed above:

MAIN   COMMON N, S(NSTAGE,NSM), IP, RET(NSTAGE,NSM),SO, DECIS (NSTAGE,NDM,NSM), FDP(NSTAGE,NSM), NTOTN, NTOTM1, IOPT

FUNC   COMMON NUMST, S(NSTAGE,NSM), IP, RET(NSTAGE,NSM),SO, DECIS(NSTAGE,NDM,NSM), FDP(NSTAGE,NSM),NTOTN, NTOTM1, IOPT

CONST   COMMON NUMST, S(NSTAGE,NSM), IP, RET(NSTAGE,NSM), SO

SCONST COMMON NUMST, S(NSTAGE,NSM), IP, RET(NSTAGE,NSM), SO

RETURS COMMON N, S(NSTAGE,NSM), IP, RET(NSTAGE,NSM), SO

5) Input Formats:

CARD TYPE	FORMAT	CONTENTS
1	(8I10)	NSTAGE, ITMAX, IPRINT, IPROB, IOPT
2	(8E10.4)	SN, SO
3	(8I10)	NDECIS(N), M, NSTEP
4	(8E10.4)	DECIS(N,I,1), I = 1, NDECIS(N)

(Repeat CARDS TYPE 3 and 4 for each stage.)

6) Output:

The program DYNAM first prints out the heading and type of problem being run.

Subroutine COMPLX prints all search values and parameters if IPRINT = 1 on Card Type 1.

If IPRINT = 2, DYNAM prints tables of values for each stage.
If IPRINT = 0 only the heading and final solution is printed.

When the solution is found, DYNAM prints the optimum return, returns at each stage, optimum decisions, the system output state value, the value of the state variable into each stage, and the system input state.

7) Summary of User Requirements:

a) Determine values for NSTAGE, ITMAX, IPRINT, IPROB, IOPT, ALPHA, BETA, GAMMA, DELTA, SN, SO, K. See Box (7) for guidelines on parameters. Also, specify NI and NO.

b) Determine NDECIS, M, NSTEP for each stage.

c) Determine the initial estimates for DECIS at each stage.

d) Adjust DIMENSION, COMMON and FORMAT statements as necessary.

e) Specify objective function, returns, and transitions by writing SUBROUTINE(S) FUNC, RETURS, TRANS.

f) Define H (upper constraints) and G (lower constraints) in SUBROUTINE CONST.

g) Define SCON (state variable limits) in SUBROUTINE SCONST.

D. Test Problem

The following test problem was taken from Wilde and Beightler (47). The calculations were performed on a CDC 6400 computer.

Objective Function: Maximize $F = \sum_{i}^{3} r_i$

Return Functions: $r_i = 5.0 d_i - i d_i^2$ $i = 1, 2, 3$

Transition Functions: $\tilde{S}_i = S_i - 0.4 d_i$ $i = 1, 2, 3$

Constraints: $0 \leq d_i \leq S_i$ $i = 1, 2, 3$

State Variable Constraints: $0 \leq S_i \leq 5.0$; $i = 1, 2, 3$

Starting Points: $d_i = 1.0$; $i = 1, 2, 3$

System Input: SN = 5.0

System Output: SO = 0.0

Parameters: NSTAGE = 3, IPROB = 1, ALPHA = 1.3, BETA = 0.01, GAMMA = 5, DELTA = 10^{-6}, K = 3, IPRINT = 0, ITMAX = 999, IOPT = +1

Stage Parameters: NDECIS = 1,1,1; M = 1,1,1; NSTEP = 100,100,100

Algorithm Answers: F = 11.43
d_1 = 2.447
d_2 = 1.190
d_3 = 0.747

Central Processor Time: 60 seconds

The listing and output for this problem are contained in the following section.

E. Program Listings and Example Output

```
C
C     MAIN LINE PROGRAM FOR DYNAMIC PROGRAMMING ALGORITHM
C
      DIMENSION NDECIS(3), X(3,1), F(3), DECMAX(3,1), NTOT(3), SMAX(3),
     1 R(3,1), G(3), H(3), XC(2), RETMAX(3)
      COMMON N, S(3,101), IP, RET(3,101), SO, DECIS(3,1,101),
     1 FDP(3,101), NTOTN, NTOTM1, IOPT
      INTEGER GAMMA
C
      NI = 50
      NO = 66
C
      READ (NI,001) NSTAGE, ITMAX, IPRINT, IPROB, IOPT
  001 FORMAT (8I10)
      IPHOLD = IPRINT
      READ (NI,002) SN, SO
  002 FORMAT (8E10.4)
C
      ALPHA = 1.3
      BETA = 0.01
      GAMMA = 5
      DELTA = 1.0E-06
      FMAX = 0.0
      WRITE (NO,015)
  015 FORMAT (1H1,10X,29HDYNAMIC PROGRAMMING PROCEDURE )
C
      GO TO (220,230,240), IPROB
  220 WRITE (NO,021) NSTAGE
  021 FORMAT (//,2X,16HOPTIMIZATION OF ,I2,28H STAGE INITIAL VALUE PROBL
     1EM  )
      GO TO 250
  230 WRITE (NO,022) NSTAGE
  022 FORMAT (//,2X,16HOPTIMIZATION OF ,I2,28H STAGE FINAL VALUE PROBLEM
     1   )
      GO TO 250
  240 WRITE (NO,023) NSTAGE
  023 FORMAT (//,2X,16HOPTIMIZATION OF ,I2,39H STAGE INITIAL AND FINAL V
     1ALUE PROBLEM
  250 CONTINUE
C
C     RECURSIVE ANALYSIS OF STAGES STARTING AT STAGE 1
C
      N = 1
   10 NS = N
   20 READ (NI,001) NDECIS(N), M, NSTEP
      K = 3*NDECIS(N)
      NDEC = NDECIS(N)
      READ (NI,002) (DECIS(N,I,1), I=1,NDEC)
C
      KODE = 0
      CALL SCONST (N,SLOW,KODE)
      KODE = 1
      CALL SCONST (N,SHIGH,KODE)
```

```
            NTOT(N) = NSTEP + 1
            STEP = (SHIGH-SLOW)/NSTEP
            NTOTN = NTOT(N)
            IF (N.GE.2) NTOTM1 = NTOT(N-1)
C
C       PREFORM SEARCH AT STAGE N FOR GIVEN STATE VALUE
C
            DO 100 IP=1,NTOTN
C
            IF (IPRINT - 1) 28, 28, 26
    26  IPRINT = 0
    28  IM1 = IP - 1
            S(N,IP)= SLOW + IM1*STEP
C
            DO 30 J=1,NDEC
    30  X(1,J) = DECIS(N,J,1)
C
            CALL COMPLX(NDEC,M,K,ITMAX,ALPHA,BETA,GAMMA,DELTA,X,F,IT,IEV2,NO,
        1 IPRINT,R,G,H,XC)
C
            FDP(N,IP) = IOPT * F(IEV2)
            DO 40 J=1,NDEC
    40  DECIS(N,J,IP) = X(IEV2,J)
            IF (IPHOLD - 1 ) 100, 45, 45
    45  IPRINT = IPHOLD
    50  IF (IP - NTOTN) 100, 55, 55
    55  WRITE (NO,003) N
   003  FORMAT (//,10X,26HTABLE OF VALUES FOR STAGE ,I2)
            WRITE (NO,005)
   005  FORMAT (/,2X,11HSTATE VALUE,7X,14HFUNCTION VALUE,11X.
        1 15HDECISION VALUES ,/)
            DO 60 II = 1,NTOTN
    60  WRITE (NO,004) S(N,II), FDP(N,II), (DECIS(N,J,II), J=1,NDEC)
   004  FORMAT (2X,1PE14.6,4X,1PE14.6,3(4X,1PE14.6))
C
   100  CONTINUE
C
            IF (N-NSTAGE) 110, 120, 120
   110  N = N + 1
            GO TO 10
C
C       SOLVING FORWARDS FOR OPTIMUM PATH
C
   120  DO 200 II=1,NSTAGE
            IBAC = NSTAGE + 1 - II
            NTOTN = NTOT(IBAC)
            NDEC = NDECIS(IBAC)
            SMAX(IBAC) = S(IBAC,NTOTN)
            RETMAX(IBAC) = RET(IBAC,NTOTN)
            DO 128 K=1,NDEC
   128  DECMAX(IBAC,K) = DECIS(IBAC,K,NTOTN)
            IF (IBAC.NE.NSTAGE) GO TO 160
            FMAX = FDP(IBAC,NTOTN)
            DO 150 J=1,NTOTN
            GO TO (122,124,122), IPROB
```

```
  122 IF (S(IBAC,J) - SN ) 150, 130, 130
  124 IF (FDP(IBAC,J) - FMAX) 150, 130, 130
  130 FMAX = FDP(IBAC,J)
      DO 140 K=1,NDEC
  140 DECMAX(IBAC,K) = DECIS(IBAC,K,J)
      SMAX(IBAC) = S(IBAC,J)
      RETMAX(IBAC) = RET(IBAC,J)
      IF (IPROB.NE.2) GO TO 200
  150 CONTINUE
      GO TO 200
  160 KODE = 0
      CALL TRANS (IBAC,SMAX,DECMAX,SO,KODE)
      DO 190 J=1,NTOTN
      IF (S(IBAC,J) - SMAX(IBAC)) 190, 170, 170
  170 SMAX(IBAC) = S(IBAC,J)
      RETMAX(IBAC) = RET(IBAC,J)
      DO 180 K=1,NDEC
  180 DECMAX(IBAC,K) = DECIS(IBAC,K,J)
      GO TO 200
  190 CONTINUE
  200 CONTINUE
C
      KODE = 1
      CALL TRANS (IBAC,SMAX,DECMAX,SO,KODE)
C
      DO 300  N=1,NSTAGE
      NDEC = NDECIS(N)
      DO 210 II=1,NDEC
  210 X(N,II)  = DECMAX (N,II)
      IP = 1
      CALL RETURS (X,N)
  300 CONTINUE
C
      WRITE (NO,011) FMAX
  011 FORMAT (///,2X,24HSYSTEM OPTIMUM RETURN = ,1PE16.8)
      WRITE (NO,012)
  012 FORMAT (//,10X,21HMAXIMUM STAGE RETURNS )
      DO 400 I=1,NSTAGE
  400 WRITE (NO,013) I, RETMAX(I)
  013 FORMAT (/,2X,6HSTAGE ,I2,10H RETURN = ,1PE16.8)
      WRITE (NO,014)
  014 FORMAT (//,10X,17HOPTIMUM DECISIONS )
      DO 500 J=1,NSTAGE
      NDEC = NDECIS(J)
      WRITE (NO,016) J, (J, I, X(J,I), I=1,NDEC)
  016 FORMAT (/,2X,6HSTAGE ,I2,6X,2(4X,2HX(,I1,1H,,I1,4H) = ,1PE16.8))
  500 CONTINUE
      WRITE (NO,018) SO
  018 FORMAT (/,2X,26HTHE SYSTEM OUTPUT STATE = ,1PE16.8)
      NSTM1 = NSTAGE - 1
      DO 600 JJ=1,NSTM1
      WRITE (NO,019) JJ, SMAX(JJ)
  019 FORMAT (/,2X,25HTHE INPUT STATE TO STAGE ,I2,3H = ,1PE16.8)
  600 CONTINUE
      SN = SMAX(NSTAGE)
      WRITE (NO,017) SN
  017 FORMAT ( /,2X,25HTHE SYSTEM INPUT STATE = ,1PE16.8)
C
      END
```

```
      SUBROUTINE COMPLX (N,M,K,ITMAX,ALPHA,BETA,GAMMA,DELTA,X,F,IT,IEV2,
     1 NO,IPRINT,R,G,H,XC)
C     COORDINATES SPECIAL PURPOSE SUBROUTINES
C
C     ARGUMENT LIST
C
C     IT    = ITERATION INDEX.
C     IEV1  = INDEX OF POINT WITH MINIMUM FUNCTION VALUE.
C     IEV2  = INDEX OF POINT WITH MAXIMUM FUNCTION VALUE.
C     I     = POINT INDEX.
C     KODE  = CONTROL KEY USED TO DETERMINE IF IMPLICIT CONSTRAINTS
C             ARE PROVIDED.
C     K1    = DO LOOP LIMIT
C
C     ALL OTHERS PREVIOUSLY DEFINED IN MAIN LINE.
C
      DIMENSION X(K,M), R(K,N), F(K), G(M), H(M), XC(N)
      INTEGER GAMMA
C
      IT = 1
      KODE = 0
      IF (M-N) 20,20,10
   10 KODE = 1
   20 CONTINUE
      DO 40 II=2,K
      DO 30 J=1,N
   30 X(II,J) = 0.0
   40 CONTINUE
C
      DO 45 II=2,K
      DO 45 JJ=1,N
C
C     GENERATE RANDOM NUMBERS FOR CALCULATING COMPLEX
C
      R(II,JJ) = RANF(Z)
C
   45 CONTINUE
C
      IF (IPRINT) 46, 48, 46
   46 WRITE (NO,001)
  001 FORMAT (//,2X,10HPARAMETERS )
      WRITE (NO,002) N, M, K, ITMAX, IC, ALPHA, BETA, GAMMA, DELTA
  002 FORMAT (//,2X,4HN = ,I2,3X,4HM = ,I2,3X,4HK = ,I2,2X,8HITMAX = ,
     1I4,2X,5HIC = ,I2,//,2X,8HALPHA = ,F5.2,5X,7HBETA = ,F10.5,3X,
     28HGAMMA = ,I2,3X,8HDELTA = ,E12.6)
      WRITE (NO,003)
  003 FORMAT (//,2X,14HRANDOM NUMBERS)
      DO 47   J=2,K
      WRITE (NO,004) (J, I, R(J,I), I=1,N)
  004 FORMAT (/,2X,3(2HR(,I2,1H,,I2,4H) = ,F6.4,2X))
   47 CONTINUE
C
C     CALCULATE COMPLEX POINTS AND CHECK AGAINST CONSTRAINTS
C
   48 DO 65 II=2,K
```

```
      DO 50 J=1,N
      I = II
      CALL CONST (N,M,K,X,G,H,I)
      X(II,J) = G(J) + R(II,J)*(H(J)-G(J))
   50 CONTINUE
      K1 = II
      CALL CHECK (N,M,K,X,G,H,I,KODE,XC,DELTA,K1)
      IF (II-2) 51, 51, 55
   51 IF (IPRINT) 52, 65, 52
   52 WRITE (NO,018)
  018 FORMAT (//,2X,30HCOORDINATES OF INITIAL COMPLEX)
      IO = 1
      WRITE (NO,019) (IO, J, X(IO,J), J=1,N)
  019 FORMAT (/,3(2X,2HX(,I2,1H,,I2,4H) = ,1PE13.6))
   55 IF (IPRINT) 56, 65, 56
   56 WRITE (NO,019) (II, J, X(II,J), J=1,N)
   65 CONTINUE
      K1 = K
      DO 70 I=1,K
      CALL FUNC (N,M,K,X,F,I)
   70 CONTINUE
      KOUNT = 1
      IA = 0
C
C     FIND POINT WITH LOWEST FUNCTION VALUE
C
      IF (IPRINT) 72, 80, 72
   72 WRITE (NO,021)
  021 FORMAT (/,2X,22HVALUES OF THE FUNCTION )
      WRITE (NO,022) (J, F(J), J=1,K)
  022 FORMAT (/,3(2X,2HF(,I2,4H) = ,1PE13.6))
   80 IEV1 = 1
      DO 100 ICM=2,K
      IF (F(IEV1)-F(ICM)) 100,100,90
   90 IEV1 = ICM
  100 CONTINUE
C
C     FIND POINT WITH HIGHEST FUNCTION VALUE
C
      IEV2 = 1
      DO 120 ICM=2,K
      IF (F(IEV2)-F(ICM)) 110,110,120
  110 IEV2 = ICM
  120 CONTINUE
C
C     CHECK CONVERGENCE CRITERIA
C
      IF (F(IEV2) - F(IEV1) - (ABS(BETA*F(IEV1)))) 140, 130, 130
  130 KOUNT = 1
      GO TO 150
  140 KOUNT = KOUNT + 1
      IF (KOUNT-GAMMA) 150,240,240
C
C     REPLACE POINT WITH LOWEST FUNCTION VALUE
C
```

```
  150 CALL CENTR (N,M,K,IEV1,I,XC,X,K1)
      DO 160  JJ=1,N
  160 X(IEV1,JJ) = (1.0+ALPHA)*(XC(JJ))-ALPHA*(X(IEV1,JJ))
      I = IEV1
      CALL CHECK (N,M,K,X,G,H,I,KODE,XC,DELTA,K1)
      CALL FUNC (N,M,K,X,F,I)
C
C     REPLACE NEW POINT IF IT REPEATS AS LOWEST FUNCTION VALUE
C
  170 IEV2 = 1
      DO 190 ICM=2,K
      IF (F(IEV2)-F(ICM)) 190,190,180
  180 IEV2 = ICM
  190 CONTINUE
      IF (IEV2-IEV1) 220,200,220
  200 DO 210 JJ=1,N
      X(IEV1,JJ)=(X(IEV1,JJ) + XC(JJ))/2.0
  210 CONTINUE
      I = IEV1
      CALL CHECK    (N,M,K,X,G,H,I,KODE,XC,DELTA,K1)
      CALL FUNC (N,M,K,X,F,I)
  220 CONTINUE
      IF (IPRINT) 230, 228, 230
  230 WRITE (NO,023) IT
  023 FORMAT (//,2X,17HITERATION NUMBER ,I5)
      WRITE (NO,024)
  024 FORMAT (/,2X,30HCOORDINATES OF CORRECTED POINT)
      WRITE (NO,019) (IEV1, JC, X(IEV1,JC), JC=1,N)
      WRITE (NO,021)
      WRITE (NO,022) (I, F(I), I=1,K)
      WRITE (NO,025)
  025 FORMAT (/,2X,27HCOORDINATES OF THE CENTROID)
      WRITE (NO,026) (JC, XC(JC), JC=1,N)
  026 FORMAT (/,2X,3(2HX(,I2,6H,C) = ,1PE14.6,4X))
  228 IT = IT + 1
      IF (IT-ITMAX) 80,80,240
  240 RETURN
      END




      SUBROUTINE CENTR (N,M,K,IEV1,I,XC,X,K1)
C
      DIMENSION X(K,M), XC(N)
C
      DO 20  J=1,N
      XC(J) = 0.0
      DO 10  IL=1,K1
   10 XC(J) = XC(J) + X(IL,J)
      RK = K1
   20 XC(J) = (XC(J)-X(IEV1,J))/(RK-1.0)
      RETURN
      END
```

```
      SUBROUTINE CHECK (N,M,K,X,G,H,I,KODE,XC,DELTA,K1)
C
C     ARGUMENT LIST
C
C     ALL ARGUMENTS DEFINED IN MAIN LINE AND CONSX
C
      DIMENSION X(K,M), G(M), H(M), XC(N)
C
   10 KT = 0
      CALL CONST (N,M,K,X,G,H,I)
C
C     CHECK AGAINST EXPLICIT CONSTRAINTS
C
      DO 50 J=1,N
      IF (X(I,J)-G(J)) 20,20,30
   20 X(I,J) = G(J) + DELTA
      GO TO 50
   30 IF (H(J)-X(I,J)) 40,40,50
   40 X(I,J) = H(J) - DELTA
   50 CONTINUE
C
      IF (KODE) 110,110,60
C
C     CHECK AGAINST THE IMPLICIT CONSTRAINTS
C
   60 NN = N + 1
      DO 100 J=NN,M
      CALL CONST (N,M,K,X,G,H,I)
      IF (X(I,J)-G(J)) 80,70,70
   70 IF (H(J)-X(I,J)) 80,100,100
   80 IEV1 = I
      KT = 1
      CALL CENTR (N,M,K,IEV1,I,XC,X,K1)
      DO 90 JJ=1,N
      X(I,JJ) = (X(I,JJ) + XC(JJ))/2.0
   90 CONTINUE
  100 CONTINUE
      IF (KT) 110, 110, 10
  110 RETURN
      END
```

```
      SUBROUTINE FUNC (N,M,K,X,F,I)
C
      DIMENSION X(K,M), F(K)
      DIMENSION SFUNC(3), XFUNC(3,1)
C
      COMMON NUMST, S(3,101), IP, RET(3,101), SO, DECIS(3,1,101),
     1 FDP(3,101), NTOTN, NTOTM1, IOPT
C
      CALL RETURS(X,I)
C
      IF (NUMST - 1) 10, 10, 20
C
   10 F(I) = RET(NUMST,IP)
      GO TO 99
C
   20 NUMT = NUMST - 1
      KODE = 0
      FSTM1 = S(NUMT,NTOTM1)
      SFUNC(NUMST) = S(NUMST,IP)
      DO 30 JF=1,N
   30 XFUNC(NUMST,JF) = X(I,JF)
      CALL TRANS (NUMT,SFUNC,XFUNC,SO,KODE)
      DO 50 IS=1,NTOTM1
      IF (S(NUMT,IS) - SFUNC(NUMT)) 50, 40, 40
   40 FSTM1 = FDP(NUMT,IS)
      GO TO 60
   50 CONTINUE
C
   60 GO TO (1,2), NUMT
C
    1 F(I) = RET(NUMST,IP) + FSTM1
      GO TO 99
C
    2 F(I) = RET(NUMST,IP) + FSTM1
C
   99 F(I) = IOPT * F(I)
      RETURN
C
      END
```

```
      SUBROUTINE SCONST (N,SCON,KODE)
C
      COMMON NUMST, S(3,101), IP, RET(3,101), SO
C
      IF (KODE) 100, 100, 200
C
  100 GO TO (1,2,3), N
C
    1 SCON = 0.00001
      GO TO 99
    2 SCON = 0.00001
      GO TO 99
    3 SCON = 0.00001
      GO TO 99
C
  200 GO TO (4,5,6), N
C
    4 SCON = 5.00
      GO TO 99
    5 SCON = 5.00
      GO TO 99
    6 SCON = 5.00
C
   99 RETURN
      END
```

```
      SUBROUTINE CONST (N,M,K,X,G,H,I)
C
      DIMENSION X(K,M), G(M), H(M)
C
      COMMON NUMST, S(3,101), IP, RET(3,101), SO
C
      GO TO (1,2,3), NUMST
C
    1 G(1) = 0.0
      H(1) = S(NUMST,IP)
      GO TO 99
    2 G(1) = 0.0
      H(1) = S(NUMST,IP)
      GO TO 99
    3 G(1) = 0.0
      H(1) = S(NUMST,IP)
C
   99 RETURN
      END
```

```
      SUBROUTINE TRANS (IBAC,SMAX,DECMAX,SO,KODE)
C
      DIMENSION SMAX(3), DECMAX(3,1)
C
      IF (KODE) 10, 10, 20
C
   10 IBACP1 = IBAC + 1
      GO TO (1,2), IBAC
C
    1 SMAX(IBAC) = SMAX(IBACP1) - 0.4 * DECMAX(IBACP1,1)
      GO TO 99
C
    2 SMAX(IBAC) = SMAX(IBACP1) - 0.4 * DECMAX(IBACP1,1)
      GO TO 99
C
   20 SO = SMAX(1) - 0.4 * DECMAX(1,1)
C
   99 RETURN
      END
```

```
      SUBROUTINE RETURS (X,I)
C
      DIMENSION X(3,1)
C
      COMMON N, S(3,101), IP, RET(3,101), SO
C
      GO TO (1,2,3), N
C
    1 RET(N,IP) = 5.0 * X(I,1) - N * X(I,1)**2.0
      GO TO 99
    2 RET(N,IP) = 5.0 * X(I,1) - N * X(I,1)**2.0
      GO TO 99
    3 RET(N,IP) = 5.0 * X(I,1) - N * X(I,1)**2.0
C
   99 RETURN
      END
```

DYNAMIC PROGRAMMING PROCEDURE

OPTIMIZATION OF 3 STAGE INITIAL VALUE PROBLEM

SYSTEM OPTIMUM RETURN = $1.14259925E+01$

MAXIMUM STAGE RETURNS

STAGE 1 RETURN = $6.21721387E+00$

STAGE 2 RETURN = $3.05545539E+00$

STAGE 3 RETURN = $2.05232569E+00$

OPTIMUM DECISIONS

STAGE 1 $X(1,1) =$ $2.44727504E+00$

STAGE 2 $X(2,1) =$ $1.18986955E+00$

STAGE 3 $X(3,1) =$ $7.47059420E-01$

THE SYSTEM OUTPUT STATE = $3.32109138E+00$

THE INPUT STATE TO STAGE 1 = $4.30000140E+00$

THE INPUT STATE TO STAGE 2 = $4.75000050E+00$

THE SYSTEM INPUT STATE = $5.00000000E+00$

Chapter 6

LEAST SQUARES OBJECTIVE FUNCTIONS

It is the goal of regression analysis or "curve fitting" to obtain values of unknown parameters in an equation utilizing experimental data. Regardless of whether the equation is linear or nonlinear in the parameters, a criterion for determining the best model parameters is required. This requirement is commonly satisfied by the least squares objective function

$$S = \sum_{i=1}^{N} (Y_i - \hat{Y}_i)^2$$

where,

$\hat{Y}_i$ = predicted or calculated value of the dependent variable for the ith observation.

Y_i = experimental value of the dependent variable for the ith observation.

N = number of experimental points.

The best values of the model parameters are obtained when the objective function is minimized.

Methods for estimating the constants may be divided into two categories, (1) linear methods, (2) nonlinear methods. The nonlinear methods are iterative in nature, i.e., starting values are picked and upgraded by the algorithm until a convergence criterion is satisfied. For linear models, the calculational procedure is direct and thus an iterative procedure is not required. In this chapter, four methods designed for least squares objective functions will be presented,

    I.   Linear Regression (LINREG ALGORITHM)

   II.   Gauss Newton (BARD ALGORITHM)

  III.   Marquardt (BSOLVE ALGORITHM)

  IV.   Powell (SSQMIN ALGORITHM)

Each of these methods can also be used for root location problems (systems of algebraic equations) by setting $Y_i = 0$ in the objective function and $\hat{Y}_i = f_i(\underline{X})$ where $f_i(\underline{X})$ is the algebraic function of the unknown independent variables $\underline{X}$. Also some of the general optimization methods described elsewhere in this text may be effective for least squares objective functions. However the methods described in this chapter have been effective for this type of problem (31, 14, 29, 15).

I. LINEAR REGRESSION (LINREG ALGORITHM)*

A. Purpose

This program solves for the coefficients in a multivariable, linear regression equation of the form

$$\hat{Y} = \hat{A}_0 + \hat{A}_1 F_1(\underline{X}) + \hat{A}_2 F_2(\underline{X}) + \ldots + \hat{A}_M F_M(\underline{X})$$

where $\hat{Y}$ is the model dependent variable

$\hat{A}_j$ are the unknown coefficients, $j = 0, 1, 2, \ldots, M$

F_j are functions of the independent variables X_i, $i = 1, 2, 3, \ldots, K$; $j = 1, 2, \ldots, M$

B. Method

The method consists of minimizing a "least squares" objective function, S, of the form,

$$S = \sum_{i=1}^{N} (Y_i - \hat{Y}_i)^2$$

where Y_i are the experimental values of the dependent variable. The algorithm proceeds as follows:

1) The "normal equations" are obtained by setting $\frac{\partial S}{\partial \hat{A}_j} = 0$, $j = 0, 1, 2, \ldots, M$ and eliminating the $\frac{\partial S}{\partial \hat{A}_0} = 0$ equation:

$$(\underline{F}^t \underline{F}) \hat{\underline{A}} = \underline{F}^t \underline{Y}$$

where

$$\underline{F} = \begin{bmatrix} (F_{1,1} - \overline{F}_1) & (F_{1,2} - \overline{F}_2) & \ldots & (F_{1,M} - \overline{F}_M) \\ (F_{2,1} - \overline{F}_1) & (F_{2,2} - \overline{F}_2) & \ldots & (F_{2,M} - \overline{F}_M) \\ \vdots & \vdots & & \vdots \\ (F_{N,1} - \overline{F}_1) & (F_{N,2} - \overline{F}_2) & \ldots & (F_{N,M} - \overline{F}_M) \end{bmatrix}$$

---

*Computer code developed by Paul S. Inglish, Arizona State University, Tempe, Arizona.

$$\underline{Y} = \begin{bmatrix} (Y_1 - \overline{Y}) \\ (Y_2 - \overline{Y}) \\ \vdots \\ (Y_N - \overline{Y}) \end{bmatrix} \qquad \underline{\hat{A}} = \begin{bmatrix} \hat{A}_1 \\ \hat{A}_2 \\ \vdots \\ \hat{A}_M \end{bmatrix}$$

$\underline{F}^t$ is the transpose of the $\underline{F}$ matrix. $\overline{Y}$ and $\overline{F}_j$ are mean values.

2) For the linear regression model, the normal equations will be linear, with the unknowns being the $\underline{\hat{A}}$ vector. Thus any appropriate linear algebraic equation solution scheme may be used to solve for the unknown coefficients, $\hat{A}_1$ to $\hat{A}_M$. $\hat{A}_o$ is obtained from

$$\hat{A}_o = \overline{Y} - \sum_{j=1}^{M} \hat{A}_j \overline{F}_j$$

3) Two tests are often performed to determine the validity of the model. First, the least squares objective function, S, is evaluated. For a perfect fit, this value would be zero. Secondly, the "multiple correlation coefficient," R^2, may be calculated,

$$R^2 = \frac{\text{Sum of Squares due to regression (SUMSR)}}{\text{Sum of Squares corrected total (SUMST)}}$$

where
$$\text{SUMSR} = \underline{\hat{A}}^t (\underline{F}^t \underline{Y}) = \sum_{i=1}^{N} (\hat{Y}_i - \overline{Y})^2$$

$$\text{SUMST} = \underline{Y}^t \underline{Y} = \sum_{i=1}^{N} (Y_i - \overline{Y})^2$$

The value of R^2 will be between 0 and 1 with $R^2 = 1$ corresponding to a perfect fit.

A flow diagram illustrating the calculational procedure is given in Figure 6.I.

C. Program Description

1) Usage:

The program consists of a short main program, the subroutine LINREG, and the subroutine SIMQ. All data is supplied through the main program and the functions of X are set up there. Subroutine LINREG

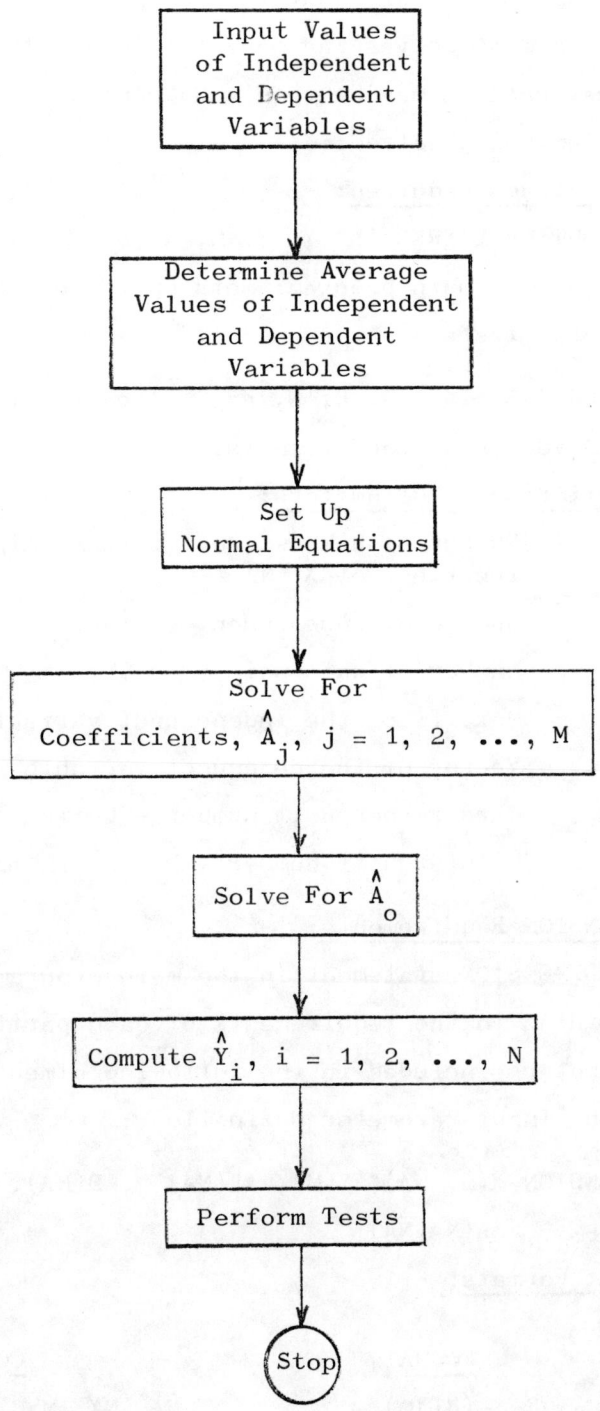

Figure 6.I. Linear Regression (LINREG ALGORITHM) Logic Diagram

sets up the normal equations and prints out the results. Subroutine SIMQ solves the normal (linear algebraic) equations. Format changes may be required, depending on the particular problem being solved.

2) <u>Subroutines Required</u>:

SUBROUTINE LINREG (X, Y, N, M, A, B, XBAR, YHAT, AA, N2, NO) called from main program; sets up the normal equations and performs tests.

SUBROUTINE SIMQ (A, B, N, KS, NS) called from LINREG, solves for the $\hat{A}$ vector of coefficients.

3) Description of Parameters:

NA	Number of unknown coefficients minus one; also number of functions of X (NA $\geq$ 2)
NX	Number of independent variables
M	Number of data points
X	Maxtrix of the independent variables
Y	Vector of the dependent variable
NI	Card reader unit number – Define in main program
NO	Printer unit number – Define in main program

4) <u>DIMENSION Requirements</u>:

The DIMENSION statement in the main program should be modified according to the requirements of each particular problem. The parameters included in the following dimension statement conform to the Input Parameter definitions:

DIMENSION X(M, NA), Y(M), A[(NA)^2], B(NA), XBAR(NA), YHAT(M), AA(NA,NA)

5) Input Formats:

CARD TYPE	FORMAT	CONTENTS
1	(8I10)	NA, NX, M
2	(8F10.5)	[X(J,I), J = 1,M]

(I = 1, NX; thus a total of NX data cards of this type are read in; repeat card or change format for M > 8.)

3	(8F10.0)	[Y(I), I = 1,M]

(If M exceeds 8, additional data cards of this type are read in.)

For multiple data sets using the same function form, repeat Cards Type 1, 2, and 3 in separate blocks.

6) **Output:**

The main program prints out the following information with the symbols translated in terms of the method description:

Output	Method
XBAR	$\overline{F}_j$, $j = 1, 2, \ldots, M$
YBAR	$\overline{Y}$
A MATRIX	$(\underline{F}^t\underline{F})$
B MATRIX	$(\underline{F}^t\underline{Y})$
AHAT	$\hat{A}_j$, $j = 1, 2, \ldots, M$
AZERO	$\hat{A}_o$
YHAT	$\hat{Y}_i$, $i = 1, 2, \ldots, N$
Y	Y_i, $i = 1, 2, \ldots, N$
SUMST	SUMST
S	S
R**2	R^2

7) **Summary of User Requirements:**

a) Determine values for NA, NX, M [X(J,I), J=1,M, I=1, NX], [Y(I), I=1,M].

b) Adjust DIMENSION and FORMAT statements as necessary.

c) Specify NI and NO in main program.

d) Specify the $F_j(\underline{X})$ forms in the main program.

D. **Test Problem**

The following test problem was constructed to test the algorithm. Calculations were performed on a CDC 6400 computer.

Model: $\hat{Y} = \hat{A}_o + \hat{A}_1 X_1 + \hat{A}_2 X_2 + \hat{A}_3 X_1^2 + \hat{A}_4 X_2^2 + \hat{A}_5 X_1 X_2$

Data: Y	X_1	X_2
2	2	11
8	5	2
6	6	3
12	8	11
3	3	6
11	7	6
7	4	3

Dimensions: NA = 5, NX = 2, M = 7

Algorithm Answers:
$\hat{A}_0 = 6.0028$
$\hat{A}_1 = -0.0998$
$\hat{A}_2 = -0.7813$
$\hat{A}_3 = 0.0957$
$\hat{A}_4 = 0.0189$
$\hat{A}_5 = 0.0833$
$S = 10.0043$
$R^2 = 0.8809$

Central Processor Time: 3 seconds

The listing and output for this problem are contained in the following section.

E. Program Listings and Example Output

```
C
C       MAIN LINE PROGRAM FOR SUBROUTINE LINREG
C

        NI = 50
        NO = 66
C
   10   READ (NI,001) NA,NX,M
  001   FORMAT (8I10)
        NN = NA*NA
        IF (NA) 999, 999, 20
   20   DO 100 I=1,NX
        READ (NI,002) (X(J,I), J=1,M)
  002   FORMAT (8F10.0)
  100   CONTINUE
C
C       SECTION 1*** SET UP FUNCTIONS***
C
        DO 200 K=1,M
        X(K,3) = X(K,1)**2
        X(K,4) = X(K,2)**2
        X(K,5) = X(K,1)*X(K,2)
C
  200   CONTINUE
C
        READ (NI,003) (Y(I), I=1,M)
  003   FORMAT (8E10.4)
C
        CALL LINREG (X,Y,NA,M,A,B,XBAR,YHAT,AA,NN,NO)
C
        GO TO 10
C
  999   STOP
        END
```

```
      SUBROUTINE LINREG (X,Y,N,M,A,B,XBAR,YHAT,AA,N2,NO)
C
      DIMENSION X(M,N), Y(M), A(N2), B(N), XBAR(N), YHAT(M), AA(N,N)
C
      WRITE (NO,001)
  001 FORMAT (1H1,10X,36HMULTIPLE LINEAR REGRESSION ALGORITHM  )
C
C     CALCULATE AVERAGE X AND Y VALUES
C
      DO 200 I=1,N
      SUMX = 0.0
      DO 100 J=1,M
  100 SUMX = SUMX + X(J,I)
  200 XBAR(I) = SUMX / FLOAT(M)
      SUMY = 0.0
      DO 300 K=1,M
  300 SUMY = SUMY + Y(K)
      YBAR = SUMY / FLOAT(M)
C
      WRITE (NO,002)
  002 FORMAT (//,2X,23HVARIABLE AVERAGE VALUES  )
      WRITE (NO,003) (II, XBAR(II), II=1,N)
  003 FORMAT (/,3(2X,5HXBAR(,I2,4H) = ,1PE14.7 ))
      WRITE (NO,004) YBAR
  004 FORMAT (/,2X,7HYBAR = ,1PE14.7 )
C
C     CALCULATE REGRESSION MATRICES
C
      KK = 1
      DO 500 I=1,N
      DO 500 J=1,N
      SUMA = 0.0
      SUMB = 0.0
      DO 400 K=1,M
      SUMA = SUMA + (X(K,I) -XBAR(I)) * (X(K,J) - XBAR(J))
  400 SUMB = SUMB + (Y(K) - YBAR) * (X(K,I) - XBAR(I))
      AA(I,J) = SUMA
      A(KK) = SUMA
      KK = KK + 1
  500 B(I) = SUMB
C
      WRITE (NO,005)
  005 FORMAT (//,10X,8HA MATRIX )
      DO 550 II=1,N
  550 WRITE (NO,006) (AA(II,JJ), JJ=1,N)
  006 FORMAT (/,8(2X,E10.4))
      WRITE (NO,007)
  007 FORMAT (//,10X,8HB MATRIX )
      WRITE (NO,006) (B(KK), KK=1,N)
C
C     SOLVE REGRESSION MATRICES FOR COEFFICIENTS
C
      CALL SIMQ (A,B,N,KS,N2)
      SUMX = 0.0
      DO 600 I=1,N
```

```
  600 SUMX = SUMX + B(I) * XBAR(I)
      AZERO = YBAR - SUMX
C
      WRITE (NO,008)
  008 FORMAT (1H1,10X,37HVALUES OF THE REGRESSION COEFFICIENTS   )
      WRITE (NO,009) (JJ, B(JJ), JJ=1,N)
  009 FORMAT (/,2(2X,5HAHAT(,I2,4H) = ,1PE16.8,8X ))
      WRITE (NO,010) AZERO
  010 FORMAT (/,2X,8HAZERO = ,1PE16.8  )
C
C     CALCULATE S AND R TEST VALUES
      STEST = 0.0
      DO 800 J=1,M
      SUMS1 = 0.0
      DO 700 K=1,N
  700 SUMS1 = SUMS1 + B(K) * X(J,K)
      YHAT(J) = AZERO + SUMS1
      DIFF = (Y(J) - YHAT(J))**2
  800 STEST = STEST + DIFF
      SUMST = 0.0
      DO 900 I=1,M
  900 SUMST = SUMST + (Y(I) - YBAR)**2
      SUMSR = SUMST - STEST
      RTEST = SUMSR / SUMST
C
      WRITE (NO,011)
  011 FORMAT (////,5X,19HEXPERIMENTAL VALUES,18X,17HREGRESSION VALUES )
      DO 1000 KK=1,M
 1000 WRITE (NO,012) KK, Y(KK), KK, YHAT(KK)
  012 FORMAT (/,2X,2HY(,I3,4H) = ,1PE16.8,10X,5HYHAT(,I3,4H) = ,1PE16.8)
C
      WRITE (NO,013) SUMST, STEST, RTEST
  013 FORMAT (///,2X,8HSUMST = ,1PE16.8,/,2X,4HS = ,1PE16.8,10X,
     1 7HR**2 = ,1PE16.8  )
C
      RETURN
      END
```

```
      SUBROUTINE SIMQ(A,B,N,KS,NS)
C
      DIMENSION  A(NS), B(N)
C
C        FORWARD SOLUTION
C
      TOL=0.0
      KS=0
      JJ=-N
      DO 65 J=1,N
      JY=J+1
      JJ=JJ+N+1
      BIGA=0
      IT=JJ-J
      DO 30 I=J,N
C
C        SEARCH FOR MAXIMUM COEFFICIENT IN COLUMN
C
      IJ=IT+I
      IF(ABS(BIGA)-ABS(A(IJ))) 20,30,30
   20 BIGA=A(IJ)
      IMAX=I
   30 CONTINUE
C
C        TEST FOR PIVOT LESS THAN TOLERANCE (SINGULAR MATRIX)
C
      IF(ABS(BIGA)-TOL) 35,35,40
   35 KS=1
      RETURN
C
C        INTERCHANGE ROWS IF NECESSARY
C
   40 I1=J+N*(J-2)
      IT=IMAX-J
      DO 50 K=J,N
      I1=I1+N
      I2=I1+IT
      SAVE=A(I1)
      A(I1)=A(I2)
      A(I2)=SAVE
C
C        DIVIDE EQUATION BY LEADING COEFFICIENT
C
   50 A(I1)=A(I1)/BIGA
      SAVE=B(IMAX)
      B(IMAX)=B(J)
      B(J)=SAVE/BIGA
C
C        ELIMINATE NEXT VARIABLE
C
      IF(J-N) 55,70,55
   55 IQS=N*(J-1)
      DO 65 IX=JY,N
      IXJ=IQS+IX
      IT=J-IX
```

```
      DO 60 JX=JY,N
      IXJX=N*(JX-1)+IX
      JJX=IXJX+IT
   60 A(IXJX)=A(IXJX)-(A(IXJ)*A(JJX))
   65 B(IX)=B(IX)-(B(J)*A(IXJ))
C
C        BACK SOLUTION
C
   70 NY=N-1
      IT=N*N
      DO 80 J=1,NY
      IA=IT-J
      IB=N-J
      IC=N
      DO 80 K=1,J
      B(IB)=B(IB)-A(IA)*B(IC)
      IA=IA-N
   80 IC=IC-1
C
      RETURN
      END
```

MULTIPLE LINEAR REGRESSION ALGORITHM

VARIABLE AVERAGE VALUES

XBAR(1) = 5.0000000E+00   XBAR(2) = 6.0000000E+00   XBAR(3) = 2.9000000E+01
XBAR(4) = 4.8000000E+01   XBAR(5) = 3.0000000E+01   XBAR(

YBAR = 7.0000000E+00

A MATRIX

```
.2800E+02  0.         .2800E+03  0.         .2520E+03
0.         .8400E+02  .8400E+02  .1140E+04  .4200E+03
.2800E+03  .8400E+02  .2884E+04  .1140E+04  .2940E+04
0.         .1140E+04  .1140E+04  .1592E+05  .5700E+04
.2520E+03  .4200E+03  .2940E+04  .5700E+04  .4584E+04
```

B MATRIX

```
.4500E+02  -.1000E+01  .4490E+03  -.5000E+01  .4180E+03
```

VALUES OF THE REGRESSION COEFFICIENTS

AHAT(1) =   -9.98883929E-02   AHAT(2) =   -7.81309186E-01
AHAT(3) =   9.57031250E-02   AHAT(4) =   1.89393939E-02
AHAT(5) =   8.33333333E-02   AHAT(

AZERO =   6.00281554E+00

EXPERIMENTAL VALUES   REGRESSION VALUES

Y(1) =   2.00000000E+00   YHAT(1) =   1.71645022E+00

Y(2) =   8.00000000E+00   YHAT(2) =   7.24242424E+00

Y(3) =   6.00000000E+00   YHAT(3) =   8.17532468E+00

Y(4) =   1.20000000E+01   YHAT(4) =   1.23593074E+01

Y(5) =   3.00000000E+00   YHAT(5) =   4.05844156E+00

Y(6) =   1.10000000E+01   YHAT(6) =   9.48701299E+00

Y(7) =   7.00000000E+00   YHAT(7) =   5.96103896E+00

SUMST =   8.40000000E+01
S =   1.00043290E+01   R**2 =   8.80900845E-01

II. GAUSS NEWTON (BARD ALGORITHM)*

A. Purpose

This program solves for the coefficients in a multivariable, nonlinear regression equation $\hat{Y} = F(X_1, X_2, \ldots, X_K; \hat{A}_1, \hat{A}_2, \ldots, \hat{A}_M)$ utilizing N data points for Y_i and $X_{k,i}$, $i = 1, 2, \ldots, N$; $k = 1, 2, \ldots, K$.

B. Method

The procedure is based on a linearization of the proposed model. A least squares objective function is utilized. The method has proved effective where good starting estimates of the unknown coefficients are available. The algorithm proceeds as follows:

1) The model is linearized by expanding $\hat{Y}_i$ in a Taylor series about current trial values for the coefficients and retaining the linear terms only,

$$\hat{Y}_i = \hat{Y}_i^* + \left[\frac{\partial \hat{Y}_i}{\partial \hat{A}_1}\right]^* \Delta \hat{A}_1 + \left[\frac{\partial \hat{Y}_i}{\partial \hat{A}_2}\right]^* \Delta \hat{A}_2 + \ldots + \left[\frac{\partial \hat{Y}_i}{\partial \hat{A}_M}\right]^* \Delta \hat{A}_M$$

where $\Delta \hat{A}_j = [\hat{A}_j - \hat{A}_j^*]$, $j = 1, 2, \ldots, M$.

The asterisk designates quantities evaluated at the initial trial values.

2) A least squares objective function is formulated,

$$\text{Minimize} \quad S = \sum_{i=1}^{N} (Y_i - \hat{Y}_i)^2.$$

3) The linearized model is substituted into the objective function and the "normal equations" formed by setting the partial derivatives of the objective function with respect to each coefficient equal to zero:

$$\frac{\partial S}{\partial \hat{A}_j} = 0, \quad j = 1, 2, \ldots, M.$$

---

*The FORTRAN program contained in this section is based on <u>Nonlinear Parameter Estimation and Programming</u>, described on page 237 of "Catalog of Programs for IBM System 360 Models 25 and Above," GC 20-1619-8; program number 360.D-13.6.003. Used by permission of International Business Machines Corporation.

The resulting normal equations will be of the form

$$(\underline{A}^t \underline{A}) \underline{\Delta \hat{A}} = \underline{A}^t (\underline{Y} - \underline{\hat{Y}}^*)$$

where

$$\underline{A} = \begin{bmatrix} \dfrac{\partial \hat{Y}_1}{\partial \hat{A}_1} & \dfrac{\partial \hat{Y}_1}{\partial \hat{A}_2} & \cdots & \dfrac{\partial \hat{Y}_1}{\partial \hat{A}_M} \\[1em] \dfrac{\partial \hat{Y}_2}{\partial \hat{A}_1} & \dfrac{\partial \hat{Y}_2}{\partial \hat{A}_2} & & \dfrac{\partial \hat{Y}_2}{\partial \hat{A}_M} \\[1em] \vdots & \vdots & & \vdots \\[1em] \dfrac{\partial \hat{Y}_N}{\partial \hat{A}_1} & \dfrac{\partial \hat{Y}_N}{\partial \hat{A}_2} & \cdots & \dfrac{\partial \hat{Y}_N}{\partial \hat{A}_M} \end{bmatrix}^*$$

$$\underline{\Delta \hat{A}} = \begin{bmatrix} (\hat{A}_1 - \hat{A}_1^*) \\ (\hat{A}_2 - \hat{A}_2^*) \\ \vdots \\ (\hat{A}_M - \hat{A}_M^*) \end{bmatrix}, \qquad (\underline{Y} - \underline{\hat{Y}}^*) = \begin{bmatrix} (Y_1 - \hat{Y}_1^*) \\ (Y_2 - \hat{Y}_2^*) \\ \vdots \\ (Y_N - Y_N^*) \end{bmatrix}$$

$\underline{A}^t$ is the transpose of the $\underline{A}$ matrix. The derivatives in the $\underline{A}$ matrix may be evaluated analytically or numerically.

4) The normal equations are a system of linear algebraic equations and are solved by an appropriate technique for $\underline{\Delta \hat{A}}$. The $\underline{\Delta \hat{A}}$ vector and S will approach zero as convergence is achieved. If convergence is achieved, the final coefficients are calculated from

$$\hat{A}_j = \hat{A}_j^* + \Delta \hat{A}_j, \qquad j = 1, 2, \ldots, M.$$

If convergence is not achieved, $\underline{\hat{A}}^*$ is updated by replacing the old values by the new values and the process repeated.

A flow sheet illustrating the above procedure is given in Figure 6.II.

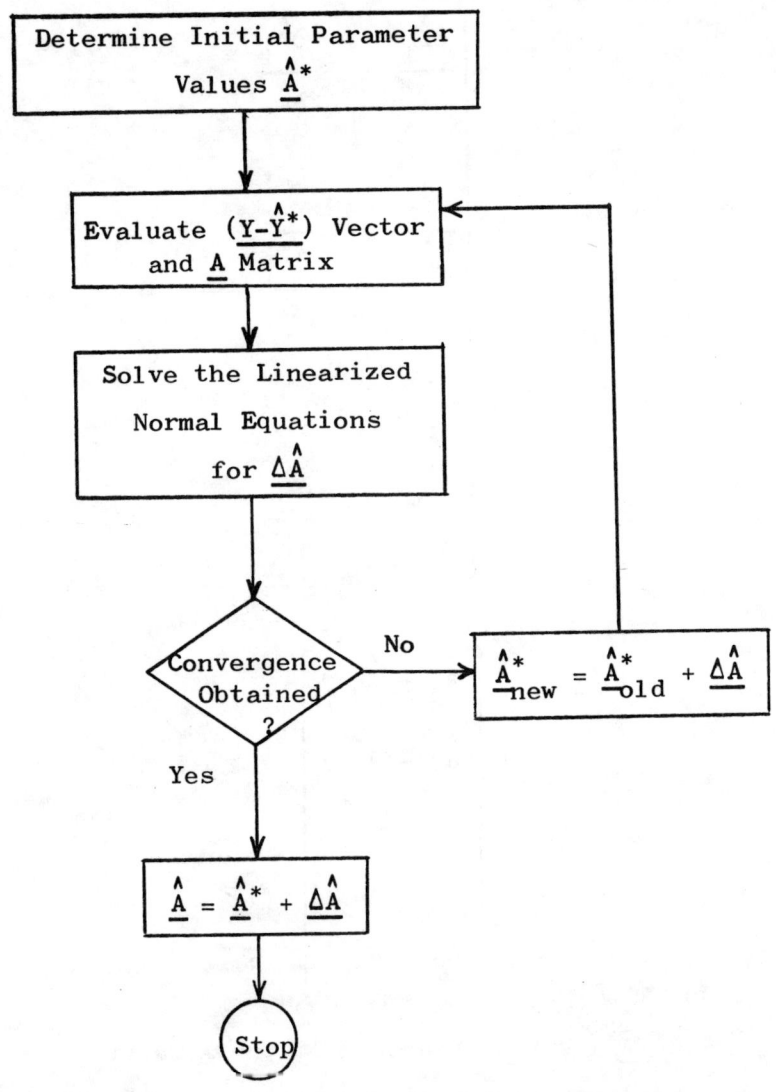

Figure 6.II. Gauss Newton (BARD ALGORITHM) Logic Diagram

C. Program Description

1) Usage:

The program consists of a main program, six subroutines and a function subprogram. The only purpose of the main program is to make a single call on the primary subroutine NLMAX. All input and output is from the subroutines. The bulk of the communication is in COMMON. Format changes may be required depending on the problem under consideration.

2) Subroutines Required:

SUBROUTINE NLMAX(NI,NO) called from main program and coordinates all calculations.

SUBROUTINE ACCUM (II,NI,NO) computes the least squares objective function.

SUBROUTINE DLSQ (II, I) calculate numerical derivatives.

SUBROUTINE EIG (N, II) computes eigenvalues and eigenvectors of a real symmetric matrix.

SUBROUTINE OUT (NO) prints out the final results.

SUBROUTINE BOUND (II, H, NI, NO) specifies upper and lower constraints on the desired coefficients (optional).

FUNCTION FUNC (C1, A, I) specifies model (user supplied).

3) Description of Parameters:

NI	Card reader unit number - Define in main program
NO	Printer unit number - Define in main program
TITLE	An eighty character "A" field used to label printout - Define in subroutine NLMAX
NTH	Number of coefficients in model - Define in subroutine NLMAX
LOUT	Printing controller for intermediate printouts. LOUT = 1 causes intermediate values to print on each iteration; LOUT = 2 suppresses printing until final solution - Define in subroutine NLMAX
NPH	Penalty function controller. NPH = 1 if penalty function is included; NPH = 2 if penalty function not used (Define in subroutine NLMAX). The penalty function modifies the objective function to prevent discontinuities near the parameter bounds (if specified)
MD	Number of model under consideration - Define in subroutine NLMAX

	C1	Initial guesses or current values of coefficients - Define in subroutine NLMAX
	M	Number of data points for model - Define in subroutine ACCUM
	NA	Number of independent plus dependent variables - Define in subroutine ACCUM
	A	Matrix of variable values - Define in subroutine ACCUM
	CLB	Lower bounds on coefficients - Define in subroutine BOUND
	CUB	Upper bounds on coefficients - Define in subroutine BOUND
	II	Defines task of subroutines - Define in subroutine NLMAX. If I = 1, the subroutine is to calculate the difference between the computed and experimental value of the dependent variable at the Ith experiment. If II = 2, in addition the derivative of the function with respect to the parameters must be calculated. If II = 3, return to the calling program occurs.
	N	Defines operation of subroutine EIG - Define in subroutine NLMAX
	I	Experimental point index - Define in subroutine ACCUM
	H	Minimum bound value - Define in subroutine BOUND

4) <u>DIMENSION Requirements:</u>

Dimensioning is performed in COMMON and should be modified according to the requirements of each particular problem. The parameters included in the following statements conform to the Input Parameter definitions, above:

COMMON C(M,M), G1(M,M), PSCA, G(M,M), F(M), Y(M), EGV(NTH), FF(NTH),
    TITLE(20), CUB(NTH), CLB(NTH), PNL(M), ..., C1(NTH), W(M),
    A(M,NTH)

5) <u>Input Formats:</u>

CARD TYPE	FORMAT	CONTENTS
1	(20A4)	TITLE
2	(16I5)	NTH, LOUT, MD
3	(4F20.5)	(C1(I), I=1, NTH)
4	(16I5)	M, NA
5	(4F20.10)	(A(I,J), J=1, NA)
		(Repeat cards for I = 1, M)
6	(4F20.5)	(CLB(I), I=1, NTH)
		(If NTH exceeds 4 an additional CARD TYPE 6 is required.)
7	(4F20.5)	(CUB(I), I=1, NTH)
		(If NTH exceeds 4 an additional CARD TYPE 7 is required.)

6) <u>Output</u>:

The initial output consists of the method label (from the main program) and model description and number (from NLMAX). The experimental observations of the independent and dependent variables are then tabulated along with the upper and lower parameter bounds (if specified), penalty function coefficients and initial parameter estimates. Intermediate printout (LOUT = 1; omitted if LOUT = 2) consists of the objective function value (negative since code maximizes objective function) and current values of the parameters for each iteration (corresponding to a new evaluation of the derivative matrix). Final output includes final values of the objective function and parameters and number of function and derivative evaluations. Also included are the gradient vector of the objective function with respect to the parameters, approximate second derivative (Hessian) matrix, eigenvalues and principal components. Finally, the residual differences between the computed and experimental values, standard deviation of residuals, standard deviation of parameters, covariance matrix of the parameter estimates, eigenvalues of the correlation matrix of parameters and the principal components of the parameter estimates with their expected values and standard deviations are printed.

7) <u>Summary of User Requirements</u>:

   a) Determine values for NTH, LOUT, MD, M, NA, CLB, CUB, NI, and NO.
   b) Determine initial guesses for coefficients, enter as (C1(I), I=1, NTH).
   c) Specify data in A(I,J); order independent variables first.
   d) Adjust the COMMON statement dimensions.
   e) Specify model by writing FUNCTION FUNC.
   f) Adjust the input and output format statements as required by the problem under consideration.

The program code presented here is part of a larger package containing a number of additional options. Further details are available from IBM (52).

D. <u>Test Problem</u>

The following problem was used to test the code (26). The calculations were performed on a CDC 6400 computer.

Model: $\hat{Y} = \hat{A}_1 + \hat{A}_2 \exp(\hat{A}_3 X)$

Data:

Y	X
127	-5
151	-3
379	-1
421	1
460	3
426	5

Starting Point:
$\hat{A}_1 = 500.$
$\hat{A}_2 = -150.$
$\hat{A}_3 = -0.20$

Constraints:
$-999. \leq \hat{A}_1 \leq 999.$
$-999. \leq \hat{A}_2 \leq 999.$
$-999. \leq \hat{A}_3 \leq 999.$

Algorithm Answers:
$\hat{A}_1 = 523.3$
$\hat{A}_2 = -156.3$
$\hat{A}_3 = -0.1997$
$S = 13,390.$

Number of Iterations: 4
Central Processor Time: 10 seconds

The listing and output for this problem are contained in the following section.

E. Program Listings and Example Output

```
C
C       MAIN LINE PROGRAM FOR SUBROUTINE NLMAX.
C
        COMMON C(20,20),G1(20,20),PSCA,G(20,20),F(20),Y(20),EGV(20),FF(20)
       1,TITLE(20),CUB(20),CLB(20),PNL(20),NCON,LOUT,F3,NTH,F6,F7,METH,NPH
       2,MD,LS,C1(20)
       3,X,XTH(20),A(200,10),M,NA
C
        NI = 50
        NO = 66
C
     10 WRITE (NO,001)
    001 FORMAT (1H1,10X,44HNONLINEAR REGRESSION BY GAUSS-NEWTON METHOD )
C
        CALL NLMAX (NI,NO)
C
        GO TO 10
C
        END




        SUBROUTINE DLSQ (II,I)
C
        COMMON C(20,20),G1(20,20),PSCA,G(20,20),F(20),Y(20),EGV(20),FF(20)
       1,TITLE(20),CUB(20),CLB(20),PNL(20),NCON,LOUT,F3,NTH,F6,F7,METH,NPH
       2,MD,LS,C1(20)
       3,X,XTH(20),A(200,10),M,NA
C
        GO TO (1,1,2),II
      2 RETURN
      1 X = FUNC(C1,A,I)
        GO TO (5,6) ,II
      6 DO 10   J=1,NTH
        C1(J) = C1(J) + .0001*C1(J)
        FOR = FUNC(C1,A,I)
        C1(J) = C1(J) - .0002*C1(J)
        REV = FUNC(C1,A,I)
        C1(J) = C1(J) + .0001*C1(J)
        XTH(J) = (FOR - REV)/(.0002*C1(J))
     10 CONTINUE
C
      5 RETURN
        END
```

```
      SUBROUTINE NLMAX (NI,NO)
C
C     ****** THIS IS DECK 1.  GAUSS-NEWTON METHOD.  ******
C
      COMMON C(20,20),G1(20,20),PSCA,G(20,20),F(20),Y(20),EGV(20),FF(20)
     1,TITLE(20),CUB(20),CLB(20),PNL(20),NCON,LOUT,F3,NTH,F6,F7,METH,NPH
     2,MD,LS,C1(20)
      DIMENSION W(20)
      EQUIVALENCE(NTH,L)
C
      METH=1
C
      READ (NI,2000) TITLE
 2000 FORMAT(20A4)
      READ (NI,2002) NTH, LOUT, MD
 2002 FORMAT(16I5)
      IF (NTH .EQ. 0 )   CALL EXIT
      WRITE (NO,2001) TITLE, MD
 2001 FORMAT (//,2X,20A4,/,2X,13HMODEL NUMBER ,I2 )
      READ (NI,2003) (C1(I), I=1,NTH)
 2003 FORMAT(4F20.5)
C
      LS=1
      CALL ACCUM (3,NI,NO)
      CALL BOUND (3,H,NI,NO)
      WRITE (NO,5000) (C1(I), I=1,L)
 5000 FORMAT(26H0PARAMETER INITIAL GUESSES/(7E16.6))
      IF(LS-3)199,907,199
  199 IPH=2
      NIN=0
      NF=0
      ND=0
      EPS=1.E-4
      EPS1=1.E-3
      DO 906 I=1,L
      FF(I)=C1(I)
  906 Y(I)=-C1(I)
      H=1.
      CALL BOUND (4,H,NI,NO)
      DO 911 I=1,L
  911 Y(I)=C1(I)*H
      NPH=1
      IF(NCON)1,899,1
  899 NPH=2
      GO TO 16
C
    1 GO TO(212,16),LOUT
  212 WRITE (NO,1001)
 1001 FORMAT(26H1PENALTY FUNCTION INCLUDED)
   16 II=2
      NRE=1
      ND=ND+1
      GO TO 100
```

```
   51   II=1
  100   NF=NF+1
        LS=1
        CALL ACCUM (II,NI,NO)
        F4=F3
        GO TO(405,68,1003),LS
  405   GO TO(401,499),NPH
  401   CALL BOUND (II,X,NI,NO)
  499   GO TO(48,409),II
  409   DO 408 I=2,L
        DO 408 J=2,I
  408   G(I,J-1)=G(J-1,I)
   48   GO TO(205,110),LOUT
  205   GO TO(208,209),II
C
  209   WRITE (NO,210) ND
  210   FORMAT(10H0ITERATIONI6)
  208   WRITE (NO,207) F3, NF, (C1(I), I=1,L)
  207   FORMAT(9H0FUNCTIONE17.7,13H   EVALUATIONI6/11H PARAMETERS/(7E17.7)
       2)
C
  110   GO TO(101,111,894),II
  101   IF(F2-F3)22,21,21
   22   GO TO(24,16),NPH
   24   GO TO(111,16),NRE
  111   DO 106 I=1,L
  106   FF(I)=C1(I)
        IF(II-1)26,26,25
   26   NRE=2
        GO TO(34,16),NRR
   34   Q=0.
        IF(NIN)14,14,16
   14   CONTINUE
        DO 27 I=1,L
   27   Q=Q+F(I)*Y(I)
        H=(Q-2.*(F3-F2))/2./(F3-F2-Q)
        IF(ABS(H)-.1)16,16,30
   30   IF(H+1.)304,304,31
   31   DO 29 I=1,L
   29   Y(I)=H*Y(I)
        GO TO 304
   25   CONTINUE
        DO 800 I=1,L
        DO 800 J=1,L
  800   C(I,J)=G(I,J)
        CALL EIG(L,2)
        LT3=2
        DO 9 I=1,L
        EGV(I)=C(I,I)
        IF(C(I,I))9,13,10
   13   C(I,I)=-1.
        GO TO 9
   10   LT3=1
```

```
            C(I,I)=-C(I,I)
    9     CONTINUE
   41     DO 11 J=1,L
            W(J)=0.
            DO 202 J1=1,L
  202     W(J)=W(J)+G1(J1,J)*F(J1)
   11     W(J)=W(J)/C(J,J)
            DO 201 I=1,L
            Y(I)=0.
            DO 201 J=1,L
  201     Y(I)=Y(I)-G1(I,J)*W(J)
  304     NRR=1
            F2=F3
            F6=F7
            H=1.
            CALL BOUND (4,H,NI,NO)
            NIN=NIN/2
            H=H/2.**NIN
            DO 604 I=1,L
  604     Y(I)=H*Y(I)
  603     J=1
            DO 700 I=1,L
            IF(ABS(Y(I))/(EPS1+ABS(FF(I)))-EPS)700,700,701
  701     J=2
  700     CONTINUE
            GO TO(33,702),J
   33     GO TO(898,16),NRE
  898     GO TO(1002,897),NPH
  897     IF(H-1.)896,1003,1003
  896     CALL BOUND (5,H,NI,NO)
            DO 895 I=1,L
  895     C1(I)=FF(I)+H*Y(I)
            CALL ACCUM (1,NI,NO)
            NF=NF+1
            II=3
            GO TO(208,894),LOUT
  894     IF(F3-F2)706,1003,1003
  702     DO 12 I=1,L
   12     C1(I)=FF(I)+Y(I)
            GO TO 51
   21     GO TO(2,6),NRE
    6     DO 3 I=1,L
    3     C1(I)=FF(I)
            GO TO 16
    2     Q=0.
            NRR=2
            DO 58 I=1,L
   58     Q=Q+F(I)*Y(I)
            H=Q/(Q+(F2-F3))*.5
   66     IF(4.*H-1.)59,59,62
   59     H=.25
   62     J=1
            NIN=NIN+1
```

```
      DO 703 I=1,L
      Y(I)=H*Y(I)
      IF(ABS(Y(I))/(EPS1+ABS(FF(I)))-EPS)703,703,704
704   J=2
703   CONTINUE
      GO TO(706,705),J
705   H=1.
      CALL BOUND (4,H,NI,NO)
      DO 23 I=1,L
      Y(I)=H*Y(I)
23    C1(I)=FF(I)+Y(I)
      GO TO 51
706   F3=F2
      F7=F6
      DO 707 I=1,L
707   C1(I)=FF(I)
19    GO TO(1002,1003),NPH
1002  IF(IPH)1004,1004,1005
1004  NPH=2
      GO TO(121,122),LOUT
C
121   WRITE (NO,214)
214   FORMAT(20H0NO PENALTY FUNCTION)
C
122   F2=-1.E30
      GO TO 16
1005  IF(ABS(F3-F4)-.1)1004,1006,1006
1006  CALL BOUND (6,H,NI,NO)
      IPH=IPH-1
      GO TO(213,16),LOUT
C
213   WRITE (NO,211)
211   FORMAT(41H0PENALTY FUNCTION REDUCED BY FACTOR OF 10)
      GO TO 16
1003  CONTINUE
      WRITE (NO,123) F3, (C1(I), I=1,L)
123   FORMAT(30H0MAXIMUM OF OBJECTIVE FUNCTIONE17.7/11H0PARAMETERS/(7E17
     2.7))
      WRITE (NO,9001) NF, ND
9001  FORMAT(21H0FUNCTION EVALUATIONSI6,25H   DERIVATIVE EVALUATIONSI5)
407   WRITE (NO,920) (F(I), I=1,L)
920   FORMAT(9H1GRADIENT/(7E17.6))
      WRITE (NO,921)
921   FORMAT(8H0HESSIAN)
      DO 922 I=1,L
922   WRITE (NO,923) (G(I,J), J=1,L)
923   FORMAT(7E17.6)
      WRITE (NO,903) (EGV(I), I=1,L)
903   FORMAT(30H0EIGENVALUES OF SCALED HESSIAN/(7E17.6))
      WRITE (NO,904)
904   FORMAT(21H0PRINCIPAL COMPONENTS)
      DO 905 I=1,L
905   WRITE (NO,37)  (G1(J,I), J=1,L)
```

```
   37 FORMAT (/7E17.6/(7E17.6))
C
      CALL BOUND (7,H,NI,NO)
      CALL OUT (NO)
C
      GO TO(217,907),LT3
  217 WRITE (NO,216)
  216 FORMAT(36H0SOLUTION IS NOT AN INTERIOR MAXIMUM)
  907 RETURN
   68 J=1
      DO 71 I=1,L
      Y(I)=.5*Y(I)
      IF(ABS(Y(I))/(EPS1+ABS(C1(I)))-EPS)71,71,925
  925 J=2
   71 C1(I)=C1(I)-Y(I)
      GO TO(909,926),J
  909 WRITE (NO,910)
  910 FORMAT(45H0FEASIBLE PARAMETER VALUES COULD NOT BE FOUND/45H ******
     2******************************************)
      RETURN
  926 CONTINUE
      WRITE (NO,924)
  924 FORMAT(8H0RESTART)
C
      IF(II-1)51,51,16
C
      END




      SUBROUTINE ACCUM (II,NI,NO)
C
      COMMON C(20,20),G1(20,20),PSCA,G(20,20),F(20),Y(20),EGV(20),FF(20)
     1,TITLE(20),CUB(20),CLB(20),PNL(20),NCON,LOUT,F3,NTH,F6,F7,METH,NPH
     2,MD,LS,C1(20)
     3,X,XTH(20),A(200,10),M,NA
C
      GO TO(100,100,101),II
C
  101 READ (NI,2000) M, NA
 2000 FORMAT(16I5)
      DO 2002 I=1,M
 2002 READ (NI,102) (A(I,J), J=1,NA)
  102 FORMAT (4F20.10)
C
      WRITE (NO,2004)
 2004 FORMAT(13H0OBSERVATIONS/)
      DO 2005 I=1,M
 2005 WRITE (NO,2006) I, (A(I,J), J=1,NA)
 2006 FORMAT(I5,7E16.6/(E21.6,6E16.6))
C
      CALL DLSQ(3,0)
      RETURN
```

```
C
 100    F3=0.
        GO TO(1,2),II
   2    DO 3 I=1,NTH
        F(I)=0.
C
C
        GO TO(13,3),METH
  13    DO 15 J=I,NTH
  15    G(I,J)=0.
C
C
   3    CONTINUE
   1    DO 4 MU=1,M
        CALL DLSQ(II,MU)
        GO TO(6,7),LS
   6    F3=F3-X*X
        GO TO(4,5),II
   5    DO 12 I=1,NTH
        F(I)=F(I)-X*XTH(I)
C
C
        GO TO(14,12),METH
  14    DO 16 J=I,NTH
  16    G(I,J)=G(I,J)-XTH(I)*XTH(J)
C
C
  12    CONTINUE
   4    CONTINUE
        GO TO(7,9),II
   9    DO 10 I=1,NTH
        F(I)=2.*F(I)
C
C
        GO TO(17,10),METH
  17    DO 18 J=I,NTH
  18    G(I,J)=2.*G(I,J)
C
C
  10    CONTINUE
   7    RETURN
C
        END




        SUBROUTINE EIG (N,II)
C
        COMMON A(20,20),V(20,20),PSCA
        DIMENSION SCA(20)
C
        PSCA=0.
        IF(N-1)107,107,108
 107    GO TO(103,109),II
```

```
      109 V(1,1)=1.
          RETURN
C
      108 VN=2.
          SUM1=0.
          DO 22 I=1,N
          IF(A(I,I))101,100,101
      100 SCA(I)=1.
          GO TO 22
      101 A1=ABS(A(I,I))
          SCA(I)=1./SQRT(A1)
          PSCA=PSCA+ALOG(A1)
          DO 102 J=1,N
          A(I,J)=A(I,J)*SCA(I)
      102 A(J,I)=A(I,J)
          A(I,I)=A(I,I)*SCA(I)
       22 CONTINUE
          DO 1 I=2,N
          DO 1 J=2,I
        1 SUM1=SUM1+A(I,J-1)*A(I,J-1)
          SUM1=SQRT(2.*SUM1)
          SUME=SUM1/10.E7
          GO TO(30,31),II
       31 DO 32 I=1,N
          DO 50 J=1,N
       50 V(I,J)=0.
       32 V(I,I)=1.
       30 IN=0
          IF(N-1)18,17,18
       18 SUM1=AMAX1(SUME,SUM1/VN)
       16 CONTINUE
          DO 3 J=2,N
          J1=J-1
          DO 3 I=1,J1
          IF(ABS(A(I,J))-SUM1)3,3,4
        4 IN=1
          Y1=-A(I,J)
          Y2=(A(I,I)-A(J,J))/2.
          OMEGA=Y1/SQRT(Y1**2+Y2**2)
          OMEGA=OMEGA*SIGN(1.,Y2)
          Y1=OMEGA/SQRT(2.+2.*SQRT(1.-OMEGA**2))
          BB1=Y1**2
          BB2=1.-BB1
          Y2=SQRT(BB2)
        9 DO 5 K=1,N
          IF (K-I) 6,5,6
        6 IF (K-J) 7,5,7
        7 Y3=A(K,I)*Y2-A(K,J)*Y1
          A(K,J)=A(K,I)*Y1+A(K,J)*Y2
          A(K,I)=Y3
          A(J,K)=A(K,J)
          A(I,K)=A(K,I)
        5 CONTINUE
          BB3=2.*Y1*Y2*A(I,J)
          Y3=A(I,I)*BB2+A(J,J)*BB1-BB3
          Y4=A(I,I)*BB1+A(J,J)*BB2+BB3
          A(I,J)=(A(I,I)-A(J,J))*Y1*Y2+A(I,J)*(BB2-BB1)
          A(J,I)=A(I,J)
```

```
      A(I,I)=Y3
      A(J,J)=Y4
      GO TO(3,20),II
   20 DO 12 K=1,N
      Y3=V(K,I)*Y2-V(K,J)*Y1
      V(K,J)=V(K,I)*Y1+V(K,J)*Y2
   12 V(K,I)=Y3
    3 CONTINUE
      IF (N-2) 17,17,21
   21 IF (IN-1) 14,15,15
   15 IN=0
      GO TO 16
   14 IF (SUM1-SUME)17,17,18
   17 GO TO(103,104),II
  104 DO 105 I=1,N
      DO 105 J=1,N
  105 V(I,J)=V(I,J)*SCA(I)
C
  103 RETURN
      END




      SUBROUTINE OUT (NO)
C
      COMMON C(20,20),G1(20,20),PSCA,G(20,20),F(20),Y(20),EGV(20),FF(20)
     1,TITLE(20),CUB(20),CLB(20),PNL(20),NCON,LOUT,F3,NTH,F6,F7,METH,NPH
     2,MD,LS,C1(20)
     3,X,XTH(20),A(200,10),M,NA
C
      WRITE (NO,003) TITLE, MD
    3 FORMAT(1H120A4/6H0MODELI5/30H0RESIDUALS (COMPUTED-OBSERVED)/)
      J=0
      DO 1 I=1,M
      J=J+1
      CALL DLSQ(1,I)
      F(J)=X
      IF(J-7)1,2,2
    2 J=0
      WRITE (NO,004) (F(K), K=1,7)
    4 FORMAT(3E16.6/3E16.6/E16.6)
    1 CONTINUE
      IF(J)5,6,5
    5 WRITE (NO,004) (F(K), K=1,J)
    6 F3=-F3
      X=F3/FLOAT(M-NTH)
      X1=SQRT(X)
      WRITE (NO,007) F3, X1
    7 FORMAT(28H0SUM OF SQUARES OF RESIDUALSE17.7/19H0STANDARD DEVIATION
     2E17.7/)
      X=2.*X
C
C
      GO TO(300,301),METH
  300 DO 8 I=1,NTH
```

```
    8   EGV(I)=X/(ABS(EGV(I))+1.E-20)
        DO 400 I=1,NTH
        DO 400 J=1,NTH
        C(I,J)=0.
        DO 400 K=1,NTH
  400   C(I,J)=C(I,J)+G1(I,K)*G1(J,K)*EGV(K)
        GO TO 303
C
C
  301   DO 302 I=1,NTH
        DO 302 J=1,NTH
  302   C(I,J)=X*G(I,J)
  303   DO 45 I=1,NTH
   45   Y(I)=SQRT(C(I,I))
        WRITE (NO,028) (C1(I), I=1,NTH)
   28   FORMAT(11H PARAMETERS//(7E17.6))
        WRITE (NO,046) (Y(I), I=1,NTH)
   46   FORMAT(//34H0STANDARD DEVIATIONS OF PARAMETERS//(7E17.6))
        WRITE (NO,038)
   38   FORMAT(//32H0COVARIANCE MATRIX OF PARAMETERS/)
        DO 34 I=1,NTH
   34   WRITE (NO,037) (C(I,J), J=1,NTH)
   37   FORMAT(/7E17.6/(7E17.6))
        CALL EIG(NTH,2)
        WRITE (NO,200) (C(I,I), I=1,NTH)
  200   FORMAT(//34H0EIGENVALUES OF CORRELATION MATRIX//(7E17.6))
        WRITE (NO,201)
  201   FORMAT(//21H0PRINCIPAL COMPONENTS/)
        DO 202 I=1,NTH
  202   WRITE (NO,037) (G1(J,I), J=1,NTH)
        DO 500 I=1,NTH
        EGV(I)=0.
        Y(I)=SQRT(C(I,I))
        DO 500 J=1,NTH
  500   EGV(I)=EGV(I)+G1(J,I)*C1(J)
        WRITE (NO,501) (EGV(I), I=1,NTH)
  501   FORMAT(//16H0EXPECTED VALUES//(7E17.6))
        WRITE (NO,502) (Y(I), I=1,NTH)
  502   FORMAT(//20H0STANDARD DEVIATIONS//(7E17.6))
C
        RETURN
        END




        FUNCTION FUNC (C1,A,I)
C
        DIMENSION C1(20),A(200,10)
C
        FUNC = C1(1) + C1(2)*EXP(C1(3)*A(I,1)) - A(I,2)
C
        RETURN
        END
```

```
      SUBROUTINE BOUND (II,H,NI,NO)
C
      COMMON C(20,20),G1(20,20),PSCA,G(20,20),F(20),Y(20),EGV(20),FF(20)
     1,TITLE(20),CUB(20),CLB(20),PNL(20),NCON,LOUT,F3,NTH,F6,F7,METH,NPH
     2,MD,LS,C1(20)
C
      GO TO(1,1,2,3,3,44,43),II
C
  44  DO 45 I=1,NTH
  45  PNL(I)=.1*PNL(I)
      RETURN
C
  1   DO 4 I=1,NTH
      AA1=C1(I)-CLB(I)
      AA2=PNL(I)/AA1
      AA3=C1(I)-CUB(I)
      AA4=PNL(I)/AA3
      F3=F3-AA2+AA4
      GO TO(4,5),II
  5   AA2=AA2/AA1
      AA4=AA4/AA3
      F(I)=F(I)+AA2-AA4
C
C
      GO TO(100,4),METH
 100  G(I,I)=G(I,I)+2.*(AA4/AA3-AA2/AA1)
C
C
  4   CONTINUE
      RETURN
C
  2   READ (NI,006) (CLB(I), I=1,NTH)
  6   FORMAT(4F20.5)
      READ (NI,006) (CUB(I), I=1,NTH)
C
      DO 20 I=1,NTH
      IF(C1(I)-CLB(I))21,21,23
  21  IF(C1(I))25,26,27
  25  CLB(I)=100.*C1(I)
      GO TO 23
  26  CLB(I)=C1(I)-1.E10
      GO TO 23
  27  CLB(I)=0.
  23  IF(C1(I)-CUB(I))20,22,22
  22  IF(C1(I))28,29,24
  28  CUB(I)=0.
      GO TO 20
  29  CUB(I)=C1(I)+1.E10
      GO TO 20
  24  CUB(I)=100.*C1(I)
  20  CONTINUE
      DO 8 I=1,NTH
  8   PNL(I)=.0001*AMIN1(.001+ABS(C1(I)),CUB(I)-CLB(I))
```

```fortran
C
      WRITE (NO,038) (I, CLB(I), CUB(I), PNL(I), I=1,NTH)
   38 FORMAT(60H0PARAMETER    LOWER BOUND       UPPER BOUND      PENALTY COEFFIC
     2IENT/(I6,2E16.6,E22.6))
C
      NCON=2*NTH
      RETURN
C
   3  HY=0.
      DO 7 I=1,NTH
   7  HY=AMIN1(Y(I)/(C1(I)-CLB(I)),Y(I)/(C1(I)-CUB(I)),HY)
      IF(II-5)40,41,43
   40 H=AMIN1(1.,-.5/HY)
      RETURN
C
   41 H=-1./HY
   43 RETURN
C
      END
```

NONLINEAR REGRESSION BY GAUSS-NEWTON METHOD

$Y = A1 + A2*EXP(A3*X1)$
MODEL NUMBER 1

OBSERVATIONS

1	-.500000E+01	0.127000E+03
2	-.300000E+01	0.151000E+03
3	-.100000E+01	0.379000E+03
4	0.100000E+01	0.421000E+03
5	0.300000E+01	0.460000E+03
6	0.500000E+01	0.426000E+03

PARAMETER	LOWER BOUND	UPPER BOUND	PENALTY COEFFICIENT
1	-.999000E+03	0.999000E+03	0.500001E-01
2	-.999000E+03	0.999000E+03	0.150001E-01
3	-.999000E+03	0.999000E+03	0.201000E-04

PARAMETER INITIAL GUESSES
   0.500000E+03   -.150000E+03   -.200000E+00

PENALTY FUNCTION INCLUDED

ITERATION   1

FUNCTION   -.1486949E+05   EVALUATION   1
PARAMETERS
   0.4999999E+03   -.1500000E+03   -.2000000E+00

FUNCTION   -.1339010E+05   EVALUATION   2
PARAMETERS
   0.5234291E+03   -.1570955E+03   -.1995004E+00

ITERATION   2

FUNCTION   -.1339010E+05   EVALUATION   3
PARAMETERS
   0.5234290E+03   -.1570955E+03   -.1995004E+00

FUNCTION   -.1339009E+05   EVALUATION   4
PARAMETERS
   0.5232434E+03   -.1568748E+03   -.1997343E+00

FUNCTION   -.1339009E+05   EVALUATION   5
PARAMETERS
   0.5232910E+03   -.1569313E+03   -.1996744E+00

ITERATION   3

FUNCTION   -.1339009E+05   EVALUATION   6
PARAMETERS
    0.5232909E+03   -.1569313E+03   -.1996744E+00

FUNCTION   -.1339009E+05   EVALUATION   7
PARAMETERS
    0.5233093E+03   -.1569523E+03   -.1996602E+00

NO PENALTY FUNCTION

ITERATION   4

FUNCTION   -.1339009E+05   EVALUATION   8
PARAMETERS
    0.5232908E+03   -.1569312E+03   -.1996743E+00

FUNCTION   -.1339009E+05   EVALUATION   9
PARAMETERS
    0.5233093E+03   -.1569523E+03   -.1996602E+00

MAXIMUM OF OBJECTIVE FUNCTION   -.1339009E+05

PARAMETERS
    0.5232908E+03   -.1569312E+03   -.1996743E+00

FUNCTION EVALUATIONS   9   DERIVATIVE EVALUATIONS   4

GRADIENT
    -.227123E-01   -.783477E-01   -.573093E+02

HESSIAN
    -.120024E+02   -.149871E+02   -.500456E+04
    -.149871E+02   -.265610E+02   -.144412E+05
    -.500456E+04   -.144412E+05   -.109476E+08

EIGENVALUES OF SCALED HESSIAN
    -.608113E-02   -.243048E+01   -.563435E+00

PRINCIPAL COMPONENTS

    0.129188E+00   -.149043E+00   0.138390E-03

    0.156409E+00   0.124236E+00   0.164547E-03

    -.205337E+00   0.862760E-03   0.212406E-03

Y = A1 + A2*EXP(A3*X1)

MODEL   1

RESIDUALS (COMPUTED-OBSERVED)

    -.295985E+02   0.866226E+02   -.473231E+02
    -.262355E+02   -.229191E+02   0.394649E+02

SUM OF SQUARES OF RESIDUALS   0.1339009E+05

STANDARD DEVIATION   0.6680842E+02

PARAMETERS

    0.523291E+03   -.156931E+03   -.199674E+00

STANDARD DEVIATIONS OF PARAMETERS

    0.158925E+03   0.180735E+03   0.170082E+00

COVARIANCE MATRIX OF PARAMETERS

    0.252570E+05   -.281959E+05   0.256477E+02

    -.281959E+05   0.326652E+05   -.301997E+02

    0.256477E+02   -.301997E+02   0.289278E-01

EIGENVALUES OF CORRELATION MATRIX

    0.681483E-02   0.294204E+01   0.511491E-01

PRINCIPAL COMPONENTS

    0.254226E-02   0.450032E-02   0.246097E+01

    -.361861E-02   0.321848E-02   -.338218E+01

    -.447610E-02   -.459010E-04   0.413199E+01

EXPECTED VALUES

    0.132710E+00   -.172333E+01   -.316015E+01

STANDARD DEVIATIONS

    0.825520E-01   0.171524E+01   0.226162E+00

III. MARQUARDT (BSOLVE ALGORITHM)*

A. Purpose

This program solves for the coefficients in a multivariable, nonlinear regression equation $\hat{Y} = F(X_1, X_2, \ldots, X_K; \hat{A}_1, \hat{A}_2, \ldots, \hat{A}_M)$ utilizing N data points for Y_i and $X_{k,i}$, $i = 1, 2, \ldots, N$; $k = 1, 2, \ldots, K$.

B. Method

The procedure was proposed by Marquardt (33) as an extension of the Gauss Newton method to allow for convergence with relatively poor starting guesses for the unknown coefficients. A least squares objective function is utilized. In this method, the Gauss Newton normal equations are modified by adding a factor λ,

$$[\underline{A}^t \underline{A} + \lambda * \underline{I}] \Delta \underline{\hat{A}} = \underline{A}^t (\underline{Y} - \underline{\hat{Y}}*)$$

where $\underline{I}$ is the identity matrix. Thus λ is added to each term of the main diagonal of the $\underline{A}^t \underline{A}$ matrix. It can be shown that when λ approaches $+\infty$, Marquardt's method is identical to Steepest Descent (31). When λ equals zero, the technique reduces to Gauss Newton. In general, a Steepest Descent procedure would be expected to converge for poor starting values but requires a lengthy solution time. Gauss Newton, on the other hand, will converge rapidly for good starting estimates. Thus in the Marquardt procedure, the initial values of λ are large and will decrease toward zero as the optimum is approached. The rules for calculating λ are discussed in the original article by Marquardt (33). All other computational details parallel the Gauss Newton procedure.

C. Program Description

1) Usage:

The program consists of a main program, a general subroutine BSOLVE, a general function subprogram ARCOS, and two user supplied subroutines, FUNC and DERIV. All input and output is through the main program. Format changes may be required depending on the problem under consideration.

---

*Computer code developed by W. E. Ball, Washington University, St. Louis, Missouri. Used by permission.

2) Subroutines Required:

SUBROUTINE BSOLVE(KK,B,NN,Z,Y,PH,FNU,FLA,TAU,EPS,PHMIN,I,ICON, FV,DV,BV,BMIN,BMAX,P,FUNC,DERIV,KD,A,AC,GAMM) called from the main program - performs primary calculations and coordinates other subroutines.

SUBROUTINE DERIV(KK,B,NN,Z,PJ,FV,DV,J,JTEST) specifies analytical derivatives if used; omit if numerical derivatives used (user supplied).

SUBROUTINE FUNC (KK,B,NN,Z,FV) specifies model (user supplied).

FUNCTION ARCOS (Z) general function subprogram internal to BSOLVE.

3) Description of Parameters:

NI	Card reader unit number - Define in main program
NO	Printer unit number - Define in main program
NN	Number of data points or number of equations (root location)
KK	Number of unknowns
B	Vector of unknowns
BMIN	Vector of minimum values of B
BMAX	Vector of maximum values of B
X	Vector of independent variable data points*- set equal to vector B for root location
Y	Vector of dependent variable - set equal to zero for root location
PH	Least squares objective function
Z	Computed values of dependent variable
BV	Code vector - set equal to 1 for numerical derivatives and -1 for analytical derivatives

4) DIMENSION Requirements:

The DIMENSION statement in the main program and subroutines should be modified according to the requirements of each particular problem. The parameters included in the following DIMENSION statement conform to the Input Parameter definitions, above:

DIMENSION P(NN*KK), A(KK,KK+2), AC(KK,KK+2), X(NN), B(KK), Z(NN),
          Y(NN), BV(KK), BMIN(KK), BMAX(KK), FV(KK), DV(KK)

---

*The program is currently set up for one independent variable. The user could easily modify the program to handle the multivariable case, say by stating the values explicitly in FUNC.

5) Input Formats:

CARD TYPE	FORMAT	CONTENTS
1	(2I10)	NN, KK
2	(8E10.4)	(B(J), J=1, KK)
3	(8E10.4)	(BMIN(J), J=1, KK)
4	(8E10.4)	(BMAX(J), J=1, KK)

(If KK exceeds 8 for CARD TYPES 2, 3, 4, additional cards will be required.)

5	(8E10.4)	(X(I), I=1, NN)
6	(8E10.4)	(Y(I), I=1, NN)

(If NN exceeds 8 for CARD TYPES 5, 6, additional cards will be required.)

6) Output:

The main program prints out the number of unknowns not satisfying the convergence criterion (ICON) and value of the objective function (PH) at each iteration, in addition to the final values of the unknowns (B).

7) Summary of User Requirements:

a) Determine values for NN, KK, B(J), BMIN(J), BMAX(J), X(I) and Y(I).

b) Specify NI and NO in main program.

c) Adjust DIMENSION statements in main program and subroutines.

d) Specify model in FUNC.

e) Specify analytical derivatives in DERIV, if used.

f) Change input and output format statements as necessary for the problem under consideration.

Further details on this program code can be found elsewhere (27).

D. Test Problem

The following problem was used to test the code (26). The calculations were performed on a CDC 6400 computer.

Model: $\hat{Y} = \hat{A}_1 + \hat{A}_2 \exp(\hat{A}_3 X)$

Data:

Y	X	Y	X
127	-5	421	1
151	-3	460	3
379	-1	426	5

Starting Point: $\hat{A}_1 = 400.$
$\hat{A}_2 = -140.$
$\hat{A}_3 = -0.13$

Constraints: $-1000. \leq \hat{A}_1 \leq 1000.$
$-1000. \leq \hat{A}_2 \leq 1000.$
$-1000. \leq \hat{A}_3 \leq 1000.$

Algorithm Answers: $A_1 = 523.3$
$A_2 = -156.9$
$A_3 = -0.1997$
$S = 13,390$

Number of Iterations: 12

Central Processor Time: 5 seconds

The listing and output for this problem are contained in the following section.

E. **Program Listings and Example Output**

```
C
C
C       MAIN LINE PROGRAM FOR SUBROUTINE BSOLVE
        DIMENSION P(50),A(10,10),AC(10,10),X(10)
        DIMENSION B(10),Z(10),Y(10),BV(10),BMIN(10),BMAX(10)
        EXTERNAL FUNC
        COMMON X
C
        NI = 50
        NO = 66
C
C       READ IN NUMBER OF DATA POINTS, UNKNOWNS.
C
        READ (NI,011) NN, KK
    011 FORMAT (8I10)
C
C       READ IN INITIAL GUESSES.
C
        READ (NI,012) (B(J), J=1,KK)
    012 FORMAT (8E10.4)
C
C       READ IN LIMITS ON VARIABLES.
C
        READ (NI,012) (BMIN(J), J=1,KK)
        READ (NI,012) (BMAX(J), J=1,KK)
C
C       READ IN INDEPENDENT VARIABLES.
C       READ IN DEPENDENT VARIABLES.
C
        READ (NI,012) (X(I), I=1,NN)
        READ (NI,012) (Y(I), I=1,NN)
C
        FNU=0.0
        FLA=0.0
        TAU=0.0
        EPS=0.0
        PHMIN=0.0
        I=0
        KD=KK
        FV=0.0
        DO 100 J=1,KK
        BV(J)=1
    100 CONTINUE
        ICON=KK
        ITER = 0
        WRITE (NO,015)
    015 FORMAT (1H1,10X,27HBSOLVE REGRESSION ALGORITHM )
C
    200 CALL BSOLVE(KK,B,NN,Z,Y,PH,FNU,FLA,TAU,EPS,PHMIN,I,ICON,FV,DV,BV,
       1BMIN,BMAX,P,FUNC,DERIV,KD,A,AC,GAMM)
```

```fortran
C
      ITER=ITER+1
      WRITE (NO,001) ICON, PH, ITER
  001 FORMAT (/,2X,6HICON = ,I3,4X, 5HPH = ,E15.8,4X, 16HITERATION NO. =
     1 ,I3)
      IF (ICON) 10, 300, 200
   10 IF (ICON+1)  20, 60, 200
   20 IF (ICON+2)  30, 70, 200
   30 IF (ICON+3)  40, 80, 200
   40 IF (ICON+4)  50, 90, 200
   50 GO TO 95
   60 WRITE (NO,004)
  004 FORMAT (//,2X,32HNO FUNCTION IMPROVEMENT POSSIBLE )
      GO TO 300
   70 WRITE (NO,005)
  005 FORMAT (//,2X, 28HMORE UNKNOWNS THAN FUNCTIONS)
      GO TO 300
   80 WRITE (NO,006)
  006 FORMAT (//,2X, 24HTOTAL VARIABLES ARE ZERO)
      GO TO 300
   90 WRITE (NO,007)
  007 FORMAT (//,2X,79HCORRECTIONS SATISFY CONVERGENCE REQUIREMENTS BUT
     1LAMDA FACTOR (FLA) STILL LARGE)
      GO TO 300
   95 WRITE (NO,008)
  008 FORMAT (//,2X, 20HTHIS IS NOT POSSIBLE)
      GO TO 300
  300 WRITE (NO,002)
  002 FORMAT (//,2X, 26HSOLUTIONS OF THE EQUATIONS)
      DO 400 J=1,KK
      WRITE (NO,003) J, B(J)
  003 FORMAT (/,2X, 2HB(,I2,4H) = ,E16.8)
  400 CONTINUE

 1000 STOP
      END




      SUBROUTINE FUNC (KK, B, NN, Z, FV)
C
      DIMENSION X(25), Z(25), B(25)
      COMMON X
C
      DO 100 JJ=1,NN
      Z(JJ) = B(1) + B(2)*(EXP(B(3)*X(JJ)))
  100 CONTINUE
C
      RETURN
      END
```

```
      SUBROUTINE BSOLVE (KK, B, NN, Z, Y, PH, FNU, FLA, TAU, EPS,
     1                   PHMIN, I, ICON, FV, DV, BV, BMIN, BMAX, P,
     2                   FUNC, DERIV, KD, A, AC, GAMM)
      DIMENSION B(10),Z(10),Y(10),BV(10),BMIN(10),BMAX(10)
      DIMENSION P(50),A(10,10),AC(10,10),X(10),FV(10),DV(10)
C
      K = KK
      N = NN
      KP1 = K + 1
      KP2 = KP1 + 1
      KBI1 = K*N
      KBI2 = KBI1 + K
      KZI = KBI2 + K
      IF( FNU .LE. 0. ) FNU = 10.0
      IF( FLA .LE. 0. ) FLA = 0.01
      IF( TAU .LE. 0. ) TAU = 0.001
      IF( EPS .LE. 0. ) EPS = 0.00002
      IF(PHMIN.LE.0.) PHMIN=0.
  120 KE=0
  130 DO 160 I1=1,K
  160 IF( BV(I1) .NE. 0. ) KE = KE + 1
      IF( KE .GT. 0 ) GO TO 170
  162 ICON = -3
  163 GO TO 2120
  170 IF( N .GE. KE ) GO TO 500
  180 ICON= -2
  190 GO TO 2120
  500 I1 = 1
  530 IF( I .GT. 0 ) GO TO 1530
  550 DO 560 J1=1,K
      J2 = KBI1 + J1
      P(J2) = B(J1)
      J3 = KBI2 + J1
  560 P(J3) = ABS(B(J1)) + 1.0E-02
      GO TO 1030
  590 IF (PHMIN .GT. PH .AND. I .GT. 1) GO TO 625
      DO 620 J1=1,K
      N1 = (J1-1)*N
      IF( BV(J1) ) 601,620,605
  601 CALL    DERIV    (K, B, N, Z, P(N1+1), FV, DV, J1, JTEST)
      IF( JTEST .NE. (-1) ) GO TO 620
      BV(J1) = 1.0
  605 DO 606 J2=1,K
      J3 = KBI1 + J2
  606 P(J3) = B(J2)
      J3 = KBI1 + J1
      J4 = KBI2 + J1
      DEN = 0.001*AMAX1(P(J4),ABS(P(J3)))
      IF (P(J3) + DEN .LE. BMAX(J1))   GO TO 55
            P(J3)    = P(J3) - DEN
            DEN      = - DEN
      GO TO 56
   55 P(J3) = P(J3) + DEN
   56 CALL   FUNC     (K, P(KBI1+1), N, P(N1+1), FV)
      DO 610 J2=1,N
      JB = J2 + N1
  610 P(JB) = (P(JB) - Z(J2))/DEN
```

```
      620 CONTINUE
C
C         SET UP CORRECTION EQUATIONS
C
      625 DO 725 J1=1,K
          N1 = (J1-1)*N
          A(J1,KP1) = 0.
          IF( BV(J1) ) 630,692,630
      630 DO 640 J2=1,N
          N2 = N1 + J2
      640 A(J1,KP1) = A(J1,KP1) + P(N2)*(Y(J2)-Z(J2))
      650 DO 680 J2=1,K
      660 A(J1,J2)=0.
      665 N2 = (J2-1)*N
      670 DO 680 J3=1,N
      672 N3 = N1 + J3
      674 N4 = N2 + J3
      680 A(J1,J2)=A(J1,J2) + P(N3)*P(N4)
          IF(A(J1,J1).GT.1.E-20) GO TO 725
      692 DO 694 J2=1,KP1
      694  A(J1,J2) = 0.
      695 A(J1,J1) = 1.0
      725 CONTINUE
          GN = 0.
          DO 729 J1=1,K
      729 GN = GN + A(J1,KP1)**2
C
C         SCALE CORRECTION EQUATIONS
C
          DO 726 J1=1,K
      726 A(J1,KP2) =   SQRT(A(J1,J1))
          DO 727 J1=1,K
          A(J1,KP1) = A(J1,KP1)/A(J1,KP2)
          DO 727 J2=1,K
      727 A(J1,J2) = A(J1,J2)/(A(J1,KP2)*A(J2,KP2))
      730 FL=FLA/FNU
          GO TO 810
      800 FL = FNU*FL
      810 DO 840 J1=1,K
      820 DO 830 J2=1,KP1
      830 AC(J1,J2)= A(J1,J2)
      840 AC(J1,J1)=AC(J1,J1) + FL
C
C         SOLVE CORRECTION EQUATIONS
C
          DO 930 L1=1,K
          L2=L1+1
          DO 910 L3=L2,KP1
      910 AC(L1,L3)=AC(L1,L3)/AC(L1,L1)
          DO 930 L3=1,K
          IF(L1-L3)920,930,920
      920 DO 925 L4=L2,KP1
      925 AC(L3,L4)=AC(L3,L4)-AC(L1,L4)*AC(L3,L1)
      930 CONTINUE
C
```

```
      DN = 0.
      DG = 0.
      DO 1028 J1=1,K
      AC(J1,KP2) = AC(J1,KP1)/A(J1,KP2)
      J2 = KBI1 + J1
      P(J2) = AMAX1(BMIN(J1),AMIN1(BMAX(J1),B(J1)+AC(J1,KP2)))
            DG    = DG + AC(J1,KP2) * A(J1,KP1) * A(J1,KP2)
      DN = DN + AC(J1,KP2)*AC(J1,KP2)
 1028 AC(J1,KP2)=P(J2)-B(J1)
      COSG     = DG/SQRT (DN*GN)
      JGAM = 0
      IF( COSG ) 1100,1110,1110
 1100 JGAM = 2
      COSG =  - COSG
 1110 CONTINUE
            COSG   = AMIN1(COSG, 1.0)
      GAMM= ARCOS(COSG)*180./(3.14159265)
      IF( JGAM .GT. 0 ) GAMM = 180. - GAMM
 1030 CALL  FUNC     (K, P(KBI1+1), N, P(KZI+1), FV)
 1500 PHI = 0.
      DO 1520 J1=1,N
      J2 = KZI + J1
 1520 PHI=PHI+(P(J2)-Y(J1))**2
      IF(PHI.LT. 1.E-10) GO TO 3000
      IF( I .GT. 0 ) GO TO 1540
 1521 ICON = K
      GO TO 2110
 1540 IF( PHI .GE. PH ) GO TO 1530
C
C     EPSILON TEST
C
 1200 ICON = 0
      DO 1220 J1=1,K
      J2=KBI1+J1
 1220 IF( ABS(AC(J1,KP2))/(TAU + ABS(P(J2))) .GT. EPS ) ICON = ICON + 1
      IF( ICON .EQ. 0 ) GO TO 1400
C
C     GAMMA LAMBDA TEST
C
      IF (FL .GT. 1.0 .AND. GAMM .GT. 90.0) ICON = -1
      GO TO 2105
C
C     GAMMA EPSILON TEST
C
 1400 IF (FL .GT. 1.0 .AND. GAMM .LE. 45.0) ICON = -4
      GO TO 2105
C
 1530 IF( I1 - 2 ) 1531,1531,2310
 1531 I1 = I1 + 1
      GO TO (530,590,800),I1
 2310 IF( FL .LT. 1.0E+8 ) GO TO 800
 1320 ICON = -1
C
 2105 FLA = FL
      DO 2091 J2=1,K
```

```
             J3 = KBI1 + J2
      2091 B(J2) = P(J3)
      2110 DO 2050 J2=1,N
             J3 = KZI + J2
      2050 Z(J2) = P(J3)
             PH = PHI
             I = I + 1
      2120 RETURN
      3000 ICON=0
             GO TO 2105
C
             END
```

```
             FUNCTION ARCOS(Z)
C
             X=Z
             KEY=0
             IF( X.LT. (-1.))  X=-1.
             IF( X.GT.  1.)   X=1.
             IF( X.GE. (-1.)  .AND.  X .LT. 0.)  KEY=1
             IF( X.LT.  0.)   X=ABS(X)
             IF( X.EQ.  0.) GO TO 10
             ARCOS=ATAN (SQRT(1.-X*X)/X)
             IF( KEY .EQ. 1)  ARCOS=3.14159265-ARCOS
             GO TO 999
         10 ARCOS=1.5707963
C
        999 RETURN
             END
```

BSOLVE REGRESSION ALGORITHM

ICON = 3	PH = .75464790E+05	ITERATION NO. =	1
ICON = 3	PH = .21277557E+05	ITERATION NO. =	2
ICON = 3	PH = .16732711E+05	ITERATION NO. =	3
ICON = 3	PH = .13570870E+05	ITERATION NO. =	4
ICON = 3	PH = .13392083E+05	ITERATION NO. =	5
ICON = 3	PH = .13390472E+05	ITERATION NO. =	6
ICON = 3	PH = .13390167E+05	ITERATION NO. =	7
ICON = 3	PH = .13390102E+05	ITERATION NO. =	8
ICON = 3	PH = .13390097E+05	ITERATION NO. =	9
ICON = 3	PH = .13390093E+05	ITERATION NO. =	10
ICON = 3	PH = .13390093E+05	ITERATION NO. =	11
ICON = 0	PH = .13390093E+05	ITERATION NO. =	12

SOLUTIONS OF THE EQUATIONS

B(1) =   .52329698E+03

B(2) =   -.15693703E+03

B(3) =   -.19967593E+00

IV. POWELL (SSQMIN ALGORITHM)*

A. Purpose

This program solves for the coefficients in a multivariable, nonlinear regression equation $\hat{Y} = F(X_1, X_2, \ldots, X_k; \hat{A}_1, \hat{A}_2, \ldots, \hat{A}_M)$ utilizing N data points for Y_i and $X_{k,i}$, $i = 1, 2, \ldots, N$; $k = 1, 2, \ldots, K$.

B. Method

The procedure was developed by M. J. D. Powell (38). The basic purpose was to modify the Gauss Newton technique to reduce the difficulties involved in solving the set of linear equations at each iteration. The procedure incorporates an iterative matrix inversion scheme discussed by Rosen (40) for symmetric matrices which changes only one row and one column of the $(\underline{A}^t\underline{A})$ matrix at each stage of the calculation. All derivatives in the Powell procedure are approximated by finite differences. The algorithm proceeds as follows:

1) A starting point is selected and an initial set of direction vector components $M_{i,j}$ ($i = 1, 2, \ldots, M$; $j = 1, 2, \ldots, M$) are selected parallel to the coordinate axes,

$$\underline{M}_1 = (1, 0, 0, \ldots, 0)$$
$$\underline{M}_2 = (0, 1, 0, \ldots, 0)$$
$$\vdots$$
$$\underline{M}_M = (0, 0, 0, \ldots, 1)$$

2) The Gauss Newton equations are set up,

$$(\underline{A}^t\underline{A}) \, \Delta\hat{\underline{A}} = \underline{A}^t (\underline{Y} - \hat{\underline{Y}}^*)$$

with $\underline{A}$ and $\underline{Y} - \hat{\underline{Y}}^*$ evaluated at the starting point. The procedure utilizes finite difference approximations for the derivatives.

3) The Gauss Newton equations are solved for $\Delta\hat{\underline{A}}$. This vector is then used to calculate a new direction vector (normalized) with the following components,

$$M_{i,\text{new}} = \frac{\Delta\hat{A}_i}{\left[\sum_{j=1}^{M} \Delta\hat{A}_j^2\right]^{1/2}} \, , \quad i = 1, 2, \ldots, M .$$

---

*Computer code developed by E. R. Beals, Lawrence Radiation Laboratory, Berkeley, California. Used by permission.

4) A one dimensional search is then conducted in the $\underline{M}_{new}$ direction (see Chapter 7) using the relationship $\hat{A}_{i(new)} = \hat{A}_{i(old)} + S\hat{M}_{i(new)}$, $i = 1, 2, \ldots, M$ where S is the distance moved in the $\underline{M}_{new}$ direction.

5) When the one dimensional minimum has been found, an overall convergence test is performed (see <u>Description of Parameters</u>). If satisfied, the procedure stops. If not, one of the previous direction vectors is replaced by the new direction vector. The vector to be replaced is the one with index corresponding to the maximum of the following quantities,

$$\left| b_i \Delta \hat{A}_i \right|, \quad i = 1, 2, \ldots, M$$

where

b_i = elements of $\underline{A}^t (\underline{Y} - \hat{\underline{Y}}^*)$

$\Delta \hat{A}_i$ = elements of $\underline{\Delta \hat{A}}$.

The values of the derivatives in the $\underline{A}$ matrix in the new direction are now found by finite differences utilizing values from the completed one dimensional search. Thus the Gauss Newton equations are now updated with respect to the new direction (one row and one column in $\underline{A}^t \underline{A}$ and one element in $\underline{A}^t (\underline{Y} - \underline{Y}^*)$). The Gauss Newton equations are now solved again for $\underline{\Delta \hat{A}}$ and the process repeated until convergence is achieved.

A logic diagram illustrating the above procedure is given in Figure 6.IV.

C. Program Description

1) <u>Usage</u>:

The program consists of a main program and three subroutines (SSQMIN, LINMIN, and CALFUN). The main program handles all input data. Output is from the main program and SSQMIN. Format changes may be required depending on the particular problem under consideration.

2) <u>Subroutines Required</u>:

SUBROUTINE SSQMIN (M, N, F, X, E, ESCALE, IPRINT, MAXFUN, FF, NI, NO) called from main program and performs or coordinates all calculations.

SUBROUTINE LINMIN (ITEST, X, F, MAXFUN, ABSACC, RELACC, XSTEP) performs one dimensional search (minimization).

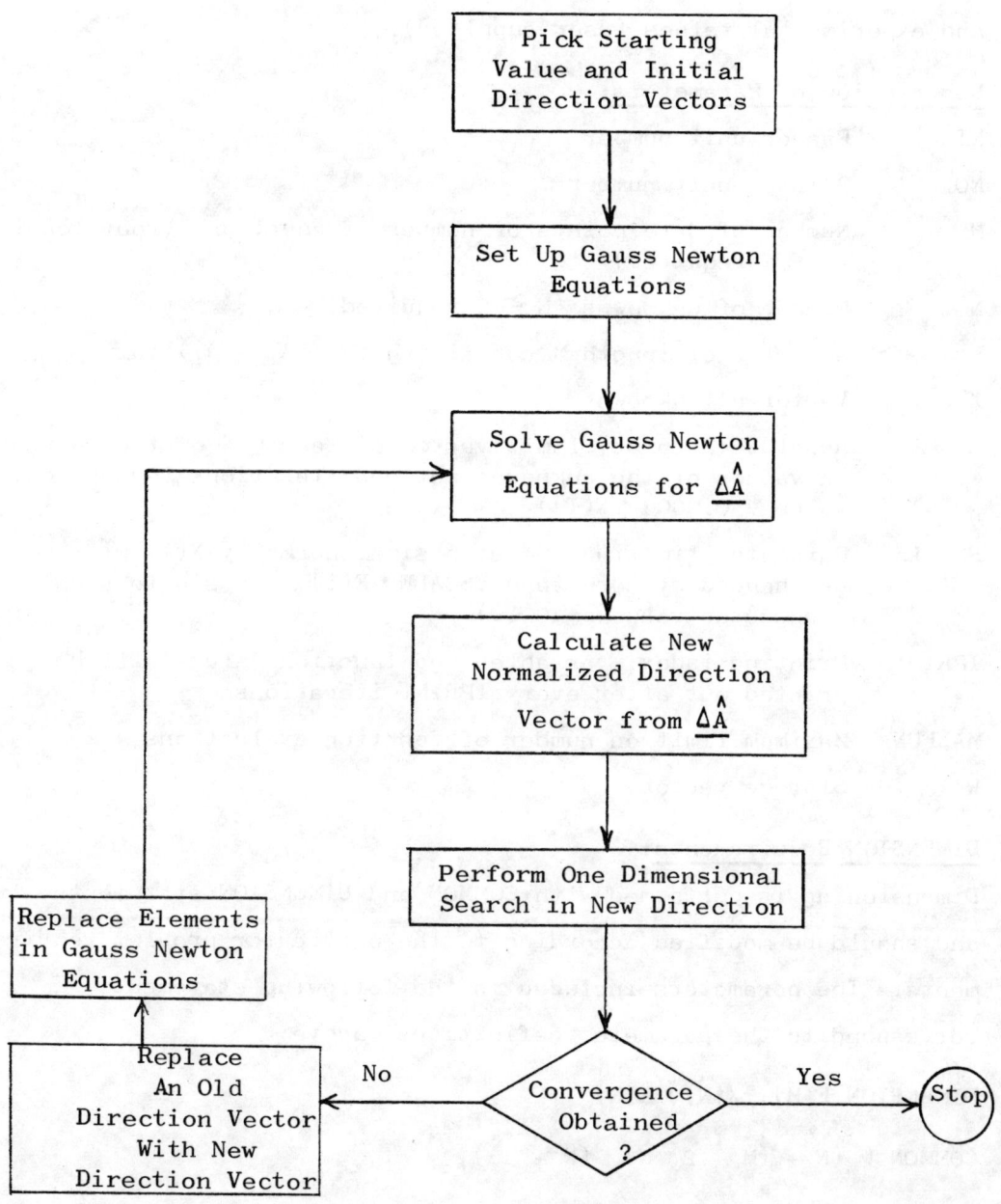

Figure 6.IV. Powell (SSQMIN ALGORITHM) Logic Diagram

SUBROUTINE CALFUN (M, N, F, X) specifies differences between model and experimental values (user supplied).

3) Description of Parameters:

- NI   Reader unit number
- NO   Printer unit number
- M   Number of data points or number of equations (root location) ($M \geq N$ required)
- N   Number of unknowns ($N \geq 2$ required)
- F   A vector of length M containing the ($\hat{Y}_i - Y_i$) values
- X   Vector of unknowns
- E   Absolute accuracy limit vector of length N on the change in values of the unknowns between iterations (suggested value = $0.0001 * X(I)$)
- ESCALE   Parameter limiting the step size; normally $X(I)$ will not be changed by more than $ESCALE * E(I)$ in a single step (suggested value = 1000.)
- IPRINT   Printing index; variables and function values will be printed out after every IPRINT iterations
- MAXFUN   Maximum limit on number of function evaluations
- W   Storage vector

4) DIMENSION Requirements:

Dimensioning is performed with COMMON and DIMENSION statements and should be modified according to the particular problem requirements. The parameters included in the following statements correspond to the parameter definitions, above:

DIMENSION F(M), X(N), E(N)

COMMON W (N + (M + 2*N) * (N + 1))

5) Input Formats:

CARD TYPE	FORMAT	CONTENTS
1	(8I10)	N,M,MAXFUN,IPRINT
2	(8E10.4)	ESCALE
3	(8E10.4)	X(I),E(I)

(Use N cards of this form)

6) Output:

The main program first prints out the program title, parameter values and initial guesses. Subject to the IPRINT specification, intermediate values of the objective function, values of the unknowns and differences between the calculated and experimental values of the dependent variables are then printed. The cumulative number of function evaluations are also listed. When convergence is achieved, a final listing of the objective function and unknowns is given.

7) <u>Summary of User Requirements</u>:
   a) Determine values of N, M, MAXFUN, IPRINT, ESCALE and E(I).
   b) Select a starting point for X(I).
   c) Specify NI and NO in main program.
   d) Adjust the DIMENSION, COMMON and FORMAT statement limits as required by the problem under consideration.
   e) Specify the $F(I) = \hat{Y}(I) - Y(I)$ equations in CALFUN.

D. <u>Test Problem</u>

The following problem was used to test the code (26). The calculations were performed on a H-G 425 computer.

Model: $\hat{Y} = \hat{A}_1 + \hat{A}_2 \exp(\hat{A}_3 X)$

Data:

Y	X
127	-5
151	-3
379	-1
421	1
460	3
426	5

Parameters: MAXFUN = 100, ESCALE = 1000.
$E(1) = 0.001$
$E(2) = 0.001$
$E(3) = 10^{-6}$

Starting Point: $\hat{A}_1 = 500.$
$\hat{A}_2 = -150.$
$\hat{A}_3 = -0.20$

Algorithm Answers: $\hat{A}_1 = 523.3$
$\hat{A}_2 = -156.9$
$\hat{A}_3 = -0.1997$
$S = 13,390.$

Number of Iterations: 8

Number of Function Evaluations: 41

Execution Time: 3 seconds

The listing and output for this problem are contained in the following section.

E. **Program Listings and Example Output**

```
C
C       MAIN LINE PROGRAM FOR SUBROUTINE SSQMIN.
C
        DIMENSION F(6), X(3), E(3)
        COMMON W(100)
C
        NI = 50
        NO = 66
C
        READ (NI,001) N, M, MAXFUN, IPRINT
  001   FORMAT (8I10)
        READ (NI,002) ESCALE
  002   FORMAT (8E10.4)
        DO 100 I=1,N
  100   READ (NI,002) X(I), E(I)
C
        WRITE (NO,003)
  003   FORMAT (1H1,10X,27HPOWELL REGRESSION ALGORITHM )
C
        CALL SSQMIN (M,N,F,X,E,ESCALE,IPRINT,MAXFUN,FF,NI,NO)
C
        WRITE (NO,004) FF
  004   FORMAT (//,2X,36HTHE SUM-OF-THE-SQUARES DIFFERENCE = ,1PE16.8 )
        WRITE (NO,005)
  005   FORMAT (//,2X,24HFINAL COEFFICIENT VALUES )
        DO 200 J=1,N
  200   WRITE (NO,006)  J, X(J)
  006   FORMAT (/,2X,2HX(,I2,4H) = ,1PE16.8)
C
        END




        SUBROUTINE CALFUN (M,N,F,X)
C
        DIMENSION F(M), X(N)
        COMMON W(100)
C
        F(1) = X(1) + X(2)*EXP(-5.*X(3)) - 127.0
        F(2) = X(1) + X(2)*EXP(-3.*X(3)) - 151.0
        F(3) = X(1) + X(2)*EXP(-1.*X(3)) - 379.0
        F(4) = X(1) + X(2)*EXP( 1.*X(3)) - 421.0
        F(5) = X(1) + X(2)*EXP( 3.*X(3)) - 460.0
        F(6) = X(1) + X(2)*EXP( 5.*X(3)) - 426.0
C
        RETURN
        END
```

```
      SUBROUTINE SSQMIN (M,N,F,X,E,ESCALE,IPRINT,MAXFUN,FF,NI,NO)
C
      DIMENSION F(M), X(N), E(N)
      COMMON W(100)
      LOGICAL STOP, MAXCAL, CONTIN, FIRST
C
      WRITE (NO,012) N, M, MAXFUN, ESCALE
   12 FORMAT (//,2X,10HPARAMETERS,//,2X,4HN = ,I2,4X,4HM = ,I3,4X,
     1 9HMAXFUN = ,I5,4X,9HESCALE = ,E10.4)
      WRITE (NO,013)
   13 FORMAT (/,2X,15HINITIAL GUESSES )
      WRITE (NO,031) (I, X(I), I=1,N)
      WRITE (NO,018)
   18 FORMAT (/,2X,21HACCURACY OF VARIABLES )
      WRITE (NO,023) (I, E(I), I=1,N)
   23 FORMAT (/,3(2X,2HE(,I2,4H) = ,E16.8))
C
C   INITIALIZE
C
      STOP=.FALSE.
      MAXCAL=.FALSE.
      IPP=IPRINT*(IPRINT-1)
      ITC=0
      IPC=0
      MPLUSN=M+N
      KST=N+MPLUSN
      NPLUS=N+1
      KINV=NPLUS*(MPLUSN+1)
      KSTORE=KINV-MPLUSN-1
      NN=N+N
      K=NN
C
C   INITIAL FUNCTION EVALUATION
C
      CALL CALFUN (M,N,F,X)
C
      MC=1
      FF=0.0
      DO 1 I=1,M
      K=K+1
      W(K)=F(I)
      FF=FF+F(I)*F(I)
    1 CONTINUE
      FOLD=FF
  100 FIRST=.TRUE.
      K=KST
      I=1
C
C   COMPUTE THE COMPONENTS OF THE GRADIENT IN THE COORDINATE DIRECTIONS
C
    2 XDUMMY=X(I)
      ISMALL=0
      DUMMY=ABS(X(I)*1.E-6)+E(I)
```

```
    5 X(I)=X(I)+DUMMY
      CALL CALFUN (M,N,F,X)
      MC=MC+1
      X(I)=XDUMMY
      DO 3 J=1,N
      K=K+1
      W(K)=0.
      W(J)=0.
    3 CONTINUE
      SUM=0.
      KK=NN
      DO 4 J=1,M
      KK=KK+1
C
C    FPLUS-FBEST
C
      F(J)=F(J)-W(KK)
      SUM=SUM+F(J)*F(J)
    4 CONTINUE
      IF (SUM .GT. FF*1.E-12) GO TO 6
C
      WRITE (NO,007) I
    7 FORMAT (5X,3HTHE,I3,58H-TH COMPONENT OF THE INITIAL STEP WAS TOO S
     1MALL  DOUBLE IT)
      DUMMY=2.0*DUMMY
C
      ISMALL=ISMALL+1
      K=K-N
      IF (ISMALL .LT. 15) GO TO 5
      ITC=0
      K=NN
      DO 8 I=1,M
      K=K+1
      F(I)=W(K)
    8 CONTINUE
      GO TO 10
C
C    SUM IS USED TO NORMALIZE G(I,K) AND D(I,J)
C
    6 SUM=1.0/SQRT(SUM)
      ISMALL=0
      J=K-N+I
C
C    W(J) IS D(I,I)   NOTE D(I,J)=0.0 J NOT EQUAL TO I
C
      W(J)=DUMMY*SUM
      DO 9 J=1,M
      K=K+1
C
C    W(K) IS G(I,K) IN THE COORDINATE DIRECTIONS
C
      W(K)=F(J)*SUM
      KK=NN+J
```

```
      DO 11 II=1,I
      KK=KK+MPLUSN
C
C   W(II) IS G*GT(I,II)
C
      W(II)=W(II)+W(KK)*W(K)
   11 CONTINUE
    9 CONTINUE
      ILESS=I-1
      IGAMAX=N+I-1
      INCINV=N-ILESS
      INCINP=INCINV+1
      IF (ILESS .GT. 0) GO TO 14
C
C   INVERSE OF G*GT(II,JJ) II,JJ=1,I  BY HOUSEHOLDER METHOD
C       RECALL (I-1)X(I-1) UPPER BLOCK ALREADY DONE
C
      W(KINV)=1.0
      GO TO 15
   14 B=1.
      DO 16 J=NPLUS,IGAMAX
      W(J)=0.
   16 CONTINUE
      KK=KINV
      DO 17 II=1,ILESS
      IIP=II+N
C
C   W(IIP)=W(N+II) IS THE SUM OF G-1(II,J)*G*GT(J,I) J=1,N
C
      W(IIP)=W(IIP)+W(KK)*W(II)
      JL=II+1
      IF (JL .GT. ILESS) GO TO 19
      DO 20 JJ=JL,ILESS
      KK=KK+1
      JJP=JJ+N
      W(IIP)=W(IIP)+W(KK)*W(JJ)
      W(JJP)=W(JJP)+W(KK)*W(II)
   20 CONTINUE
C
C   B IS G*GT(I,I)-SUM OF G*GT(I,II)*G-1(II,JJ)*G*GT(JJ,I)
C       WHICH IS A0
C
   19 B=B-W(II)*W(IIP)
      KK=KK+INCINP
   17 CONTINUE
      B=1./B
      KK=KINV
      DO 21 II=NPLUS,IGAMAX
      BB=-B*W(II)
      DO 22 JJ=II,IGAMAX
C
C   W(KK) IS G-1(II,JJ)   WHICH EQUALS A1-1+A1-1*A2*A0-1*A3*A1-1
C
```

```
          W(KK)=W(KK)-BB*W(JJ)
          KK=KK+1
       22 CONTINUE
C   W(KK) IS G-1(I,II) WHICH EQUALS -A0-1*A3      WHICH EQUALS G-1(II,I)
          W(KK)=BB
          KK=KK+INCINV
       21 CONTINUE
C
C   W(KK) IS G-1(I,I) WHICH EQUALS A0-1
C
          W(KK)=B
       15 IF ( .NOT. FIRST) GO TO 27
          I=I+1
          IF ( I .LE. N) GO TO 2
C
C   0-TH ITERATION INITIALIZATION
C
          FIRST=.FALSE.
          ISAME=0
          FF=0.
          KL=NN
          DO 26 I=1,M
          KL=KL+1
          F(I)=W(KL)
          FF=FF+F(I)*F(I)
       26 CONTINUE
          CONTIN=.TRUE.
       27 IPC=IPC-IPRINT
          IF (IPC .GE. 0) GO TO 29
C
C   ITERATION PRINTOUT
C
       28 WRITE (NO,030) ITC, MC, FF
       30 FORMAT (//,2X,9HITERATION,I3,4X,33HNUMBER OF FUNCTION EVALUATIONS
         1= ,I4,//,2X,21HSUM-OF-THE-SQUARES = ,E16.8,//,2X,12HCOEFFICIENTS)
          WRITE (NO,031) (I, X(I), I=1,N)
       31 FORMAT (/,3(2X,2HX(,I2,4H) = ,E16.8))
          WRITE (NO,032)
       32 FORMAT (/,2X,15HFUNCTION VALUES )
          WRITE (NO,035) (J, F(J), J=1,M)
       35 FORMAT (/,3(2X,2HF(,I2,4H) = ,E16.8))
          IPC=IPP
          IF (STOP) GO TO 33
C
C   CONVERGENCE TESTS
C   1 N+1 VALUES OF F ARE THE SAME
C   2 MAXIMUM OF STEP(I)/E(I)  LESS THAN OR EQUAL TO 1.0   (CONTIN FALSE)
C   3 MAXIMUM OF THE I-TH COMPONENT OF THE ACTUAL STEP TAKEN / E(I)
C         LESS THAN OR EQUAL TO 1.0          CHANGE LESS THAN OR EQUAL TO 1
C
          CHANGE = 0.0
       29 IF (CHANGE .NE. 0.0) ISAME=0
          ISAME=ISAME+1
```

```
      IF (ISAME .LE. N) GO TO 291
      IF (IPRINT .LE. 0) GO TO 33
      WRITE (NO,295)
  295 FORMAT (//5X,28HN+1 VALUES OF F ARE THE SAME)
      IF (FF .GE. FOLD) GO TO 10
      FOLD=FF
      K=NN
      DO 293 I=1,M
      K=K+1
      W(K)=F(I)
  293 CONTINUE
      GO TO 100
  291 IF (CONTIN) GO TO 34
      IF (CHANGE .GT. 1.0) GO TO 36
   10 IF (IPRINT .LE. 0) GO TO 33
C
C     TERMINAL PRINTOUT
C
      WRITE (NO,038)
   38 FORMAT (//5X,46HSSQMIN FINAL VALUES OF FUNCTIONS AND VARIABLES)
      STOP=.TRUE.
      GO TO 28
   33 RETURN
C
   36 CONTIN=.TRUE.
C
C     START NEXT ITERATION
C
   34 ITC=ITC+1
      K=N
      KK=KST
C
C     CALCULATION OF P
C
      DO 39 I=1,N
      K=K+1
      W(K)=0.
      KK=KK+N
      W(I)=0.
      DO 40 J=1,M
      KK=KK+1
C
C     W(I) IS THE SUM OF G(I,K)*F(K) WHICH IS -P(I)
C
      W(I)=W(I)+W(KK)*F(J)
   40 CONTINUE
   39 CONTINUE
      DM=0.
      K=KINV
C
C     CALCULATION OF Q
C
      DO 41 II=1,N
```

```
      IIP=II+N
C
C   W(IIP)=W(N+II) IS THE SUM OF G-1(II,J)*(-P(J)) 4=1,N   WHICH IS -Q(I)
C
      W(IIP)=W(IIP)+W(K)*W(II)
      JL=II+1
      IF (JL .GT. N) GO TO 43
      DO 44 JJ=JL,N
      JJP=JJ+N
      K=K+1
      W(IIP)=W(IIP)+W(K)*W(JJ)
      W(JJP)=W(JJP)+W(K)*W(II)
   44 CONTINUE
      K=K+1
C
C   MAXIMUM OF P(I)*Q(I)     KL INDEX OF THE DIRECTION OF D(I,J)
C          TO BE REPLACED BY STEP(J)
C
   43 IF (DM .GE. ABS(W(II)*W(IIP))) GO TO 41
      DM=ABS(W(II)*W(IIP))
      KL=II
   41 CONTINUE
      II=N+MPLUSN*KL
      CHANGE=0.
      CHANGE=0.
      DO 46 I=1,N
      JL=N+I
      W(I)=0.
      DO 47 J=NPLUS,NN
      JL=JL+MPLUSN
C
C   W(I) IS THE SUM OF (-Q(J))*D(J,I) J=1,N   WHICH IS -STEP(I)
C
      W(I)=W(I)+W(J)*W(JL)
   47 CONTINUE
      II=II+1
C
C   INTERCHANGING KL AND N ROWS OF D(I,J) PUT XBEST IN D(N,J)
C
      W(II)=W(JL)
      W(JL)=X(I)
C
C   CHANGE IS THE MAXIMUM OF ABS(STEP(I)/E(I)))
C
      IF (ABS(E(I)*CHANGE) .GT. ABS(W(I))) GO TO 46
      CHANGE=ABS(W(I)/E(I))
   46 CONTINUE
      DO 49 I=1,M
      II=II+1
      JL=JL+1
C
C   INTERCHANGING KL AND N ROWS OF G   PUT FBEST IN G(N,K)
C
```

```
            W(II)=W(JL)
            W(JL)=F(I)
         49 CONTINUE
            FC=FF
            ACC=0.1/CHANGE
            IT=3
            XC=0.
            XL=0.
            IS=3
            XSTEP=-AMIN1(0.5,ESCALE/CHANGE)
            IF (CHANGE .LE. 1.0) CONTIN=.FALSE.
C
C     LINEAR SEARCH
C
         51 CALL LINMIN (IT,XC,FC,6,ACC,0.1,XSTEP)
            IF (IT .NE. 1) GO TO 53
            MC=MC+1
            IF (MC .LE. MAXFUN) GO TO 54
            WRITE (NO,056) MAXFUN
         56 FORMAT (5X,I6,16H CALLS OF CALFUN)
            MAXCAL=.TRUE.
            GO TO 53
         54 XL=XC-XL
            DO 57 J=1,N
            X(J)=X(J)+XL*W(J)
         57 CONTINUE
            XL=XC
            CALL CALFUN (M,N,F,X)
            FC=0.
            DO 58 J=1,M
            FC=FC+F(J)*F(J)
         58 CONTINUE
            IF (IS .NE. 3) GO TO 59
            K=N
C
C     DETERMINATION OF SECOND BEST POINT
C
            IF (FC-FF) 61,51,62
         61 IS=2
            FMIN=FC
            FSEC=FF
            GO TO 63
         62 IS=1
            FMIN=FF
            FSEC=FC
            GO TO 63
         59 IF (FC .GE. FSEC) GO TO 51
            K=KSTORE
            IF (IS .EQ. 2) GO TO 74
            K=N
         74 IF (FC-FMIN) 65,51,66
         66 FSEC=FC
            GO TO 63
```

```
   65 IS=3-IS
      FSEC=FMIN
      FMIN=FC
   63 DO 67 J=1,N
      K=K+1
      W(K)=X(J)
   67 CONTINUE
      DO 68 J=1,M
      K=K+1
      W(K)=F(J)
   68 CONTINUE
      GO TO 51
   53 K=KSTORE
      KK=N
C
C  IF IS=2  XBEST AND FBEST LIE IN W(N+   ) SECOND BEST X AND X LIE IN
C  W(KSTORE+    )=D(N,J)   AND G(N,K)
C  IF IS IS NOT 2 XBEST AND FBEST LIE IN W(KSTORE+    )AND THE SECOND BES
C         LIE IN W(N+    )
C
      IF (IS .NE. 2) GO TO 69
      K=N
      KK=KSTORE
   69 SUM=0.
      DM=0.
      JJ=KSTORE
      DO 71 J=1,N
      K=K+1
      KK=KK+1
      JJ=JJ+1
C
C  XBEST INTO X
C  XBEST-XSECOND INTO D(V,J)
C
      X(J)=W(K)
      W(JJ)=W(K)-W(KK)
   71 CONTINUE
      DO 72 J=1,M
      K=K+1
      KK=KK+1
      JJ=JJ+1
C
C  FBEST INTO F
C  FBEST-FSECOND   INTO G(N,K)
C
      F(J)=W(K)
      W(JJ)=W(K)-W(KK)
      SUM=SUM+W(JJ)*W(JJ)
      DM=DM+F(J)*W(JJ)
   72 CONTINUE
      IF (MAXCAL) GO TO 10
      J=KINV
      KK=NPLUS-KL
```

```
              DO 76 I=1,KL
              K=J+KL-I
              J=K+KK
C
C     INTERCHANGE KL AND N ROWS OF G-1
C
              W(I)=W(K)
              W(K)=W(J-1)
           76 CONTINUE
              IF (KL .GE. N) GO TO 78
              KL=KL+1
              JJ=K
              DO 79 I=KL,N
              K=K+1
              J=J+NPLUS-I
              W(I)=W(K)
              W(K)=W(J-1)
           79 CONTINUE
              W(JJ)=W(K)
              B=1./W(KL-1)
              W(KL-1)=W(N)
              GO TO 88
           78 B=1./W(N)
           88 K=KINV
C
C     DETERMINE A1-1 FROM G-1  FOR USE IN CALCULATING NEW G-1
C
              DO 80 I=1,ILESS
              BB=B*W(I)
              DO 81 J=I,ILESS
C
C     W(K) IS G-1(I,J) WHICH IS A1-1=B1-B2*B4-1*B3
C
              W(K)=W(K)-BB*W(J)
              K=K+1
           81 CONTINUE
              K=K+1
           80 CONTINUE
              IF (FMIN .LT. FF) GO TO 82
              CHANGE=0.0
              GO TO 84
           82 FF=FMIN
C
C     CHANGE IS THE MAXIMUM OF THE COMPONENTS OF THE ACTUAL STEP TAKEN
C           DIVIDED BY THE COMPONENTS OF E
C     SUM IS USED TO NORMALIZE G(N,K) AND D(N,J)
C
              CHANGE=ABS(XC)*CHANGE
           84 XL=-DM/FMIN
              SUM=1.0/SQRT(SUM+DM*XL)
              K=KSTORE
              DO 85 I=1,N
              K=K+1
```

```
C
C     W(K) IS D(N,J) THE STEP TAKEN PROPERLY NORMALIZED
C
      W(K)=SUM*W(K)
      W(I)=0.
   85 CONTINUE
      DO 86 I=1,M
      K=K+1
C
C     W(K) IS G(N,K) WHICH IS (FBEST-FSECOND+(SUM OF (FBEST-FSECOND)*FBEST/
C            FMIN)*FBEST)   NORMALIZED
C
      W(K)=SUM*(W(K)+XL*F(I))
      KK=NN+I
      DO 87 J=1,N
      KK=KK+MPLUSN
C
C     W(J) IS THE N-TH ROW OF G*GT
C
      W(J)=W(J)+W(KK)*W(K)
   87 CONTINUE
   86 CONTINUE
      GO TO 14
C
      END
```

```
      SUBROUTINE LINMIN(ITEST,X,F,MAXFUN,ABSACC,RELACC,XSTEP)
C
      GO TO (1,2,2),ITEST
C
    2 IS=6-ITEST
      ITEST=1
      IINC=1
      XINC=XSTEP+XSTEP
      MC=IS-3
      IF (MC) 4,4,15
    3 MC=MC+1
      IF (MAXFUN .GE. MC) GO TO 15
      ITEST=4
   43 X=DB
      F=FB
      IF (FB .LE. FC) GO TO 15
      X=DC
      F=FC
   15 RETURN
C
    1 GO TO (5,6,7,8),IS
    8 IS=3
    4 DC=X
      FC=F
      X=X+XSTEP
      GO TO 3
    7 IF (FC-F) 9,10,11
   10 X=X+XINC
      XINC=XINC+XINC
      GO TO 3
    9 DB=X
      FB=F
      XINC=-XINC
      GO TO 13
   11 DB=DC
      FB=FC
      DC=X
      FC=F
   13 X=DC+DC-DB
      IS=2
      GO TO 3
    6 DA=DB
      DB=DC
      FA=FB
      FB=FC
   32 DC=X
      FC=F
      GO TO 14
    5 IF (FB .LT. FC) GO TO 16
      IF (F .GE. FB) GO TO 32
      FA=FB
      DA=DB
   19 FB=F
```

```
         DB=X
         GO TO 14
      16 IF (FA .LE. FC) GO TO 21
         XINC=FA
         FA=FC
         FC=XINC
         XINC=DA
         DA=DC
         DC=XINC
      21 XINC=DC
         IF ((D-DB)*(D-DC) .LT. 0.0) GO TO 32
         IF (F .GE. FA) GO TO 24
         FC=FB
         DC=DB
         GO TO 19
      24 FA=F
         DA=X
      14 IF (FB .GT. FC) GO TO 29
         IINC=2
         XINC=DC
         IF (FB .EQ. FC) GO TO 45
      29 D=(FA-FB)/(DA-DB)-(FA-FC)/(DA-DC)
         IF (D*(DB-DC) .LT. 0.0) GO TO 33
         D=0.5*(DB+DC-(FB-FC)/D)
         IF ((ABS(D-X) .GT. ABS(ABSACC)) .AND. (ABS(D-X) .GT. ABS(D*RELACC
        1)) GO TO 36
         ITEST=2
         GO TO 43
      36 IS=1
         X=D
         IF ((DA-DC)*(DC-D)) 3,26,38
      38 IS=2
         GO TO (39,40),IINC
      33 IS=2
         GO TO (41,42),IINC
      39 IF (ABS(XINC) .GE. ABS(DC-D)) GO TO 3
      41 X=DC
         GO TO 10
      40 IF (ABS(XINC-X) .GT. ABS(X-DC)) GO TO 3
      42 X=0.5*(XINC+DC)
         IF ((XINC-X)*(X-DC) .GT. 0.0) GO TO 3
         GO TO 26
      45 X=0.5*(DB+DC)
         IF ((DB-X)*(X-DC) .GT. 0.0) GO TO 3
      26 ITEST=3
         GO TO 43
C
         END
```

POWELL REGRESSION ALGORITHM

PARAMETERS

N = 3   M = 6   MAXFUN =   100   ESCALE = 0.1000E+04

INITIAL GUESSES

X(1) =   0.50000000E+03   X(2) =   -.15000000E+03   X(3) =   -.20000000E+00

ACCURACY OF VARIABLES

E(1) =   0.10000000E-02   E(2) =   0.10000000E-02   E(3) =   0.10000000E-05

ITERATION   0   NUMBER OF FUNCTION EVALUATIONS =   4

SUM-OF-THE-SQUARES =   0.14869486E+05

COEFFICIENTS

X(1) =   0.50000000E+03   X(2) =   -.15000000E+03   X(3) =   -.20000000E+00

FUNCTION VALUES

F(1) =   -.34742274E+02   F(2) =   0.75682180E+02   F(3) =   -.62210414E+02

F(4) =   -.43809613E+02   F(5) =   -.42321745E+02   F(6) =   0.18818084E+02

ITERATION   1   NUMBER OF FUNCTION EVALUATIONS =   10

SUM-OF-THE-SQUARES =   0.13390102E+05

COEFFICIENTS

X(1) =   0.52343094E+03   X(2) =   -.15709583E+03   X(3) =   -.19950076E+00

FUNCTION VALUES

F(1) =   -.29535177E+02   F(2) =   0.86612068E+02   F(3) =   -.47350575E+02

F(4) =   -.26252479E+02   F(5) =   -.22914308E+02   F(6) =   0.39494169E+02

(Iterations 2, 3, 4, 6, and 7 are omitted.)

```
ITERATION  5     NUMBER OF FUNCTION EVALUATIONS =   24
SUM-OF-THE-SQUARES =    0.13390093E+05
COEFFICIENTS
X( 1) =   0.52330593E+03   X( 2) =   -.15694810E+03   X( 3) =   -.19966452E+00
FUNCTION VALUES
F( 1) =   -.29608204E+02   F( 2) =   0.86615526E+02   F( 3) =   -.47326608E+02
F( 4) =   -.26235419E+02   F( 5) =   -.22915744E+02   F( 6) =   0.39471022E+02

ITERATION  8     NUMBER OF FUNCTION EVALUATIONS =   41
SUM-OF-THE-SQUARES =    0.13390093E+05
COEFFICIENTS
X( 1) =   0.52330621E+03   X( 2) =   -.15694856E+03   X( 3) =   -.19966395E+00
FUNCTION VALUES
F( 1) =   -.29607956E+02   F( 2) =   0.86615457E+02   F( 3) =   -.47326783E+02
F( 4) =   -.26235591E+02   F( 5) =   -.22915866E+02   F( 6) =   0.39470966E+02

    SSQMIN FINAL VALUES OF FUNCTIONS AND VARIABLES

ITERATION  8     NUMBER OF FUNCTION EVALUATIONS =   41
SUM-OF-THE-SQUARES =    0.13390093E+05
COEFFICIENTS
X( 1) =   0.52330621E+03   X( 2) =   -.15694856E+03   X( 3) =   -.19966395E+00
FUNCTION VALUES
F( 1) =   -.29607956E+02   F( 2) =   0.86615457E+02   F( 3) =   -.47326783E+02
F( 4) =   -.26235591E+02   F( 5) =   -.22915866E+02   F( 6) =   0.39470966E+02

THE SUM-OF-THE-SQUARES DIFFERENCE =     1.3390093E+04

FINAL COEFFICIENT VALUES
X( 1) =    5.2330621E+02
X( 2) =   -1.5694856E+02
X( 3) =   -1.9966395E-01
```

PART TWO:

SEARCH METHODS

Chapter 7

SINGLE VARIABLE UNCONSTRAINED METHODS

The simplest type of search problems are those involving only a single independent variable not subject to constraints. Few practical applications exist in this category. However a large number of multivariable search methods (see Chapters 9, 10) require an unconstrained single variable search building block. Many variations exist for this type of problem (5, 8, 14). All start from some base point and move toward the optimum based on sequential improvement in the value of the objective function. The step size may be fixed or accelerated and decelerated subject to a set of rules. In this chapter, only one technique will be presented, that of Coggin (COGGIN ALGORITHM). Other methods are incorporated into other multivariable codes presented in this text (see SSQMIN, BOTM, FMCG, FMFP, CONMIN).

I. COGGINS (COGGIN ALGORITHM)*

A. Purpose

This program finds the unconstrained maximum of a single variable, nonlinear function:

Maximize   F(X)

B. Method

The procedure used is based on that of Coggins (8). The method is a combination of the single variable techniques proposed by Davies, Swann, and Campey (8) and Powell (37). Unimodality is assumed and thus the use of multiple starting points is recommended if a multimodal function is suspected. The algorithm proceeds as follows:

1) A starting point is chosen and the objective function evaluated.

2) The independent variable is incremented a distance ΔX and the objective function evaluated again. If a function improvement is obtained the step size is doubled for the next function evaluation. If a function improvement is not obtained on the first step, the direction is reversed and the next point located a distance $-\Delta X$ from the starting point.

3) After the first step, the step size is doubled if a function improvement is obtained and halved if a worse function evaluation is obtained.

4) When a local optimum is encountered, the procedure will yield three points (X_k, X_{k-1}, X_{k-2}) straddling the optimum. An additional point, X_{k+1}, is then located:

$$X_{k+1} = X_{k-1} + \frac{\Delta X}{2}$$

where ΔX is the current step size. The best three points are then retained (call X_1, X_2, X_3).

5) A quadratic equation, f, is then curve fitted to the three retained points. The optimum location, X^*, located by setting $\frac{\partial f}{\partial X} = 0$, is

---

*Computer code developed by Marc Voorhees, Arizona State University, Tempe, Arizona.

$$X^* = \frac{1}{2} \left[\frac{(X_2^2 - X_3^2)F(X_1) + (X_3^2 - X_1^2)F(X_2) + (X_1^2 - X_2^2)F(X_3)}{(X_2 - X_3)F(X_1) + (X_3 - X_1)F(X_2) + (X_1 - X_2)F(X_3)} \right]$$

6) The objective function at X^* is then compared with the best previous point subject to a convergence limit,

$$\left| X^* - X_i \text{ (best)} \right| \leq \text{limit}$$

If the above criterion is satisfied, the procedure stops. If not, the worst point is replaced by X^* and a new quadratic surface fitted and local optimum obtained. This process is repeated until the convergence criterion is satisfied.

A flow sheet illustrating the above procedure is given in Figure 7.I.

C. Program Description

1) Usage:

The program consists of a main program and a user supplied subroutine (FUNC). The starting point, the initial increment and the convergence limit are all read into the main program. Output printing is from the main program. Format changes may be required depending on the problem under consideration.

2) Subroutines Required:

SUBROUTINE FUNC (XX, YY) called from the main program; specifies objective function (user supplied).

3) Description of Parameters:

XLIM	Convergence criteria limit - Define in main program
DELX	Independent variable step size - Define in main program
XX	Independent variable
YY	Objective function
X1	Starting value of independent variable - Define in main program
Y1	Objective function at the starting value
NI	Card reader unit number - Define in main program
NO	Printer unit number - Define in main program

4) DIMENSION Requirements:

The DIMENSION statement in the main program is fixed by the method as follows:

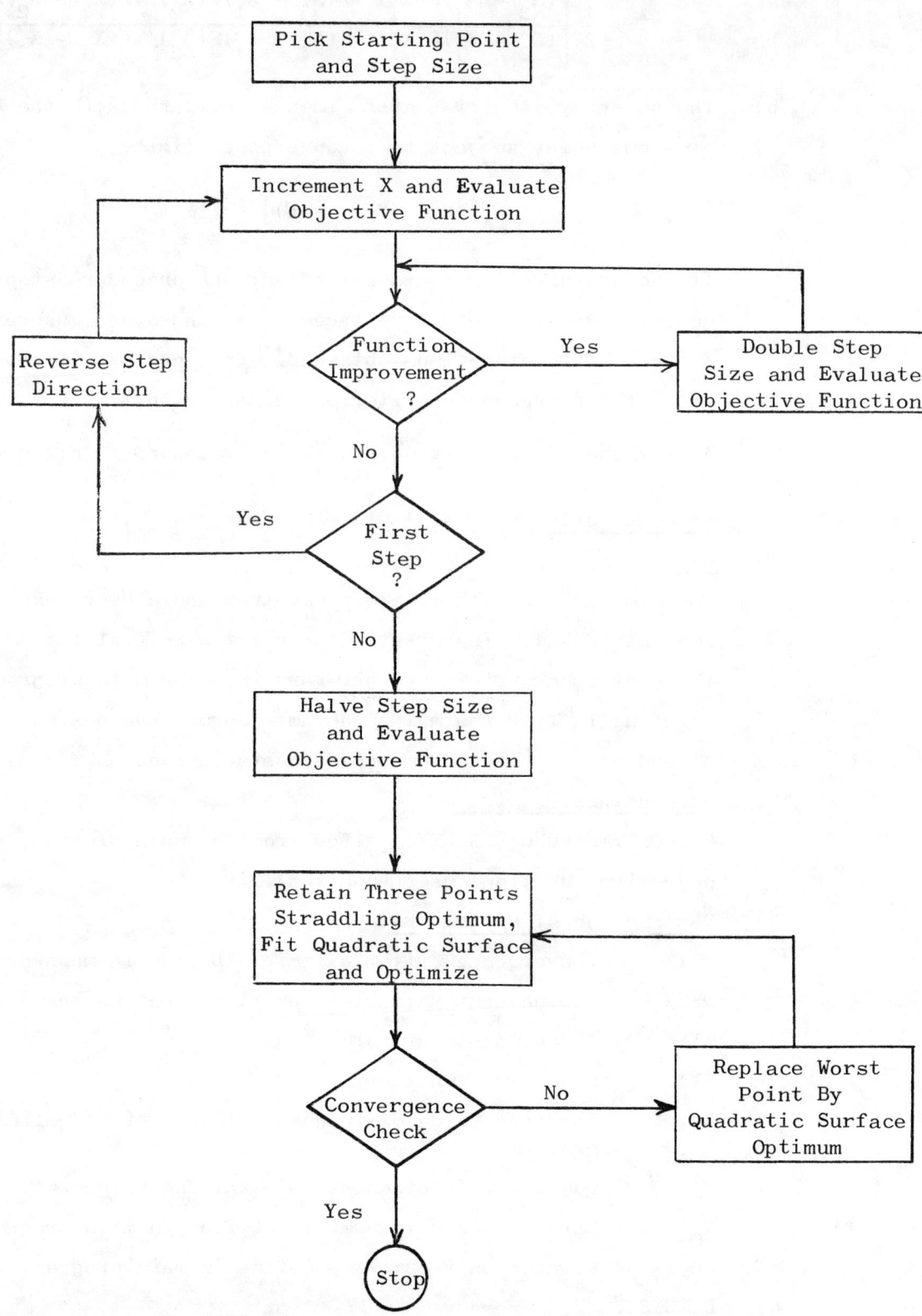

Figure 7.I. Coggins (COGGIN ALGORITHM) Logic Diagram

DIMENSION ZZ(3), WW(3), YF(3)

No DIMENSION statements are required in SUBROUTINE FUNC.

5) <u>Input Formats:</u>

CARD TYPE	FORMAT	CONTENTS
1	(3E10.4)	XLIM, DELX, X1

6) <u>Output:</u>

The main program first prints out the method title, XLIM, DELX, and X1. The current values of the independent variables (XX), objective function (YY), and step size are then printed under the heading, DAVIES, SWANN, AND CAMPEY ALGORITHM. When the optimum is straddled, the X^* value and $F(X^*)$ values from the quadratic fits are printed out as XMAX and YMAX under the heading POWELL ALGORITHM. When the convergence criteria is satisfied, the final value of the independent variable and objective function are printed.

7) <u>Summary of User Requirements:</u>

   a) Determine values for XLIM, DELX, and X1.
   b) Specify objective function in SUBROUTINE FUNC.
   c) Supply NI and NO in main program.
   d) Change input and output format statements as necessary for the particular problem under study.

D. <u>Test Problem</u>

The following test problem was posed to test the code. The calculations were performed on a CDC 6400 computer.

Function: $F = -X^4 + 12X^3 - 15X^2 - 56X + 60$

Starting Point: X = 0

Parameters: DELX = .01, XLIM = .001

Algorithm Answers: F = 88.9
                     X = -0.87

Number of Function Evaluations: 13

Central Processor Time: 2 seconds

The listing and output for this problem are given in the following section.

E. Program Listings and Example Output

```
C
C         METHOD OF COGGINS
C
C
C
      DIMENSION ZZ(3), WW(3), YF(3)
C
      NI = 50
      NO = 66
C
C        INPUT VALUES OF LIMIT, STEP SIZE, AND INITIAL GUESS FOR X
C
      READ (NI,001) XLIM, DELX, X1
    1 FORMAT (8E10.4)
C
      WRITE (NO,002)
  002 FORMAT (1H1,10X,41HCOGGINS METHOD OF SINGLE VARIABLE SEARCH )
C
      WRITE (NO,201) XLIM
  201 FORMAT(     //,15X,37HTHE ABSOLUTE LIMIT ON THE INDEPENDENT,
     1 12H VARIABLE IS,2X,E14.7)
      WRITE (NO,202) DELX
  202 FORMAT(//,15X,24HTHE INITIAL STEP SIZE IS,23X,E14.7)
      WRITE (NO,203) X1
  203 FORMAT(//,15X,36HTHE INITIAL VALUE OF THE INDEPENDENT,
     1 12H VARIABLE IS,2X,E14.7)
C
C        DAVIES, SWANN, AND CAMPEY ALGORITHM
C
C
C
C        EVALUATE THE FUNCTION FOR THE INITIAL VALUE OF THE INDEPENDENT
C        VARIABLE
C
      CALL FUNC(X1,Y1)
      WRITE (NO,210)
  210 FORMAT(     //,15X,35HDAVIES, SWANN, AND CAMPEY ALGORITHM)
      WRITE (NO,204)
  204 FORMAT(     //,21X,2HXX,14X,2HYY,8X,9HSTEP SIZE)
      WRITE (NO,205) X1, Y1, DELX
  205 FORMAT(/,15X,E14.7,2X,E14.7,2X,E14.7)
C
C        INCREMENT THE INDEPENDENT VARIABLE AND EVALUATE THE FUNCTION
C
```

```
      X2=X1+DELX
      II=0
    8 CALL FUNC(X2,Y2)
      WRITE (NO,205) X2, Y2, DELX
C
C        SEE WHICH OF THE FUNCTION EVALUATIONS IS THE GREATEST
C
    9 IF(Y2-Y1) 10,12,12
   10 IF(II-1)14,14,16
   14 II=II+1
C
C        IF THE FUNCTION EVALUATION IS LESS AFTER THE INDEPENDENT
C        VARIABLE HAS BEEN INCREMENTED, SWITCH DIRECTION
C        OF THE STEP
C
      DELX=-DELX
      X2=X1+DELX
      GO TO 8
C
C        IF THE MAXIMUM IS BRACKETED,  TRANSFER CONTROL TO THE POWELL
C        ALGORITHM.
C
   16 GO TO 80
   12 CONTINUE
      X3=X2+DELX
      GO TO 13
   17 X3=X4
   13 CALL FUNC(X3,Y3)
      WRITE (NO,205) X3, Y3, DELX
C
C        IF THE FUNCTION EVALUATION IS GREATER THAN THE PREVIOUS
C        VALUE,  DOUBLE THE STEP SIZE.
C
      DELX=2.*DELX
      X4=X3+DELX
      CALL FUNC(X4,Y4)
      IF(Y4-Y3) 20,22,22
   20 GO TO 90
   22 GO TO 17
C
C        WHEN THE OPTIMUM IS STRADDLED,  EVALUATE THREE POINTS
C        ABOUT THE MAXIMUM.
C
C
C        POWELL ALGORITHM
C
   80 ZZ(1)=X1
      ZZ(2)=X1+DELX/2.
      ZZ(3)=X2
      GO TO 99
   90 ZZ(1)=X3
      ZZ(2)=X3+DELX/2.
      ZZ(3)=X4
      GO TO 99
C
```

```
C         EVALUATE THE FUNCTION AT THESE THREE POINTS
C
   99 DO 97 I=1,3
      CALL FUNC (ZZ(I),YF(I))
   97 CONTINUE
C
C         FIT A QUADRATIC CURVE TO THESE THREE POINTS
C
      WRITE (NO,211)
  211 FORMAT(   ///,15X,16HPOWELL ALGORITHM)
      WRITE (NO,212)
  212 FORMAT(//,20X,4HXMAX,12X,4HYMAX)
   98 CONTINUE
      A=ZZ(2)-ZZ(3)
      B=ZZ(3)-ZZ(1)
      C=ZZ(1)-ZZ(2)
      D=ZZ(2)**2-ZZ(3)**2
      E=ZZ(3)**2-ZZ(1)**2
      F=ZZ(1)**2-ZZ(2)**2
C
C         ANALYTICALLY EVALUATE THE MAXIMUM VALUE OF THE FITTED CURVE
C
      ZZT=.5*(D*YF(1)+E*YF(2)+F*YF(3))
      ZZB=A*YF(1)+B*YF(2)+C*YF(3)
      ZZM=ZZT/ZZB
C
C         EVALUATE THE FUNCTION VALUE AT THIS POINT
C
      CALL FUNC(ZZM,YFM)
      WRITE (NO,205) ZZM, YFM
C
C         CHECK TO SEE IF ANY OF THE POINTS ARE WITHIN THE
C         DESIRED ACCURACY
C
      DO 100 J=1,3
      WW(J)=ABS(ZZ(J)-ZZM)
      IF(WW(J)-XLIM) 105,105,106
  105 XMAX=ZZ(J)
      GO TO 200
  106 CONTINUE
  100 CONTINUE
C
C         SEE WHICH FUNCTION VALUE IS THE SMALLEST
C         AND REPLACE BY THE INTERPOLATED   MAXIMUM POINT
C
      IF(YF(2)-YF(1)) 110,111,111
  111 IF(YF(3)-YF(1)) 110,113,113
  110 IF(YF(3)-YF(2)) 115,114,114
  113 JK=1
      GO TO 117
  114 JK=2
      GO TO 117
  115 JK=3
  117 ZZ(JK)=ZZM
      YF(JK)=YFM
```

```
      C
      C       FIT A QUADRATIC TO THE NEW POINTS
      C
            GO TO 98
      C
      C       EVALUATE THE MAXIMUM FUNCTION VALUE
      C
        200 CALL FUNC(XMAX,YMAX)
            WRITE (NO,300) XMAX
        300 FORMAT(   ///,15X,26HTHE MAXIMUM VALUE OF XX IS,E14.7)
            WRITE (NO,400) YMAX
        400 FORMAT(//,15X,26HTHE MAXIMUM VALUE OF YY IS,E14.7)
      C
            END




            SUBROUTINE FUNC(XX,YY)
      C
            YY=XX**4-12.*XX**3+15.*XX**2+56.*XX-60.
            YY=-YY
      C
            RETURN
            END
```

COGGINS METHOD OF SINGLE VARIABLE SEARCH

THE ABSOLUTE LIMIT ON THE INDEPENDENT VARIABLE IS   .1000000E-02

THE INITIAL STEP SIZE IS   .1000000E-01

THE INITIAL VALUE OF THE INDEPENDENT VARIABLE IS -0.

DAVIES, SWANN, AND CAMPEY ALGORITHM

XX	YY	STEP SIZE
-0.	.6000000E+02	.1000000E-01
.1000000E-01	.5943851E+02	.1000000E-01
-.1000000E-01	.6055849E+02	-.1000000E-01
-.2000000E-01	.6111390E+02	-.1000000E-01
-.4000000E-01	.6221523E+02	-.2000000E-01
-.8000000E-01	.6437782E+02	-.4000000E-01
-.1600000E+00	.6852619E+02	-.8000000E-01
-.3200000E+00	.7598030E+02	-.1600000E+00
-.6400000E+00	.8638250E+02	-.3200000E+00

POWELL ALGORITHM

XMAX	YMAX
-.8590926E+00	.8888534E+02
-.8671530E+00	.8889110E+02
-.8699886E+00	.8889157E+02
-.8701784E+00	.8889157E+02

THE MAXIMUM VALUE OF XX IS -.8699886E+00

THE MAXIMUM VALUE OF YY IS .8889157E+02

Chapter 8

SINGLE VARIABLE CONSTRAINED METHODS

The most widely used single variable search technique subject to inequality constraints is that of Fibonacci. This is an interval elimination procedure, i.e., the region in which the optimum lies is sequentially reduced by the search procedure. Several variations of this method exist (5). In this chapter, the basic Fibonacci method is presented (FIBON ALGORITHM). A variation is used as a building block elsewhere in this book (see SUMT).

I. FIBONACCI (FIBON ALGORITHM)*

A. Purpose

This program finds the minimum of a single variable, nonlinear function subject to constraints:

Minimize $\quad F(X)$

Subject to $\quad a_1 \leq X \leq b_1$

The upper and lower bounds, b_1 and a_1, are constants.

B. Method

The procedure is an interval elimination search method. Thus, starting with the original boundaries on the independent variable, the interval in which the optimum value of the function occurs is reduced to some final value, the magnitude of which depends on the desired accuracy. The location of points for function evaluations is based on the use of positive integers known as the Fibonacci numbers (5). No derivatives are required. A specification of the desired accuracy will determine the number of function evaluations. A unimodal function is assumed. Thus the use of multiple starting points is recommended if a multimodal function is suspected. The algorithm proceeds as follows:

1) Designate the original search interval as L_1 with boundaries a_1 and b_1.

2) Predetermine the desired accuracy α and thus the number, N, of required Fibonacci numbers (equals number of required function evaluations).

$$\alpha = \frac{1}{F_N}$$

$$F_0 = F_1 = 1$$

$$F_n = F_{n-1} + F_{n-2}, \; n \geq 2$$

where F_n is called a Fibonacci number.

---

*Computer code developed by Marc Voorhees, Arizona State University, Tempe, Arizona.

3) Place the first two points, X_1 and X_2 ($X_1 < X_2$) within L_1 at a distance ℓ_1 from each boundary,

$$\ell_1 = \frac{F_{N-2}}{F_N} L_1$$

$$X_1 = a_1 + \ell_1$$

$$X_2 = b_1 - \ell_1$$

4) Evaluate the objective function at X_1 and X_2. Designate the functions as $F(X_1)$ and $F(X_2)$. Narrow the search interval as follows:

$$a_1 \leq X^* \leq X_2 \quad \text{for} \quad F(X_1) < F(X_2)$$
$$X_1 \leq X^* \leq b_1 \quad \text{for} \quad F(X_1) > F(X_2)$$

where X^* is the location of the optimum. The new search interval is given by

$$L_2 = \frac{F_{N-1}}{F_N} \cdot L_1 = L_1 - \ell_1$$

with boundaries a_2 and b_2.

5) Place the third point in the new L_2 subinterval, symmetric about the remaining point,

$$\ell_2 = \frac{F_{N-3}}{F_{N-1}} L_2$$

$$X_3 = a_2 + \ell_2 \text{ or } b_2 - \ell_2$$

6) Evaluate the objective function $F(X_3)$, compare with the function for the point remaining in the interval and reduce the interval to

$$L_3 = \frac{F_{N-2}}{F_N} L_1 = L_2 - \ell_2$$

7) The process is continued per the preceeding rules for N evaluations. The general equations are

$$\ell_k = \frac{F_{N-(k+1)}}{F_{N-(k-1)}} L_k$$

$$X_{k+1} = a_k + \ell_k \text{ or } b_k - \ell_k \quad \text{(symmetric about mid point)}$$

$$L_k = \frac{F_{N-(k-1)}}{F_N} L_1 = L_{k-1} - \ell_{k-1}$$

After N-1 evaluations and discarding the appropriate interval at each step, the remaining point will be precisely in the center of the remaining interval. Thus point N is also at the midpoint and is replaced by a point perturbed some small distance ϵ to one side or the other of the midpoint. The objective function is then evaluated and the final interval where the optimum is located is thus determined.

A flow sheet illustrating the procedure is given in Figure 8.I.

C. Program Description

1) Usage:

The program consists of a main program and one user supplied subroutine (FUNC). The constraints and desired accuracy are read into the main program. All output is from the main program. Format changes may be required depending on the problem under consideration.

2) Subroutines Required:

SUBROUTINE FUNC (X, Y) specifies objective function (user supplied).

3) Description of Parameters:

X	Independent variable
ALPHA	Desired accuracy specified as a fraction of the original search interval - Define in main program
A	Lower constraint - Define in main program
B	Upper constraint - Define in main program
FIB	Fibonacci number - Defined in main program
Y	Objective function - Define in Subroutine FUNC
NO	Printer unit number - Define in main program
NI	Card reader unit number - Define in main program

4) DIMENSION Requirements:

The DIMENSION statement in the main program should be modified according to the requirements of each particular problem. The only dimensioned quantity is FIB. The value depends upon the desired accuracy and corresponds to the number of Fibonacci

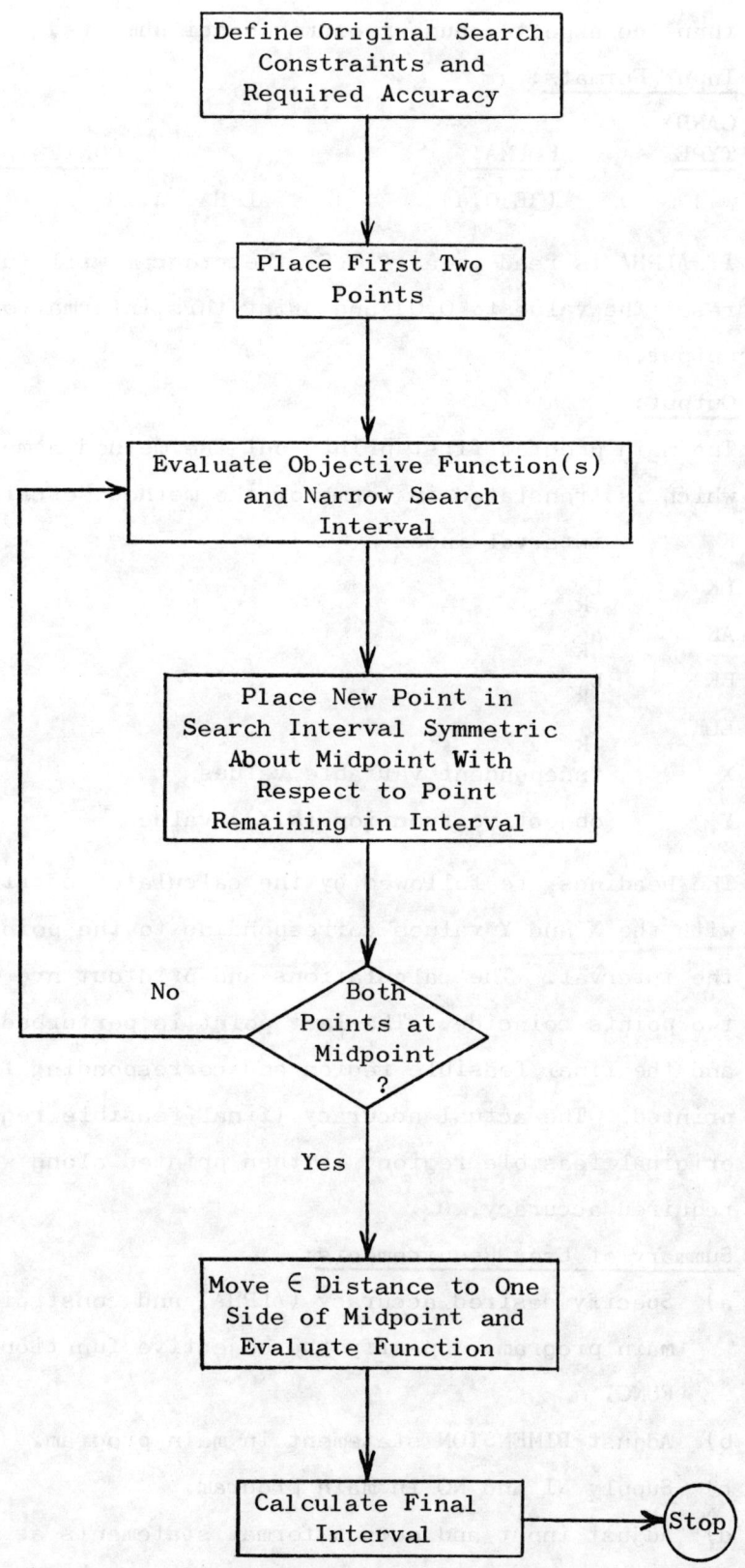

Figure 8.1. Fibonacci (FIBON ALGORITHM) Logic Diagram

numbers expected. Thus set the dimension equal to or greater than the expected number of Fibonacci numbers.

5) Input Formats:

CARD TYPE	FORMAT	CONTENTS
1	(3E10.4)	ALPHA, A, B

If ALPHA is read in as ≥ 0.5, the program will automatically reset the value to 0.01 and print this information in the output.

6) Output:

The main program first prints out the method name and then a heading which is translated in terms of the method description as follows:

K	interval index
LK	L_k
AK	a_k
BK	b_k
LLK	ℓ_k
X	independent variable values
Y	objective function (F(X)) values

The headings are followed by the calculated entries for each step with the X and Y values corresponding to the points remaining in the interval. The calculations and printout are terminated when two points coincide. The last point is perturbed by $\epsilon = X_N \cdot 10^{-3}$ and the final feasible region and corresponding function values printed. The actual accuracy (final feasible region divided by original feasible region) is then printed along with the original required accuracy, α.

7) Summary of User Requirements:

a) Specify desired accuracy (ALPHA) and constraints (A,B) in the main program. Specify the objective function (Y) in Subroutine FUNC.

b) Adjust DIMENSION statement in main program.

c) Supply NI and NO in main program.

d) Adjust input and output format statements as required by the particular problem under investigation.

D. Test Problem

The following test problem was chosen (10) to test the code. The calculations were performed on a CDC 6400 computer.

Function: $F(x) = X^2 - 6X + 2$

Constraints: $0 \leq X \leq 10$

Parameters: ALPHA = 0.03

Algorithm Answers: $-7.00 \leq F \leq -6.94$

$2.94 \leq X \leq 3.24$

Number of Function Evaluations: 9

Central Processor Time: 2 seconds

The listing and output for this problem are contained in the following sections.

E. <u>Program Listings and Example Output</u>

```fortran
C
C       MAIN LINE PROGRAM FOR FIBONACCI PROCEDURE
C
        DIMENSION FIB(50)
C
        NI = 50
        NO = 66
C
        READ (NI,016) ALPHA, A, B
    016 FORMAT (8E10.4)
        DEL = B - A
        WRITE (NO,001)
    001 FORMAT (1H1,10X,35HFIBONACCI SINGLE-VARIABLE PROCEDURE )
C
C       DEFINE THE FIRST THREE FIBONACCI NUMBERS
C
        FIB0 = 1.0
        FIB(1) = 1.0
        FIB(2) = 2.0
C
C          CALCULATE THE REMAINING FIBONACCI NUMBERS
C
      5 BB = 1.0/ALPHA
        IF (BB - 2.0)   10, 10, 11
     10 GO TO 14
     11 CONTINUE
        JJ=2
     12 JJ=JJ+1
        FIB(JJ)=FIB(JJ-1)+FIB(JJ-2)
        CC=FIB(JJ)
        IF(CC-BB) 13,15,15
     13 GO TO 12
     14 WRITE (NO,002)
    002 FORMAT (///,10X,42HACCURACY SPECIFIED IN FUNC NOT SUFFICIENT. ,
       1 //,10X,33HPROGRAM RESET ALPHA, ALPHA = 0.01 )
        ALPHA = 0.01
        GO TO 5
C
C          FIRST STEP IN THE TABLEAU
C
```

```
   15 I=0
      KK=JJ-2
      IK=JJ-2
      BL=B-A
      ALL=FIB(IK)*BL/FIB(JJ)
      W=A+ALL
      V=B-ALL
      CALL FUNC(W,T)
      CALL FUNC(V,U)
      JK=1
      WRITE (NO,003)
  003 FORMAT (//,1X, 1HK,5X,2HLK,10X,2HAK,11X,2HBK,09X,3HLLK,11X,1HX,
     1 12X,1HY )
      WRITE (NO,004) JK, BL, A, B, ALL, W, T
  004 FORMAT (/,1X, I1,2X,E10.4,2X,E10.4,2X,E10.4,2X,E10.4,
     1 2X,E10.4,2X,E10.4)
  006 FORMAT(52X,E10.4,2X,E10.4)
C
C     SUCCEEDING STEPS IN THE TABLEAU
C
      IK=IK-1
      JJ=JJ-1
      DO 70 I=1,KK
      IF(U-T) 20,20,22
   20 A=A+ALL
      BL=B-A
      W=V
      CALL FUNC(W,T)
      ALL=FIB(IK)*BL/FIB(JJ)
      V=B-ALL
      CALL FUNC(V,U)
      II=I+1
      IK=IK-1
      JJ=JJ-1
      IF(IK-1) 28,29,29
   28 IK=1
   29 CONTINUE
      WRITE (NO,004) II, BL, A, B, ALL, W, T
      WRITE (NO,006) V, J
      GO TO 70
   22 B=B-ALL
      BL=B-A
      V=W
      CALL FUNC(V,U)
      ALL=FIB(IK)*BL/FIB(JJ)
      W=A+ALL
      CALL FUNC(W,T)
      II=I+1
      IK=IK-1
      JJ=JJ-1
      IF(IK-1) 30,31,31
   30 IK=1
   31 CONTINUE
```

```
      WRITE (NO,004) II, BL, A, B, ALL, V, U
      WRITE (NO,006) W, T
      GO TO 70
   70 CONTINUE
C
C     CALCULATION OF THE FINAL RANGE OF THE DEPENDENT VARIABLE
C
      EPS = 0.001 * W
      DL=W+EPS
      CALL FUNC(DL,YL)
      IF(YL-T) 80,80,81
   80 CALL FUNC(B,BF)
      WRITE (NO,007) W, B
  007 FORMAT(///         ,25HTHE FINAL FEASIBLE REGION,2X,2HX=,
     1 E10.4,2X,2HX=,E10.4)
      WRITE (NO,008) T, BF
  008 FORMAT(/      ,20HWITH FUNCTION VALUES,7X,2HY=,E10.4,2X,2HY=,E10.4)
      GO TO 87
   81 CALL FUNC(A,AF)
      WRITE (NO,009) W, A
  009 FORMAT(///         ,25HTHE FINAL FEASIBLE REGION,2X,2HX=,
     1 E10.4,2X,2HX=,E10.4)
      WRITE (NO,017) T, AF
  017 FORMAT(/      ,20HWITH FUNCTION VALUES,7X,2HY=,E10.4,2X,2HY=,E10.4)
   87 ACC=(W-A)/(DEL)
      WRITE (NO,018) ACC
  018 FORMAT(/    ,     15HTHE ACCURACY IS,12X,E10.4)
      WRITE (NO,019) ALPHA
  019 FORMAT(/       ,25HTHE REQUIRED ACCURACY WAS,2X,E10.4)
  999 CONTINUE
C
      END




      SUBROUTINE FUNC(X,Y)
C
      Y=X**2-6.*X+2.
C
      RETURN
      END
```

FIBONACCI SINGLE-VARIABLE PROCEDURE

K	LK	AK	BK	LLK	X	Y
1	0.1000E+02	0.0000E+00	0.1000E+02	0.3824E+01	0.3824E+01	-.6322E+01
2	0.6176E+01	0.0000E+00	0.6176E+01	0.2353E+01	0.3824E+01	-.6322E+01
					0.2353E+01	-.6581E+01
3	0.3824E+01	0.0000E+00	0.3824E+01	0.1471E+01	0.2353E+01	-.6581E+01
					0.1471E+01	-.4661E+01
4	0.2353E+01	0.1471E+01	0.3824E+01	0.8824E+00	0.2353E+01	-.6581E+01
					0.2941E+01	-.6997E+01
5	0.1471E+01	0.2353E+01	0.3824E+01	0.5882E+00	0.2941E+01	-.6997E+01
					0.3235E+01	-.6945E+01
6	0.8824E+00	0.2353E+01	0.3235E+01	0.2941E+00	0.2941E+01	-.6997E+01
					0.2647E+01	-.6875E+01
7	0.5882E+00	0.2647E+01	0.3235E+01	0.2941E+00	0.2941E+01	-.6997E+01
					0.2941E+01	-.6997E+01

THE FINAL FEASIBLE REGION   X=0.2941E+01   X=0.3235E+01
WITH FUNCTION VALUES   Y=-.6997E+01   Y=-.6945E+01
THE ACCURACY IS   0.2941E-01
THE REQUIRED ACCURACY WAS   0.3000E-01

Chapter 9

MULTIVARIABLE UNCONSTRAINED METHODS

Search methods for use on multivariable unconstrained problems have rapidly increased in number and sophistication in recent years. While all realistic problems are constrained, an unconstrained building block is often required (see Chapter 10). The unconstrained methods are normally divided into two categories:
1) derivative free methods
2) gradient methods.

The gradient methods require function and derivative evaluations while the derivative free methods require function evaluations only. In general, one would expect the gradient methods to be more effective, due to the added information provided. However, if analytical derivatives are available, the question of whether a search technique should be used at all is presented. If numerical derivative approximations are utilized, the efficiency of the gradient methods should be approximately the same as that of the derivative free methods. Gradient methods incorporating numerical derivatives would be expected to present some numerical problems in the vicinity of the optimum, i.e., the approximations would become very small.

For this chapter, several of the most widely used techniques in each category are presented:

   I. Nelder and Mead (NELDER ALGORITHM)
  II. Hooke and Jeeves (HOOKE ALGORITHM)
 III. Rosenbrock (ROSENB ALGORITHM)
  IV. Powell (BOTM ALGORITHM)
   V. Fletcher-Reeves (FMCG ALGORITHM)
  VI. Fletcher-Powell (FMFP ALGORITHM)

I. NELDER AND MEAD (NELDER ALGORITHM)*

A. **Purpose**

This program finds the minimum of a multivariable unconstrained, nonlinear function:

$$\text{Minimize} \quad F(X_1, X_2, \ldots, X_N)$$

B. **Method**

The procedure is an extension of the simplex method by Spendley, Hext, and Himsworth (43). Both methods utilize a regular geometric figure (called a simplex) consisting of N+1 vertices. This method accelerates the simplex method and makes it more general. The procedue is based on the work by J. A. Nelder and R. Mead (35). This simplex method adapts itself to the local landscape, using reflected, expanded, and contracted points to locate the minimum. Unimodality is assumed and thus several sets of starting points should be considered. Derivatives are not required. The algorithm proceeds as follows:

1) A starting point, $\underline{X}_1$, is selected.
2) A starting "simplex" is constructed consisting of the starting point and the following additional points:

$$\underline{X}_j = \underline{X}_1 + \underline{\xi}_j, \qquad j = 2, 3, \ldots, N+1$$

where $\underline{\xi}_j$ is determined from the following table,

j	$\xi_{1,j}$	$\xi_{2,j}$	...	$\xi_{N-1,j}$	$\xi_{N,j}$
2	p	q	...	q	q
3	q	p	...	q	q
⋮	⋮	⋮		⋮	⋮
N	q	q	...	p	q
N+1	q	q	...	q	p

---

*Computer code developed by H. T. Bates, Kansas State University, Manhattan, Kansas. Used by permission.

$$N = \text{total number of variables}$$

$$a = \text{side length of simplex}$$

$$p = \frac{a}{N\sqrt{2}}(\sqrt{N+1} + N-1)$$

$$q = \frac{a}{N\sqrt{2}}(\sqrt{N+1} - 1)$$

3) Once the simplex is formed, the objective function is evaluated at each point. The worst point (highest value of objective function) is replaced by a new point. Three operations are used--reflection, contraction, and expansion. A reflected point is located first as follows:

$$X_{i,j}(\text{reflected}) = \bar{X}_{i,c} + \alpha(\bar{X}_{i,c} - X_{i,j}(\text{worst}))$$

$$i = 1, 2, \ldots, N$$

where α is a positive constant.

$\bar{X}_{i,c}$ are the centroid coordinates of all points excluding the worst point and are calculated from the following:

$$\bar{X}_{i,c} = \frac{1}{K-1}\left[\sum_{j=1}^{K} X_{i,j} - X_{i,j}(\text{worst})\right], \quad i = 1, 2, \ldots, N$$

$$K = N+1$$

4) If the reflected point has the worst objective function value of the current points, a contracted point is located as follows:

$$X_{i,j}(\text{contracted}) = \bar{X}_{i,c} - \beta(\bar{X}_{i,c} - X_{i,j}(\text{worst})), \quad i = 1, 2, \ldots, N$$

where β lies between 0 and 1.

If the reflected point is better than the worst point but is not the best point, a contracted point is calculated from the reflected point as follows:

$$X_{i,j}(\text{contracted}) = \bar{X}_{i,c} - \beta(\bar{X}_{i,c} - X_{i,j}(\text{reflected})),$$

$$i = 1, 2, \ldots, N$$

The objective function is now evaluated at the contracted point. If an improvement over the current points is achieved, the process is restarted. If an improvement is not achieved, the points are moved one half the distance toward the best point:

$$X_{i,j} \text{ (new)} = (X_{i,j} \text{ (best)} + X_{i,j} \text{ (old)})/2$$

$$i = 1, 2, \ldots, N$$

The process is then restarted.

5) If the reflected point calculated in step 3) is the best point, an expansion point is calculated as follows:

$$X_{i,j} \text{ (expansion)} = \bar{X}_{i,c} + \gamma (X_{i,j} \text{ (reflected)} - \bar{X}_{i,c})$$

$$i = 1, 2, \ldots, N$$

where γ is a positive constant. If the expansion point is an improvement over the reflected point, the reflected point is replaced by the expansion point and the process restarted. If the expansion point is not an improvement over the reflected point, the reflected point is retained and the process restarted.

6) The procedure is terminated when the convergence criterion is satisfied or a specified number of iterations has been exceeded (see Description of Parameters).

A flow sheet illustrating the procedure is given in Figure 9.I.

C. Program Description

1) Usage:

The program consists of a main program and a user supplied subroutine FUNC. An initial estimate, side length of simplex, the number of variables, the three operator coefficients, and the convergence criteria limit are read into the main program. Subroutine FUNC contains the objective function to be minimized. All output is from the main program.

2) Subroutine Description:

SUBROUTINE FUNC (I,X,Z,N,NP1) called from the main program; specifies objective function (user supplied).

3) Description of Parameters:

N	Number of independent variables
A	Side length of simplex
ALFA	Reflection coefficient, suggested = 1.0
BETA	Contraction coefficient, suggested = 0.5
GAM	Expansion coefficient, suggested = 2.0
X	Independent variables

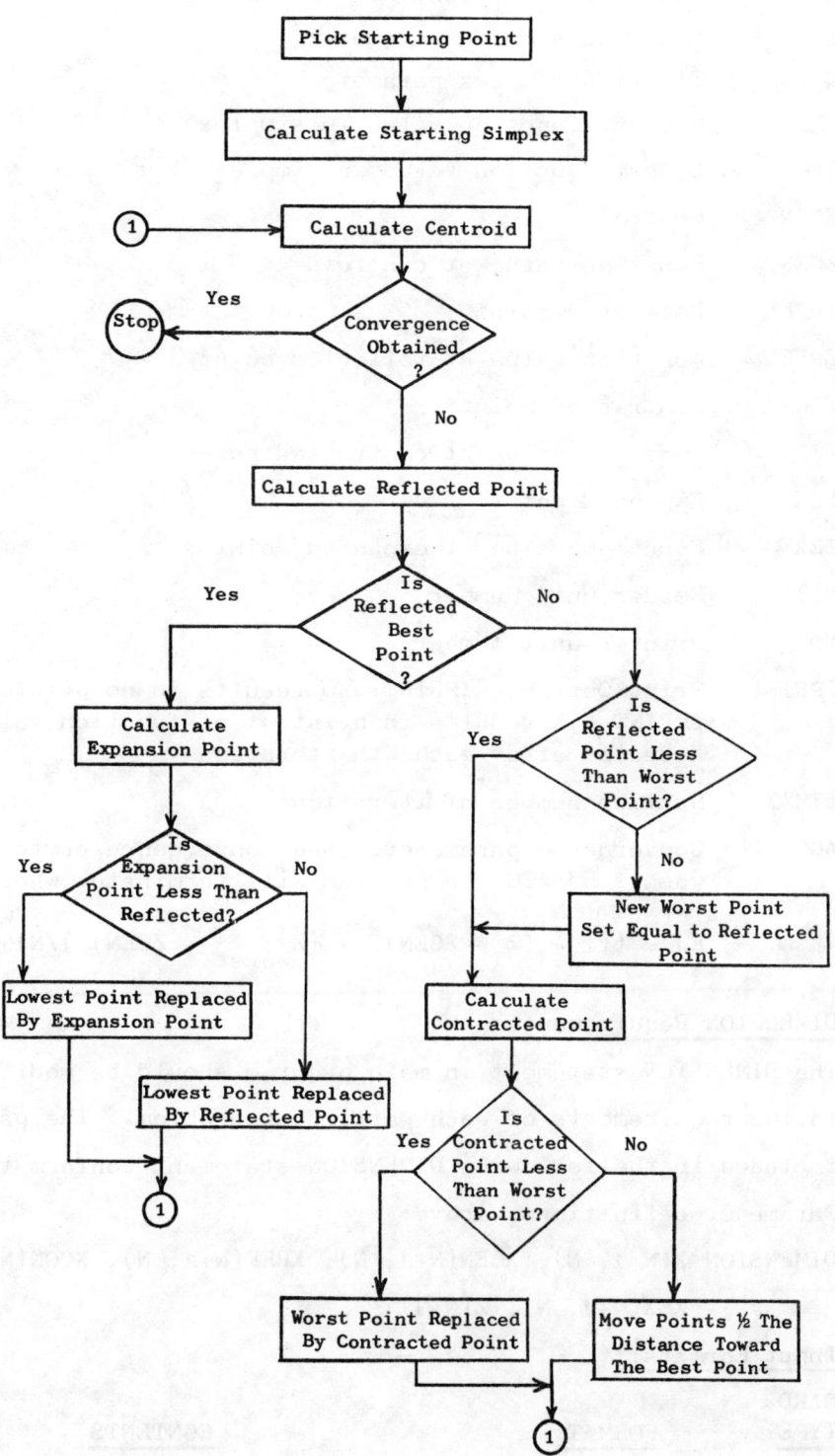

Figure 9.I. Nelder and Mead (NELDER ALGORITHM) Logic Diagram

	Z	Objective function - Defined in FUNC
	P	Starting simplex parameter
	Q	Starting simplex parameter
	ZHI	Highest function value in simplex
	ZLO	Lowest function value in simplex
	XCEN	Centroid
	ZCEN	Function value at centroid
	XREF	Reflected point
	ZREF	Function value at reflected point
	XCON	Contracted point
	ZCON	Function value at contracted point
	XEX	Expanded point
	ZEX	Function value at expanded point
	NI	Reader unit number
	NO	Printer unit number
	IPRINT	Print Option. IPRINT = 0 results in no printout. IPRINT = K results in printout of function value and X vector after each Kth iteration
	ITMAX	Maximum number of iterations
	ACC	Convergence parameter. The convergence criterion used was if EJ<ACC the program will terminate, where

$$EJ = (((\sum_{i=1}^{N}(Z_i - ZCEN)^2 - (Z_{worst} - ZCEN)^2)/N)^{1/2}$$

4) **DIMENSION Requirements:**

The DIMENSION statement in main program should be modified according to the requirements of each particular problem. The parameters included in the following DIMENSION statement conform to the Input Parameter definitions, above:

DIMENSION X(N+1, N), XCEN(N+1, N), XREF(N+1, N), XCON(N+1, N),
          XEX(N+1, N), Z(N+1)

5) **Input Formats:**

CARD TYPE	FORMAT	CONTENTS
1	(3I10)	N, ITMAX, IPRINT
2	(5E10.4)	ALFA, BETA, GAM, ACC, A
3	(8E10.4)	(X(1,J), J = 1, N)

(If N exceeds 8, additional CARD TYPE 3's are required.)

6) **Output:**

The main program prints out values of parameters and initial X vector.

Intermediate results are printed after each group of iterations as specified by IPRINT.

7) <u>Summary of User Requirements</u>:
   a) Determine values for N, ITMAX, IPRINT, A, ALFA, BETA, GAM, NI, NO, and ACC.
   b) Determine initial estimates for optimum values of independent variables; enter as (X(1,J), J + 1,N).
   c) Adjust DIMENSION statement in
   d) Specify objective function by writing SUBROUTINE FUNC.
   e) Adjust FORMAT statements as required.

D. <u>Test Problem</u>

The following problem was suggested by Derringer (13). The computations were performed on a CDC 6400 computer.

1) $F = -3803.84 - 138.08X_1 - 232.92X_2 + 123.08X_1^2 + 203.64X_2^2 + 182.25X_1X_2$

Starting Point: $X_1 = 1.0, X_2 = 0.5$

Parameters: ALFA = 1.0, BETA = 0.5, GAM = 2.0, ACC = 0.0001, IPRINT = 1

Algorithm Answers: $F = -3873.9$
$X_1 = 0.20519$
$X_2 = 0.47983$

No. of Iterations: 16

Central Processor Time: 4 seconds

The listing and output for this problem are contained in the following section.

E. Program Listings and Example Output

```
C
C       MAIN LINE PROGRAM FOR NELDER AND MEAD MINIMIZATION ROUTINE.
C
        DIMENSION X(3,2), XCEN(3,2), XREF(3,2), XCON(3,2), XEX(3,2), Z(3)
C
        NI = 50
        NO = 66
C
        READ(NI,001) N, ITMAX, IPRINT
    001 FORMAT (8I10)
        NP1 = N + 1
        READ (NI,002) ALFA, BETA, GAM, ACC, A
    002 FORMAT (8E10.4)
        READ (NI,002) (X(1,J), J=1,N)
C
        Q=(A/N*(2.**.5))*((N+1.)**.5-1.)
        P=(A/N*(2.**.5))*((N+1)**.5+N-1.)
        M=N+1
        DO 130 I=2,M
        AP=1.0
        DO 120 J=1,N
        AP=AP+1
        IF(I.EQ.AP)GO TO 135
        X(I,J)=X(1,J)+Q
        GO TO 120
    135 X(I,J)=X(1,J)+P
    120 CONTINUE
    130 CONTINUE
C
        IF(ALFA.EQ.0.)ALFA=1.
        IF(BETA.EQ.0.)BETA=.5
        IF(GAM.EQ.0.)GAM=2.
        IF(ACC.EQ.0.)ACC=.0001
C
        WRITE (NO,003)
    003 FORMAT (1H1,10X,28HNELDER AND MEAD OPTIMIZATION )
        WRITE (NO,004)
    004 FORMAT (/,2X,10HPARAMETERS )

        WRITE (NO,005) N, ACC, ALFA, BETA, GAM
    005 FORMAT (/,2X,4HN = ,I2,4X,11HACCURACY = ,E10.4,/,2X,8HALPHA = ,
       1E10.4,4X,7HBETA = ,E10.4,4X,8HGAMMA = ,E10.4)
        WRITE (NO,007)
    007 FORMAT (//,10X,16HSTARTING SIMPLEX )
        DO 140 I=1,NP1
        WRITE (NO,006) (I, J, X(I,J), J=1,N)
    006 FORMAT (/,2(2X,2HX(,I2,1H,,I2,4H) = ,1PE12.5))
    140 CONTINUE
        ITR=0
    150 DO 155 I=1,NP1
        CALL FUNC (I,X,Z,N,NP1)
```

```
    155 CONTINUE
        ITR=ITR+1
        IF (ITR.GE.ITMAX) GO TO 145
        IF (IPRINT) 158, 162, 158
    158 WRITE (NO,008) ITR
    008 FORMAT (//,2X,17HITERATION NUMBER ,I3)
        DO 160 J=1,NP1
    160 WRITE (NO,006) (J, I, X(J,I), I=1,N)
        WRITE (NO,009) (I, Z(I), I=1,NP1)
    009 FORMAT (/,3(2X,2HF(,I2,4H) = ,E16.8))
    162 ZHI=AMAX1(Z(1),Z(2),Z(3))
        ZLO=AMIN1(Z(1),Z(2),Z(3))
        DO 165 I=1,NP1
        IF(ZHI.EQ.Z(I))GO TO 170
    165 CONTINUE
    170 K=I
        EN=N
        DO 180 J=1,N
        SUM=0.
        DO 175 I=1,NP1
        IF(K.EQ.I)GO TO 175
        SUM=SUM+X(I,J)
    175 CONTINUE
    180 XCEN(K,J)=SUM/EN
        I=K
        CALL FUNC (I,XCEN,Z,N,NP1)
        ZCEN=Z(I)
        SUM=0.
        DO 185 I=1,NP1
        IF(K.EQ.I)GO TO 185
        SUM=SUM+(Z(I)-ZCEN)*(Z(I)-ZCEN)/EN
    185 CONTINUE
        EJ=SQRT(SUM)
        IF (EJ.LT.ACC) GO TO 998
        DO 190 J=1,N
        XREF(K,J)=XCEN(K,J)+ALFA*(XCEN(K,J)-X(K,J))
    190 CONTINUE
        I=K
        CALL FUNC (I,XREF,Z,N,NP1)
        ZREF=Z(I)
        DO 200 I=1,NP1
        IF(ZLO.EQ.Z(I))GO TO 205
    200 CONTINUE

    205 L=I
        IF(ZREF.LE.Z(L))GO TO 240
        DO 207 I=1,NP1
        IF(ZREF.LT.Z(I))GO TO 208
    207 CONTINUE
        GO TO 215
    208 DO 210 J=1,N
    210 X(K,J)=XREF(K,J)
        GO TO 150
    215 DO 220 J=1,N
```

```
  220 XCON(K,J)=XCEN(K,J)+BETA*(X(K,J)-XCEN(K,J))
      I=K
      CALL FUNC (I,XCON,Z,N,NP1)
      ZCON=Z(I)
      IF(ZCON.LT.Z(K))GO TO 230
      DO 225 J=1,N
      DO 225 I=1,NP1
  225 X(I,J)=(X(I,J)+X(L,J))/2.
      GO TO 150
  230 DO 235 J=1,N
  235 X(K,J)=XCON(K,J)
      GO TO 150
  240 DO 245 J=1,N
  245 XEX(K,J)=XCEN(K,J)+GAM*(XREF(K,J)-XCEN(K,J))
      I=K
      CALL FUNC (I,XEX,Z,N,NP1)
      ZEX=Z(I)
      IF(ZEX.LT.Z(L))GO TO 255
      DO 250 J=1,N
  250 X(K,J)=XREF(K,J)
      GO TO 150
  255 DO 260 J=1,N
  260 X(K,J)=XEX(K,J)
      GO TO 150
  145 WRITE (NO,011) ITMAX
  011 FORMAT (///,10X,20HDID NOT CONVERGE IN ,I5,11HITERATIONS. )
  998 WRITE (NO,012) ZLO
  012 FORMAT (//,2X,21HOPTIMUM VALUE OF F = ,E16.8)
      WRITE (NO,013)
  013 FORMAT (//,2X,27HOPTIMUM VALUES OF VARIABLES )
      DO 300 I=1,N
  300 WRITE (NO,014) I, X(NP1,I)
  014 FORMAT (/,2X,2HX(,I2,4H) = ,1PE16.8)
      END


      SUBROUTINE FUNC (I,X,Z,N,NP1)
C
      DIMENSION X(NP1,N), Z(NP1)
C
      X1 = X(I,1)
      X2 = X(I,2)
      X12 = X1**2
      X22 = X2**2
      Z(I) = 3803.84 + 138.08*X1 + 232.92*X2 - 123.08*X12 - 203.64*X22
     1 - 182.25*X1*X2
      Z(I) = - Z(I)
C
      RETURN
      END
```

NELDER AND MEAD OPTIMIZATION

PARAMETERS

N = 2   ACCURACY = .1000E-03
ALPHA = .1000E+01   BETA = .5000E+00   GAMMA = .2000E+01

STARTING SIMPLEX

X(1, 1) =   1.00000E+00   X(1, 2) =   5.00000E-01

X(2, 1) =   1.19319E+00   X(2, 2) =   5.51764E-01

X(3, 1) =   1.05176E+00   X(3, 2) =   6.93185E-01

ITERATION NUMBER   1

X(1, 1) =   1.00000E+00   X(1, 2) =   5.00000E-01

X(2, 1) =   1.19319E+00   X(2, 2) =   5.51764E-01

X(3, 1) =   1.05176E+00   X(3, 2) =   6.93185E-01

F(1) =   -.37932650E+04   F(2) =   -.37399017E+04   F(3) =   -.37436496E+04

(Every 5th iteration is included.)

ITERATION NUMBER   5

X(1, 1) =   1.24661E-01   X(1, 2) =   5.79255E-01

X(2, 1) =   4.07968E-01   X(2, 2) =   6.32753E-01

X(3, 1) =   2.79023E-01   X(3, 2) =   4.86130E-01

F(1) =   -.38725716E+04   F(2) =   -.38584887E+04   F(3) =   -.38731693E+04

ITERATION NUMBER 19

X(1, 1) =   2.08196E-01   X(1, 2) =   4.72756E-01

X(2, 1) =   1.88901E-01   X(2, 2) =   4.84396E-01

X(3, 1) =   2.24314E-01   X(3, 2) =   4.91083E-01

F(1) =   -.38739158E+04   F(2) =   -.38738986E+04   F(3) =   -.38738169E+04

ITERATION NUMBER 15

X(1, 1) =   2.07402E-01   X(1, 2) =   4.80248E-01

X(2, 1) =   2.06395E-01   X(2, 2) =   4.79102E-01

X(3, 1) =   2.08608E-01   X(3, 2) =   4.79520E-01

F(1) =   -.38739230E+04   F(2) =   -.38739235E+04   F(3) =   -.38739226E+04

ITERATION NUMBER 16

X(1, 1) =   2.07402E-01   X(1, 2) =   4.80248E-01

X(2, 1) =   2.06395E-01   X(2, 2) =   4.79102E-01

X(3, 1) =   2.05189E-01   X(3, 2) =   4.79830E-01

F(1) =   -.38739230E+04   F(2) =   -.38739235E+04   F(3) =   -.38739235E+04

OPTIMUM VALUE OF F =   -.38739235E+04

OPTIMUM VALUES OF VARIABLES

X(1) =   2.05188700E-01

X(2) =   4.79829703E-01

II. HOOKE AND JEEVES (HOOKE ALGORITHM)*

A. Purpose

This program finds the minimum of a multivariable, unconstrained, nonlinear function:

Minimize $\quad F(X_1, X_2, \ldots, X_N)$

B. Method

The procedure is based on the direct search method proposed by Hooke and Jeeves (30). No derivatives are required. The procedure assumes a unimodal function; therefore, if more than one minimum exists or the shape of the surface is unknown, several sets of starting values are recommended. The algorithm proceeds as follows:

1) A base point is picked and the objective function evaluated.
2) Local searches are made in each direction by stepping X_i a distance S_i to each side and evaluating the objective function to see if a lower function value is obtained.
3) If there is no function decrease, the step size is reduced and searches are made from the previous best point.
4) If the value of the objective function has decreased, a "temporary head", $X_{i,o}^{(k+1)}$, is located using the two previous base points $X_i^{(k+1)}$ and $X_i^{(k)}$:

$$X_{i,o}^{(k+1)} = X_i^{(k+1)} + \alpha(X_i^{(k+1)} - X_i^{(k)})$$

where i is the variable index = 1, 2, 3, ..., N

    o denotes the temporary head

    k is stage index (a stage is the end of N searches)

    α is an acceleration factor, $\alpha \geq 1$.

5) If the temporary head results in a lower function value, a new local search is performed about the temporary head, a new head is located and the value of F checked. This expansion continues as long as F decreases.

---

*Computer code developed by A. I. Johnson, University of Western Ontario, Canada. Used by permission.

6) If the temporary head does not result in a lower function value, a search is made from the previous best point.

7) The procedure terminates when the convergence criterion is satisfied (see Description of Parameters).

A flow sheet illustrating the above procedure is given in Figure 9.II.

C. Program Description

1) Usage:

The program consists of a short main program, the main subroutine HOOKE and the user supplied functional evaluation subroutine OBJECT. Initial values of the independent variables, step sizes, and solution parameters are supplied through the main program. Subroutine HOOKE performs all searches and provides all printout.

2) Subroutine Required:

SUBROUTINE HOOKE (RK, EPS, NSTAGE, MAXK, NKAT, EPSY, ALPHA, BETA, QD, Q, QQ, W, IPRINT) called from main program, performs all searches.

SUBROUTINE OBJECT (SUMN, AKE, NSTAGE) function evaluation subroutine (user supplied).

3) Description of Parameters:

NSTAGE	Number of decision variables to be used
RK	Vector of initial guesses for decision variables
EPS	Vector of initial step size to be used for each of the variables
ITMAX	Maximum number of times the objective function is called (=MAXK)
NKAT	Maximum number of times the initial step size is to be reduced
EPSY	Error in objective function to be reached before program terminates (difference between current value and previous stage value)
ALPHA	Factor for extending the size of the initial steps, greater than or equal to 1.0
BETA	Factor for reducing the initial step size, $0.0 \leq BETA \leq 1.0$
QD	Optimum value of the function resulting from the search
AKE	Vector of independent variables in subroutine OBJECT
SUMN	Objective function to be minimized - define in OBJECT
IPRINT	Print control. IPRINT = 0 results in no intermediate output. IPRINT = 1 results in output on each iteration

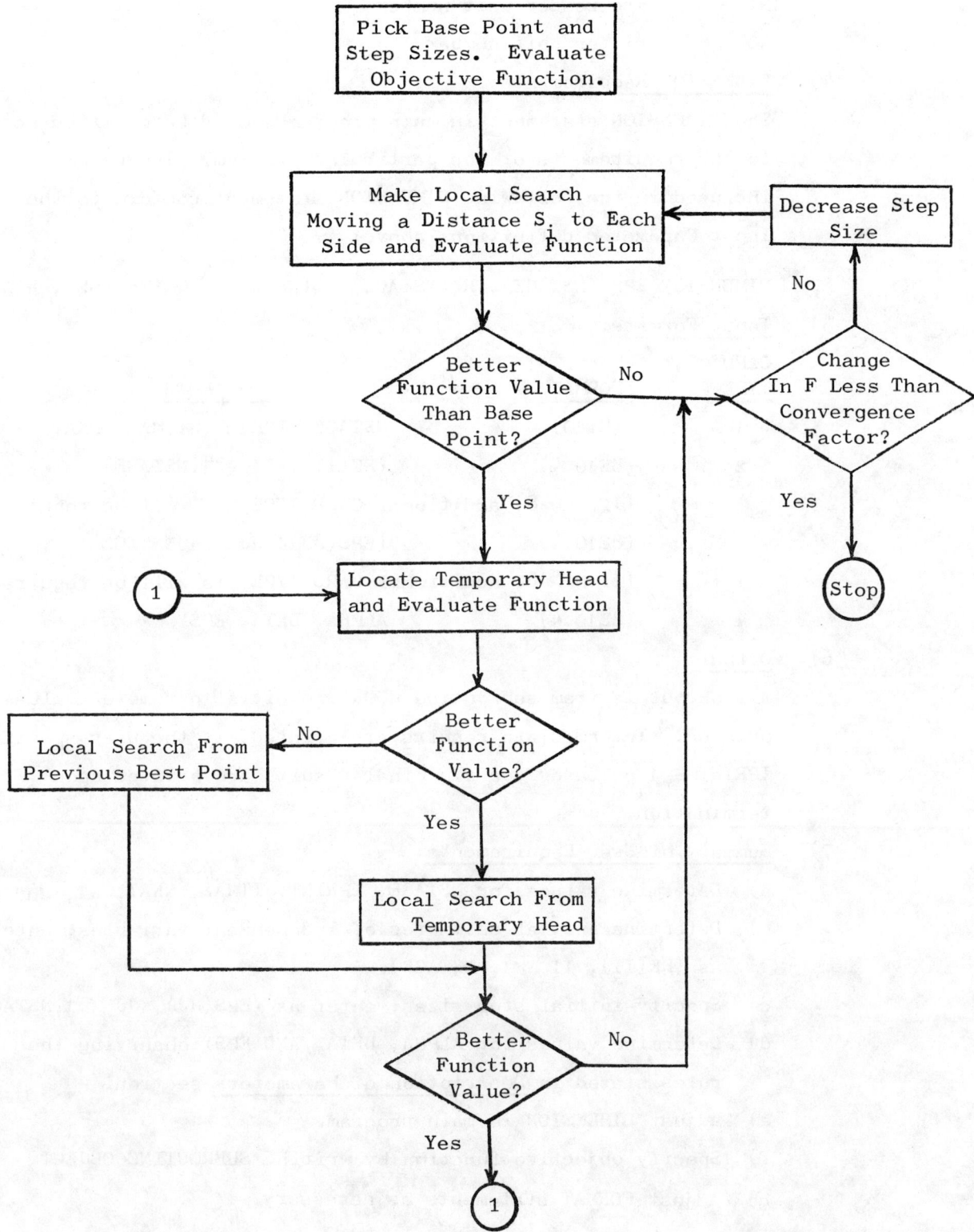

Figure 9.II. Hooke and Jeeves (HOOKE ALGORITHM) Logic Diagram

NI   Card reader unit number

NO   Printer unit number

4) <u>DIMENSION Requirements</u>:

The DIMENSION statement in main program should be modified according to the requirements of the particular problem. The parameters included in the following DIMENSION statement conform to the Input Parameter definitions above:

DIMENSION EPS (NSTAGE), RK(NSTAGE), Q(NSTAGE), QQ(NSTAGE), W(NSTAGE)

5) <u>Input Formats</u>:

CARD TYPE	FORMAT	CONTENTS
1	(8I10)	NSTAGE, IPRINT, ITMAX, NKAT
2	(8E10.4)	(RK(II), II = 1,NSTAGE)
	(If N > 8, additional CARD TYPE 2's will be required.)	
3	(8E10.4)	(EPS(JJ), JJ = 1,NSTAGE)
	(If N > 8, additional CARD TYPE 3's will be required.)	
4	(8E10.4)	ALPHA, BETA, EPSY

6) <u>Output</u>:

All output is from subroutine HOOKE. Initial parameter values are printed. Intermediate results are printed, if the user specifies IPRINT = 1 on Card Type 1. Final results are printed upon termination.

7) <u>Summary of User Requirements</u>:

a) Determine values for NSTAGE, IPRINT, ITMAX, NKAT, NI, and NO.

b) Determine initial estimates of independent variables; enter as (RK(II), II = 1, NSTAGE).

c) Specify initial step sizes; enter as (EPS(JJ), JJ = 1,NSTAGE).

d) Determine values for ALPHA, BETA, AND EPSY observing the rules stated in <u>Description of Parameters</u> section.

e) Adjust DIMENSION in main program.

f) Specify objective function by writing SUBROUTINE OBJECT

g) Adjust FORMAT statements as necessary.

D. <u>Test Problem</u>

The following test program was taken from the literature (13). Calculations were performed on a CDC 6400 computer.

Function: $F = -3803.84 - 138.08X_1 - 232.92X_2 + 123.08X_1^2 + 203.64X_2^2 + 182.25X_1X_2$

Starting Point: $X_1 = 1.0$, $X_2 = 0.5$

Parameters: N = 2, ITMAX = 500, NKAT = 20, EPSY = 0.00001, ALPHA = 1.0, BETA = 0.5

Initial Step Sizes: EPS(1) = 0.10, EPS(2) = 0.10

Algorithm Answers: $F = -3873.9$
$X_1 = 0.20576$
$X_2 = 0.47979$

Number of Function Evaluations: 110
Central Processor Time: 3 seconds

The listing and output for this problem are contained in the following section.

E. <u>Program Listings and Example Output</u>

```fortran
C
C
C     MAIN LINE PROGRAM FOR SUBROUTINE HOOKE.
C
      DIMENSION EPS(2), RK(2), Q(2), QQ(2), W(2)
      COMMON NI,NO
C
      NI = 50
      NO = 66
C
      READ (NI,001) NSTAGE, IPRINT, ITMAX, NKAT
  001 FORMAT (8I10)
      READ (NI,002) (RK(II), II=1,NSTAGE)
  002 FORMAT (8E10.4)
      READ (NI,002) (EPS(JJ), JJ=1,NSTAGE)
C
      READ (NI,003) ALPHA, BETA, EPSY
  003 FORMAT (8E10.4)
      QD = 0.0
C
      CALL HOOKE (RK,EPS,NSTAGE,ITMAX,NKAT,EPSY,ALPHA,BETA,QD,Q,QQ,W,
     1 IPRINT)
C
      END




      SUBROUTINE HOOKE (RK,EPS,NSTAGE,MAXK,NKAT,EPSY,ALPHA,BETA,QD,
     1 Q,QQ,W,IPRINT)
      DIMENSION RK(NSTAGE), EPS(NSTAGE), Q(NSTAGE), QQ(NSTAGE),
     1 W(NSTAGE)
```

```
      COMMON NI,NO
C
C
      WRITE (NO,001)
  001 FORMAT (1H1,10X,37HHOOKE AND JEEVES OPTIMIZATION ROUTINE)
      WRITE (NO,002) ALPHA, BETA, MAXK, NKAT
  002 FORMAT (//,2X,10HPARAMETERS,/,2X,8HALPHA = ,F5.2,4X,
     1 7HBETA = ,F5.2,4X,8HITMAX = ,I4,4X,7HNKAT = ,I3)
      WRITE (NO,003) NSTAGE
  003 FORMAT (/,2X,22HNUMBER OF VARIABLES = ,I3)
      WRITE (NO,004)
  004 FORMAT (/,2X,18HINITIAL STEP SIZES)
      DO 6  I=1,NSTAGE
      WRITE (NO,005)  I, EPS(I)
  005 FORMAT (/,2X,4HEPS(,I2,4H) = ,E16.8)
    6 CONTINUE
      WRITE (NO,007) EPSY
  007 FORMAT (/,2X,43HERROR IN FUNCTION VALUES FOR CONVERGENCE = ,E16.8)
      KFLAG = 0
      DO 601 I=1,NSTAGE
      Q(I) =RK(I)
      W(I) = 0.0
  601 CONTINUE
      KAT =0.0
      KK1 =0
   70 KCOUNT =0
      WBEST = W(NSTAGE)
      CALL OBJECT (SUM,RK,NSTAGE)
      KK1= KK1+ 1
      BO =SUM
      IF (KK1.EQ. 1)  QD = SUM
      IF (KK1.EQ. 1)   GO TO 201
      IF(BO.GT.QD)   KFLAG = 1
      IF (BO.LT.QD)  QD = BO
C
C      ESTABLISHING THE SEARCH PATTERN
C
  201 DO   55   I = 1,NSTAGE
      QQ(I)=RK(I)
      TSRK = RK(I)
      RK(I) = RK(I) + EPS(I)
      CALL OBJECT (SUM,RK,NSTAGE)
      KK1= KK1+ 1
      W(I) = SUM
      IF (W(I) .LT.QD)   GO TO 58
      RK(I) = RK(I) - 2.0*EPS(I)
      CALL OBJECT (SUM,RK,NSTAGE)
      KK1= KK1+ 1
      W(I) = SUM
      IF (W(I) .LT.QD)    GO TO 58
      RK(I) = TSRK
      IF (I.EQ. 1) GO TO 513
      W(I) =W(I-1)
```

```
          GO TO 613
   513 W(I) =BO
   613 CONTINUE
       KCOUNT =1+ KCOUNT
       GO TO 55
    58 QD= W(I)
       QQ(I) =RK(I)
    55 CONTINUE
       IF (IPRINT) 60, 65, 60
    60 WRITE (NO,100) KK1
C
C         RECORD RESPONSES AND LOCATION
C
C
       WRITE(NO,102)
       WRITE(NO,207) (RK(I), I=1,NSTAGE), QD
C
C         TEST TO DETERMINE TERMINATION OF PROGRAM
C
    65 IF (KK1.GT.MAXK) GO TO 94
       IF (KAT .GE. NKAT)   GO TO 94
       IF (ABS(W(NSTAGE)-WBEST).LE.EPSY) GO TO 94
C
C         IF ALL AXES FAIL REDUCE STEP SIZE
C
       IF (KCOUNT .GE. NSTAGE ) GO TO 28
       DO 26    I = 1,NSTAGE
       RK(I) =RK(I) + ALPHA*(RK(I) - Q(I))
    26 CONTINUE
       DO 25    I = 1,NSTAGE
       Q(I) =QQ(I)
    25 CONTINUE
       GO TO 70
C
C         REDUCE STEP SIZE
C
    28 KAT = KAT + 1
       IF (KFLAG .EQ. 1)   GO TO 202
       GO TO 204
   202 KFLAG = 0
       DO 203   I = 1,NSTAGE
       RK(I) = Q(I)
   203 CONTINUE
   204 DO 80    I=1,NSTAGE
       EPS(I) =EPS(I) *BETA
    80 CONTINUE
       IF (IPRINT) 85, 70, 85
    85 WRITE (NO,101) KAT
       GO TO 70
    94 WRITE (NO,460) (EPS(I), I=1,NSTAGE)
       WRITE (NO,461) (RK(I), I=1,NSTAGE)
       WRITE (NO,462) QD
       DO 104 I=1,NSTAGE
   104 WRITE (NO,103) I, RK(I)
```

```
      WRITE (NO,100) KK1
100   FORMAT (//,2X,33HNUMBER OF FUNCTION EVALUATIONS = ,I8)
101   FORMAT (/,2X,18HSTEP SIZE REDUCED ,I2,6H TIMES)
102   FORMAT(1X,26HEND OF EACH PATTERN SEARCH/)
103   FORMAT (//,2X,8HFINAL X(,I2,4H) = ,1PE16.8)
207   FORMAT(1X,18HVARIABLES AND SUMN,3X,9E12.4//)
465   FORMAT (10X,3HSUM,3X,E14.5)
460   FORMAT(1X, 18H THE FINAL EPS ARE,    4F20.8/)
461   FORMAT (1X, 18H THE FINAL RK ARE ,   5F20.8/)
462   FORMAT (1X, 24H THE MINIMUM RESPONSE IS,    F20.8/)
      RETURN
      END




      SUBROUTINE OBJECT (SUMN,AKE,NSTAGE)
C
      DIMENSION AKE(NSTAGE)
C
      X1 = AKE(1)
      X2 = AKE(2)
      X12 = (X1**2)
      X22 = (X2**2)
      SUMN = 3803.84 + 138.08*X1 + 232.92*X2 - 123.08*X1**2 - 203.64
     1 *X2**2 - 182.25*X1*X2
      SUMN = - SUMN
C
      RETURN
      END
```

HOOKE AND JEEVES OPTIMIZATION ROUTINE

PARAMETERS
ALPHA = 1.00   BETA = 0.50   ITMAX = 500   NKAT = 20

NUMBER OF VARIABLES = 2

INITIAL STEP SIZES

EPS(1) =   0.10000000E+00

EPS(2) =   0.10000000E+00

ERROR IN FUNCTION VALUES FOR CONVERGENCE =   0.10000000E-04

NUMBER OF FUNCTION EVALUATIONS =   5
END OF EACH PATTERN SEARCH

VARIABLES AND SUMN   0.9000E+00   0.4000E+00   -.3823E+04

(9 intervening printouts are omitted.)

NUMBER OF FUNCTION EVALUATIONS =   50
END OF EACH PATTERN SEARCH

VARIABLES AND SUMN   0.2250E+00   0.4750E+00   -.3874E+04

STEP SIZE REDUCED   4 TIMES

(11 intervening printouts are omitted.)

```
   NUMBER OF FUNCTION EVALUATIONS =           106
   END OF EACH PATTERN SEARCH

   VARIABLES AND SUMN        0.2059E+00  0.4797E+00  -.3874E+04

   STEP SIZE REDUCED 10 TIMES


   NUMBER OF FUNCTION EVALUATIONS =           110
   END OF EACH PATTERN SEARCH

   VARIABLES AND SUMN        0.2058E+00  0.4798E+00  -.3874E+04
   THE FINAL EPS ARE         0.00009766              0.00009766
   THE FINAL RK ARE          0.20576172              0.47978516
   THE MINIMUM RESPONSE IS      -3873.92354660


   FINAL X( 1) =     2.0576172E-01

   FINAL X( 2) =     4.7978516E-01

   NUMBER OF FUNCTION EVALUATIONS =           110
```

III. ROSENBROCK (ROSENB ALGORITHM)*

A. Purpose

This program finds the minimum of a multivariable, unconstrained, nonlinear function:

$$\text{Minimize} \quad F(X_1, X_2, \ldots, X_N)$$

B. Method

The procedure is based on the direct search method proposed by H. H. Rosenbrock (41). No derivatives are required. The procedure assumes a unimodal function; therefore, several sets of starting values for the independent variables should be used if it is known that more than one minimum exists or if the shape of the surface is unknown. The algorithm proceeds as follows:

1) A starting point and initial step sizes, S_i, $i = 1, 2, \ldots, N$, are picked and the objective function evaluated.

2) The first variable X_1 is stepped a distance S_1 parallel to the axis, and the function evaluated. If the value of F decreased, the move is termed a success and S_1 increased by a factor α, $\alpha \geq 1.0$. If the value of F increased the move is termed a failure and S_1 decreased by a factor β, $0 < \beta \leq 1.0$, and the direction of movement reversed.

3) The next variable, X_i, is in turn stepped a distance S_i parallel to the axis. The same acceleration or deceleration and reversal procedure is followed for all variables in consecutive repetitive sequences until a success (decrease in F) and failure (increase in F) have been encountered in all N directions.

4) The axes are then rotated by the following equations. Each rotation of the axes is termed a <u>stage</u>.

$$M_{i,j}^{(k+1)} = \frac{D_{i,j}^{(k)}}{\left[\sum_{\ell=1}^{N} (D_{\ell,j}^{(k)})^2\right]^{1/2}}$$

---

*Computer code developed by A. I. Johnson, University of Western Ontario, London, Ontario, Canada. Used by permission.

where

$$D_{i,1}^{(k)} = A_{i,1}^{(k)}$$

$$D_{i,j}^{(k)} = A_{i,j}^{(k)} - \sum_{\ell=1}^{j-1}\left[(\sum_{n=1}^{j} M_{n,\ell}^{(k+1)} \cdot A_{n,j}^{(k)}) \cdot M_{i,\ell}^{(k+1)}\right], \; j = 2, 3, \ldots, N$$

$$A_{i,j}^{(k)} = \sum_{\ell=j}^{N} d_\ell^{(k)} \cdot M_{i,\ell}^{(k)}$$

where

    i = variable index = 1, 2, 3, ..., N

    j = direction index = 1, 2, 3, ..., N

    k = stage index

    d_i = sum of distances moved in the i direction since last rotation of axes

    $M_{i,j}$ = direction vector component (normalized).

5) Search is made in each of the X directions using the new coordinate axes:

$$\text{new } X_i^{(k)} = \text{old } X_i^{(k)} + S_j^{(k)} M_{i,j}^{(k)}$$

6) The procedure terminates when the convergence criterion is satisfied (see <u>Description of Parameters</u>).

A flow sheet illustrating the above procedure is given in Figure 9.III.

C. <u>Program Description</u>

  1) <u>Usage</u>:

The program consists of a small main program, the general subroutine ROSENB, and the user supplied function evaluation subroutine OBJECT. Initial values for independent variables, step sizes, solution parameters, and limits are supplied through the main program. Intermediate function and independent variable values and final values are printed out in ROSENB.

  2) <u>Subroutines Required</u>:

SUBROUTINE ROSENB (AKE, EPS, KM, MAXK, MKAT, MCYC, ALPHA, BETA, V, NSTEP, EPSY, D, BL, BLEN, AJ, E, AL, AFK, NI, NO) called from main program, performs search.

SUBROUTINE OBJECT (AKE, SUMN, KM) specifies objective function (user supplied).

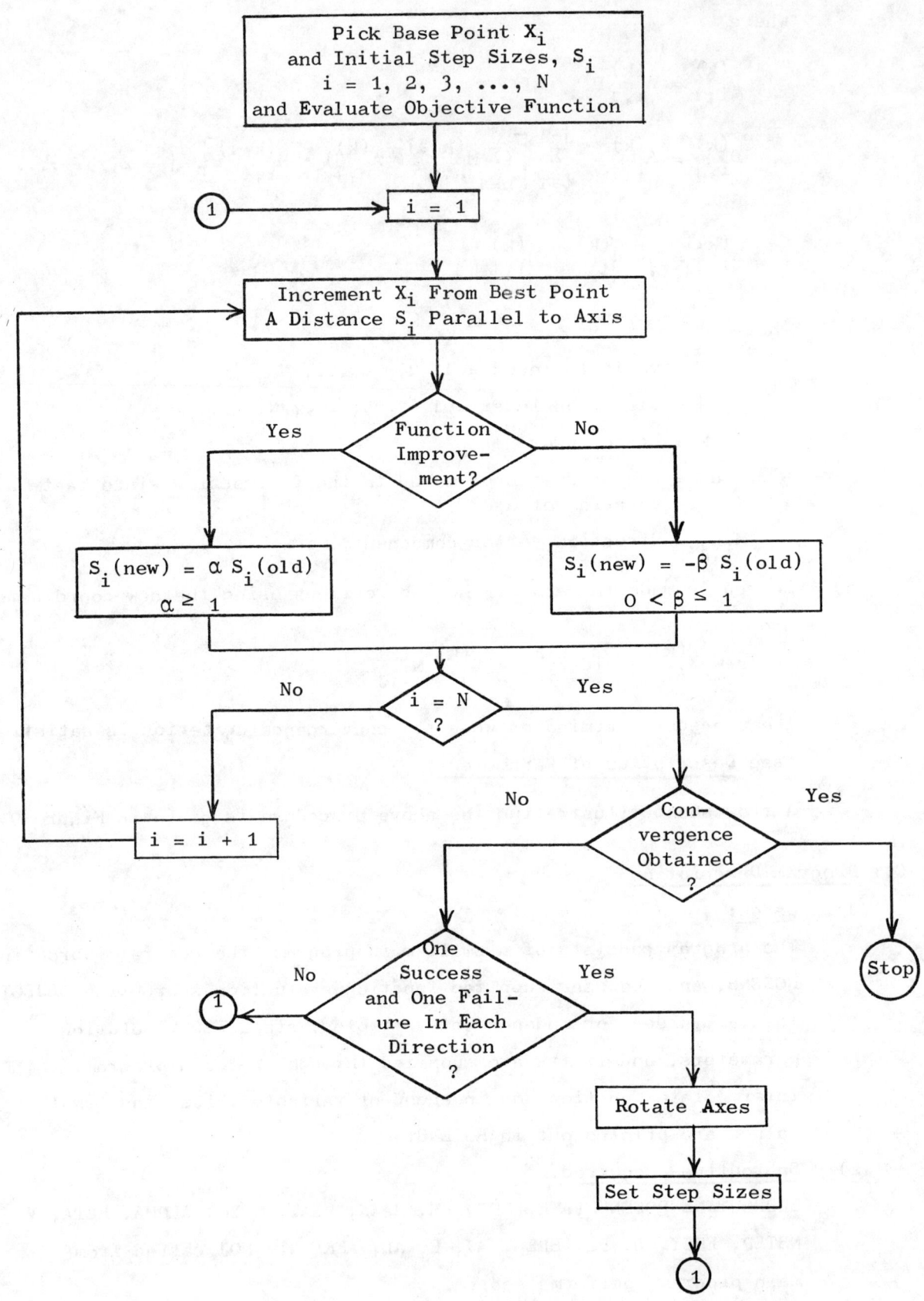

Figure 9.III. Rosenbrock (ROSENB ALGORITHM) Logic Diagram

3) Description of Parameters:

KM	Number of independent variables
AKE	Vector of initial guesses for independent variables
EPS	Vector of initial step sizes for the variables
MAXK	Maximum number of times program is to evaluate objective function
MKAT	Maximum number of times axes are to be rotated
MCYC	Number of successive failures encountered in all directions before termination
EPSY	Error in objective function to be reached before program terminates (difference between current value and previous stage value)
NSTEP	Control variable determining step size to be used after each rotation of the axes; NSTEP = 1 if initial step size to be used; NSTEP = 2 if final step size on previous stage to be used
ALPHA	Scaling factor for step size increase
BETA	Scaling factor for step size reduction
NI	Card reader unit number -- Define in main program
NO	Printer unit number -- Define in main program

4) DIMENSION Requirements:

The DIMENSION statement in the main program should be modified according to the requirements of each particular problem. The parameters included in the following DIMENSION statement conform to the Input Parameter definitions.

DIMENSION EPS(KM), AKE(KM), V(KM,KM), D(KM), BL(KM,KM), BLEN(KM),
          AJ(KM), E(KM), AL(KM,KM), AFK(KM)

5) Input Formats:

CARD TYPE	FORMAT	CONTENTS
1	(5I10)	KM, MAXK, MKAT, MCYC, NSTEP
2	(8E10.4)	(AKE(I), I=1, KM)
3	(8E10.4)	(EPS(J), J=1, KM)
4	(8E10.4)	EPSY, ALPHA, BETA

(If KM exceeds 8, additional CARD TYPE 2 and 3 are required.)

6) Output:

The main program prints out values of the input parameters. Subroutine ROSENB provides function and independent variable values for each stage. When convergence is achieved, the total number of function evaluations is also given.

7) <u>Summary of User Requirements</u>:

a) Determine values for KM, MAXK, MKAT, MCYC, NSTEP, EPSY, ALPHA, BETA, [AKE(I), I = 1, KM], [EPS(J), J = 1, KM].

b) Specify NI and NO in main program.

c) Adjust DIMENSION and FORMAT statements as necessary.

d) Specify objective function, SUMN, in OBJECT.

D. <u>Test Problem</u>

The following test problem was taken from the literature (13). The calculations were performed on a CDC 6400 computer.

Function: $F = -3803.84 - 138.08X_1 - 232.92X_2 + 123.08X_1^2 + 203.64X_2^2 + 182.25X_1X_2$

Starting Point: $X_1 = 1.0$, $X_2 = 0.5$

Dimensions: KM = 2

Parameters: MAXK = 1000, MKAT = 30, MCYC = 50, EPSY = 0.00001, ALPHA = 2.0, BETA = 0.50, NSTEP = 2

Initial Step Size: EPS(1) = 0.1, EPS(2) = 0.1

Algorithm Answers: $F = -3873.9$
$X_1 = 0.20609$
$X_2 = 0.47961$

Number of Function Evaluations: 62

Central Processor Time: 3 seconds

The listing and output for this problem are contained in the following section.

E. Program Listings and Example Output

```
C
C      MAIN LINE PROGRAM FOR SUBROUTINE ROSENB.
C
       DIMENSION EPS(2), AKE(2), V(2,2), D(2), BL(2,2), BLEN(2), AJ(2),
      1 E(2), AL(2,2), AFK(2)
C
       NI = 50
       NO = 66
C
       READ (NI,001) KM, MAXK, MKAT, MCYC, NSTEP
   001 FORMAT (5I10)
       READ(NI,02)(AKE(I),I=1,KM)
   002 FORMAT (8E10.4)
       READ(NI,02)(EPS(J),J=1,KM)
       READ (NI,002) EPSY, ALPHA, BETA
C
       CALL ROSENB (AKE,EPS,KM,MAXK,MKAT,MCYC,ALPHA,BETA,V,NSTEP,EPSY,D,
      1 BL,BLEN,AJ,E,AL,AFK,NI,NO)
C
   999 STOP
       END




       SUBROUTINE ROSENB (AKE,EPS,KM,MAXK,MKAT,MCYC,ALPHA,BETA,V,NSTEP,
      1 EPSY,D,BL,BLEN,AJ,E,AL,AFK,NI,NO)
C
       DIMENSION AKE(KM), D(KM), V(KM,KM), BL(KM,KM), BLEN(KM), EPS(KM),
      1 AJ(KM), E(KM), AL(KM,KM), AFK(KM)
C
       WRITE (NO,099)
    99 FORMAT (1H1,10X,34HROSENBROCK MINIMIZATION PROCEDURE   )
       WRITE (NO,1004) MAXK, MKAT, MCYC, NSTEP, ALPHA, BETA, EPSY
  1004 FORMAT (//,2X,10HPARAMETERS ,/,2X,7HMAXK = ,I4,4X,7HMKAT = ,I2,4X,
      1 7HMCYC = ,I2,4X,8HNSTEP = ,I2,//,2X,8HALPHA = ,F5.2,4X,7HBETA = ,
      2 F5.2,4X,7HEPSY = ,1PE12.4 )
C
       KAT = 1
       DO 98 II=1,KM
       DO 98 JJ=1,KM
       V(II,JJ) = 0.0
       IF(II-JJ) 98,97,98
    97 V(II,JJ) = 1.0
    98 CONTINUE
       CALL OBJECT (AKE,SUMN,KM)
       SUMO = SUMN
       DO 812 K = 1,KM
       AFK(K) = AKE(K)
```

```
    812 CONTINUE
        KK1 = 1
        IF ( NSTEP - 1 ) 701,700,701
    700 GO TO 1000
    701 CONTINUE
        DO 350 I=1,KM
        E(I) = EPS(I)
    350 CONTINUE
   1000 DO 250 I=1,KM
        FBEST = SUMN
        AJ(I) = 2.0
        IF ( NSTEP - 1 ) 702,703,702
    702 GO TO 250
    703 CONTINUE
        E(I) = EPS(I)
    250 D(I) = 0.0
        III = 0
    397 III = III + 1
    258 I = 1
    259 DO 251 J = 1,KM
    251 AKE(J) = AKE(J) + E(I)*V(I,J)
        CALL OBJECT (AKE,SUMN,KM)
        KAT = KAT + 1
        SUMDIF = FBEST - SUMN
        IF (ABS(SUMDIF)-EPSY) 704,704,705
    704 GO TO 1001
    705 CONTINUE
        IF ( KAT - MAXK ) 706,707,707
    707 GO TO 1001

    706 CONTINUE
        IF ( SUMN - SUMO ) 708,708,709
    708 GO TO 253
    709 CONTINUE
        DO 254 J=1,KM
    254 AKE(J) = AKE(J) - E(I)*V(I,J)
        E(I) = -BETA*E(I)
        IF ( AJ(I) - 1.5 ) 710,711,711
    710 AJ(I) = 0.0
    711 CONTINUE
        GO TO 255
    253 D(I) = D(I) + E(I)
        E(I) = ALPHA*E(I)
        SUMO = SUMN
        DO 813 K=1,KM
    813 AFK(K) = AKE(K)
        IF ( AJ(I) - 1.5 ) 712,712,713
    713 AJ(I) = 1.0
    712 CONTINUE
    255 DO 256 J=1,KM
        IF ( AJ(J) - 0.5 ) 256,256,715
    715 GO TO 299
    256 CONTINUE
        GO TO 257
    299 IF ( I - KM ) 717,716,717
    716 GO TO 399
```

```
    717 CONTINUE
        I = I + 1
        GO TO 259
    399 DO 398 J=1,KM
        IF ( AJ(J) - 2. ) 718,398,398
    718 GO TO 258
    398 CONTINUE
        IF ( III - MCYC ) 720,721,721
    720 GO TO 397
    721 CONTINUE
        GO TO 1001
    257 CONTINUE
        DO 290 I=1,KM
        DO 290 J=1,KM
    290 AL(I,J) = 0.0
C
C       PRINT VALUES OF STAGE, FUNCTION, INDEPENDENT VARIABLES
C
        WRITE (NO,280) KK1
    280 FORMAT (//,2X,13HSTAGE NUMBER ,I2)
        WRITE (NO,281) SUMO
    281 FORMAT (/,7X,34HVALUE OF THE OBJECTIVE FUNCTION = ,E16.8)
        WRITE (NO,282)
    282 FORMAT (/,7X,35HVALUES OF THE INDEPENDENT VARIABLES,/)
        DO 284 IX = 1,KM
        WRITE (NO,283) IX, AKE(IX)
    283 FORMAT (/,7X,2HX(,I2,4H) = ,E16.8)
    284 CONTINUE
C
C       ROTATE AXES
C
        DO 260 I=1,KM
        KL = I
        DO 260 J=1,KM
        DO 261 K=KL,KM
    261 AL(I,J) = D(K)*V(K,J) + AL(I,J)
    260 BL(I,J) = AL(I,J)
        BLEN(1) = 0.0
        DO 351 K=1,KM
        BLEN(1) = BLEN(1) + BL(1,K)*BL(1,K)
    351 CONTINUE
        BLEN(1) = SQRT(BLEN(1))
        DO 352 J=1,KM
        V(1,J) = BL(1,J)/BLEN(1)
    352 CONTINUE
        DO 263 I=2,KM
        II = I - 1
        DO 263 J=1,KM
        SUMAVV = 0.0
        DO 264 KK=1,II
        SUMAV = 0.0
        DO 262 K=1,KM
    262 SUMAV = SUMAV + AL(I,K)*V(KK,K)
    264 SUMAVV = SUMAV*V(KK,J) + SUMAVV
```

```
  263 BL(I,J) = AL(I,J) - SUMAVV
      DO 266 I=2,KM
      BLEN(I) = 0.0
      DO 267 K=1,KM
  267 BLEN(I) = BLEN(I) + BL(I,K)*BL(I,K)
      BLEN(I) = SQRT(BLEN(I))
      DO 266 J=1,KM
  266 V(I,J) = BL(I,J)/BLEN(I)
      KK1 = KK1 + 1
      IF ( KK1 - MKAT ) 723,722,722
  722 GO TO 1001
  723 GO TO 1000
 1001 WRITE (NO,1002) KK1
 1002 FORMAT (///,2X,25HTOTAL NUMBER OF STAGES = ,I2)
      WRITE (NO,1003) KAT
 1003 FORMAT (/,2X,38HTOTAL NUMBER OF FUNCTION EVALUATIONS  ,I5)
      WRITE (NO,1005) SUMO
 1005 FORMAT (/,2X,33HFINAL VALUE OF OBJECT FUNCTION = ,1PE16.8)
      DO 1007 IX=1,KM
      WRITE (NO,1006) IX, AKE(IX)
 1006 FORMAT (/,2X,2HX(,I2,4H) = ,E16.8)
 1007 CONTINUE
C
      RETURN
      END




      SUBROUTINE OBJECT (AKE,SUMN,KM)
C
      DIMENSION AKE(KM)
C
      X1 = AKE(1)
      X2 = AKE(2)
      SUMN = 3803.84 + 138.08*X1 + 232.92*X2 - 123.08*X1**2 - 203.64
     1 *X2**2 - 182.25*X1*X2
      SUMN = - SUMN
C
      RETURN
      END
```

ROSENBROCK MINIMIZATION PROCEDURE

PARAMETERS
MAXK = 1000   MKAT = 30   MCYC = 50   NSTEP = 2
ALPHA = 2.00   BETA = 0.50   EPSY = 1.000E-05

STAGE NUMBER 1

    VALUE OF THE OBJECTIVE FUNCTION =   -.38737413E+04
    VALUES OF THE INDEPENDENT VARIABLES

    X(1) =   0.25000000E+00
    X(2) =   0.45000000E+00

STAGE NUMBER 2

    VALUE OF THE OBJECTIVE FUNCTION =   -.38738958E+04
    VALUES OF THE INDEPENDENT VARIABLES

    X(1) =   0.21133583E+00
    X(2) =   0.46621401E+00

STAGE NUMBER 3

    VALUE OF THE OBJECTIVE FUNCTION =   -.38739205E+04
    VALUES OF THE INDEPENDENT VARIABLES

    X(1) =   0.21040619E+00
    X(2) =   0.48015848E+00

STAGE NUMBER 4

    VALUE OF THE OBJECTIVE FUNCTION =   -.38739232E+04
    VALUES OF THE INDEPENDENT VARIABLES

    X(1) =   0.20739205E+00
    X(2) =   0.47839157E+00

STAGE NUMBER 5

    VALUE OF THE OBJECTIVE FUNCTION = $-.38739235E+04$

    VALUES OF THE INDEPENDENT VARIABLES

    X(1) = $0.20480070E+00$

    X(2) = $0.48049486E+00$

STAGE NUMBER 6

    VALUE OF THE OBJECTIVE FUNCTION = $-.38739235E+04$

    VALUES OF THE INDEPENDENT VARIABLES

    X(1) = $0.20577213E+00$

    X(2) = $0.47983217E+00$

TOTAL NUMBER OF STAGES = 7

TOTAL NUMBER OF FUNCTION EVALUATIONS   62

FINAL VALUE OF OBJECT FUNCTION = $-3.8739235E+03$

X(1) = $0.20609482E+00$

X(2) = $0.47961203E+00$

  STOP

IV. POWELL (BOTM ALGORITHM)*

A. Purpose

This program finds the minimum of an unconstrained, multivariable, nonlinear function:

Minimize $\quad F(X_1, X_2, \ldots, X_N)$

B. Method

The procedure is based on the method of M. J. D. Powell (37). Derivatives are not required. Trials with several starting points are recommended if a multimodal function is suspected. The algorithm proceeds as follows:

1) A starting point, $\underline{X}_o^{(o)}$, is selected. The initial search directions, $\underline{M}_i^{(o)}$, $i = 1, 2, \ldots, N$, are parallel to the original coordinate axes.

2) A sequence of single variable searches are made in the N initial directions, using quadratic approximation (see Chapter 7).

3) The following points are then located:

   $\underline{X}_N^{(k)}$ = the last point from the sequence of single variable searches

   $\underline{X}_M^{(k)}$ = the point of greatest function improvement between successive single variable searches

   $\underline{X}_t^{(k)} = 2\underline{X}_N^{(k)} - \underline{X}_o^{(k)}$ = expanded point

   $\underline{X}_o^{(k)}$ = starting point for the iteration

   where k is the stage index which is incremented for each new set of search directions.

4) A check is made to see if the value of the objective function at the expanded point, $\underline{X}_t^{(k)}$, is better than that at the starting point, $\underline{X}_o^{(k)}$. If there is no improvement, the last point, $\underline{X}_N^{(k)}$, becomes the new starting point and a new sequence of single variable searches are made in the same directions as before,

---

*Computer code developed by M. J. D. Powell, Atomic Energy Research Establishment, Harwell, Berkshire, England. Used by permission.

$$\underline{X}_o^{(k+1)} = \underline{X}_N^{(k)}$$

$$\underline{M}_i^{(k+1)} = \underline{M}_i^{(k)}, \quad i = 1, 2, \ldots, N.$$

If the objective function, $F_t^{(k)}$, at the expanded point is an improvement over the starting point function, $F_o^{(k)}$, then the following test is performed,

$$[F_o^{(k)} - 2F_N^{(k)} + F_t^{(k)}][F_o^{(k)} - F_N^{(k)} - \Delta]^2 \geq \frac{\Delta(F_o^{(k)} - F_t^{(k)})^2}{2}$$

where $\Delta = |F_M^{(k)} - F_{M-1}^{(k)}|$.

The test determines whether the function in this region is a valley, i.e., $F_t^{(k)}$ is an improvement but the surface is rising. If the test is satisfied, the old search directions are retained and a new sequence of single variable searches started as above. If the test is not satisfied, a single variable search is performed in the $\underline{\mu}^{(k)}$ direction,

$$\underline{\mu}^{(k)} = \underline{X}_N^{(k)} - \underline{X}_o^{(k)}$$

until the best value, $\underline{X}_o^{(k+1)}$ is found. New search directions are then chosen as follows:

$$\underline{M}_i^{(k+1)} = \underline{M}_i^{(k)} \quad \text{for} \quad i = 1, 2, \ldots, M-1$$

$$\underline{M}_i^{(k+1)} = \underline{M}_{i+1}^{(k)} \quad \text{for} \quad i = M, \ldots, N-1$$

$$\underline{M}_N^{(k+1)} = \underline{\mu}^{(k)}$$

A new sequence of single variable searches is then started.

5) Convergence is assumed when the values for the independent variables between successive iterations are less than preset limits,

$$|x_i^{(k)} - x_i^{(k-1)}| < \epsilon, \quad i = 1, 2, \ldots, N$$

A flow sheet illustrating the procedure is given in Figure 9.IV.

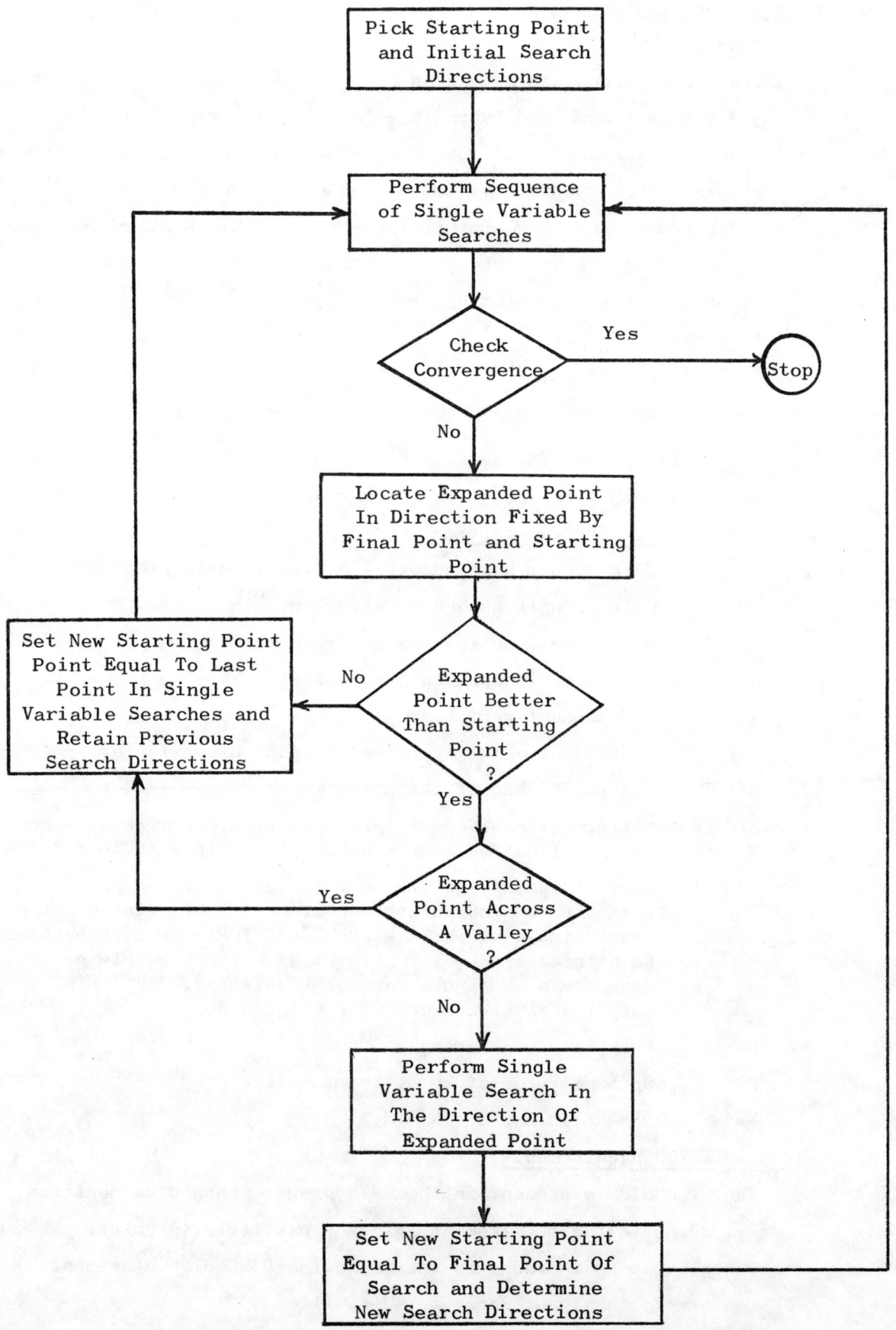

Figure 9.IV. Powell (BOTM ALGORITHM) Logic Diagram

C. Program Description

1) Usage:

The program consists of a main program, the general subroutine BOTM and the user supplied subroutine CALCFX. The starting point, number of variables and solution parameters are read into the main program. Intermediate values of the objective function and independent variables are printed out in BOTM. Final values are printed from the main program. Format changes may be required depending on the problem under consideration.

2) Subroutines Required:

SUBROUTINE BOTM (X, E, N, EF, ESCALE, IPRINT, MAXIT, W, NI, NO, NW) called from main program. Performs single variable searches.

SUBROUTINE CALCFX (N, X, F)

specifies objective function (user supplied).

3) Description of Parameters:

NI	Card reader unit number - Define in main program
NO	Printer unit number - Define in main program
F	Objective function - specify in CALCFX
N	Number of independent variables
X	Independent variables
E	Convergence limit for the independent variables
MAXIT	Maximum number of iterations
ESCALE	Maximum step size multiplier in single variable search - X(I) will not be incremented by more than ESCALE* E(I)
IPRINT	Controls printing. IF IPRINT = 1, the independent variables and objective function will be printed after every single variable search. IF IPRINT = 2, they will be printed after every iteration only (a complete sequence of N single variable searches). IF IPRINT = 3, only the final answers will be printed.
W	Working vector area
EF	Optimum value of objective function
NW	Dimension for W = N* (N+3)

4) DIMENSION Requirements:

The DIMENSION statement in the main program should be modified according to the requirements of each particular problem. The parameters included in the following DIMENSION statement conform to the Input Parameter definitions,

DIMENSION X(N), W(NW), E(N)

5) Input Formats:

CARD TYPE	FORMAT	CONTENTS
1	(3I10, 1F10.5)	N, IPRINT, MAXIT, ESCALE
2	(8E10.4)	(X(II), II=1,N)
3	(8E10.4)	(E(JJ), JJ=1,N)

(Additional CARD TYPES 2 and 3 will be required if N exceeds 8.)

6) Output:

The input values are printed out first. The remaining output is controlled by IPRINT. If IPRINT = 1, the independent variables, function value, and number of cumulative function evaluations will be printed for each single variable search. For IPRINT = 2, the values of the independent variables and objective function will be printed for each iteration. For IPRINT = 3, final values only will be printed.

7) Summary of User Requirements:

a) Specify NI, NO, N, IPRINT, MAXIT, ESCALE, X and E.

b) Adjust the DIMENSION and FORMAT statements as required.

c) Specify the objective function in SUBROUTINE CALCFX.

D. Test Problem

The following test problem was obtained from the literature (13). The calculations were performed on a CDC 6400 computer.

Function: $F = -3803.84 - 138.08X_1 - 232.92X_2 + 123.08X_1^2 + 203.64X_2^2 + 182.25X_1X_2$

Starting Point: $X_1 = 1.0$, $X_2 = 0.5$

Dimensions: N = 2

Parameters: ESCALE = 0.5, IPRINT = 1, MAXIT = 999, E(1) = 0.1
E(2) = 0.1.

Algorithm Answers: $F = -3873.0$
$X_1 = 0.20566$
$X_2 = 0.47986$

Number of Iterations: 3

Number of Function Evaluations: 19

Central Processor Time: 4 seconds

The listing and output for this problem are contained in the following section.

E. Program Listings and Example Output

```
C
C       MAIN LINE PROGRAM FOR SUBROUTINE BOTM
C
        DIMENSION X(2), W(10), E(2)
C
        NI = 50
        NO = 66
C
        READ (NI,005) N, IPRINT, MAXIT, ESCALE
  005   FORMAT (3I10,2F10.5)
        READ (NI,006) (X(II),II=1,N)
  006   FORMAT (8E10.4)
        READ (NI,006) (E(JJ), JJ=1,N)
C
        NW = N * (N+3)
C
        CALL BOTM (X,E,N,EF,ESCALE,IPRINT,MAXIT,W,NI,NO,NW)
C
        WRITE (NO,001)
  001   FORMAT (//,5X,23HVALUES OF THE VARIABLES)
        DO 100 J=1,N
        WRITE (NO,002) J, X(J)
  002   FORMAT (/,5X,2HX(,I2,4H) = ,E16.8)
  100   CONTINUE

        WRITE (NO,003) EF
  003   FORMAT (//,5X,21HOPTIMUM VALUE OF F = ,E16.8)
C
        END




        SUBROUTINE CALCFX (N,X,F)
C
        DIMENSION X(N)
C
        X1 = X(1)
        X2 = X(2)
        X12 = X1**2
        X22 = X2**2
        F = 3803.84 + 138.08*X1 + 232.92*X2 - 123.08*X12 - 203.64*X22
       1  - 182.25*X1*X2
C
        F= -F
C
        RETURN
        END
```

```
      SUBROUTINE BOTM (X,E,N,EF,ESCALE,IPRINT,MAXIT,W,NI,NO,NW)
C
      DIMENSION X(N), W(NW), E(N)
C
      WRITE (NO,001)
  001 FORMAT (1H1,10X,32HPOWELL-BOTM OPTIMIZATION ROUTINE )
      WRITE (NO,002) N, MAXIT, ESCALE, (I, X(I), I=1,N), (J, E(J), J=1,
     1 N)
  002 FORMAT (//,2X,10HPARAMETERS,//,2X,4HN = ,I2,4X,8HMAXIT = ,I4,4X,
     1 9HESCALE = ,F5.2,//,2X,15HINITIAL GUESSES,//,2(2X,2HX(,I2,4H) = ,
     2 1PE16.8),//,2X,31HACCURACY REQUIRED FOR VARIABLES ,//,2(2X,2HE(,
     3 I2,4H) = ,E16.8) )
C
      DDMAG=0.1*ESCALE
      SCER=0.05/ESCALE
      JJ=N*(N+1)
      JJJ=JJ+N
      K=N+1
      NFCC=1
      IND=1
      INN=1
      DO 4 I=1,N
      W(I)=ESCALE
      DO 4 J=1,N
      W(K)=0.
      IF(I-J)4,3,4
C
    3 W(K)=ABS (E(I))
    4 K=K+1
      ITERC=1
      ISGRAD=2
      CALL CALCFX(N,X,F)
      FKEEP=2.*ABS (F)
    5 ITONE=1
      FP=F
      SUM=0.
      IXP=JJ
      DO 6 I=1,N
      IXP=IXP+1
    6 W(IXP)=X(I)
      IDIRN=N+1
      ILINE=1
    7 DMAX=W(ILINE)
      DACC=DMAX*SCER
      DMAG=AMIN1 (DDMAG,0.1*DMAX)
      DMAG=AMAX1(DMAG,20.*DACC)
      DDMAX=10.*DMAG
      GO TO (70,70,71),ITONE
C
   70 DL=0.
      D=DMAG
      FPREV=F
      IS=5
      FA=FPREV
      DA=DL
    8 DD=D-DL
      DL=D
```

```
   58 K=IDIRN
      DO 9 I=1,N
      X(I)=X(I)+DD*W(K)
    9 K=K+1
      CALL CALCFX(N,X,F)
      NFCC=NFCC+1
      GO TO (10,11,12,13,14,96),IS
   14 IF(F-FA)15,16,24
C
   16 IF (ABS (D)-DMAX) 17,17,18
   17 D=D+D
      GO TO 8
   18 WRITE (NO,019)
   19 FORMAT(5X,38HMAXIMUM CHANGE DOES NOT ALTER FUNCTION)
      GO TO 20
C
   15 FB=F
      DB=D
      GO TO 21
   24 FB=FA
      DB=DA
      FA=F
      DA=D
   21 GO TO (83,23),ISGRAD
   23 D=DC+DB-DA

      IS=1
      GO TO 8
   83 D=0.5*(DA+DB-(FA-FB)/(DA-DB))
      IS=4
      IF((DA-D)*(D-DB))25,8,8
   25 IS=1
      IF(ABS (D-DB)-DDMAX)8,8,26
   26 D=DB+SIGN (DDMAX,DB-DA)
      IS=1
      DDMAX=DDMAX+DDMAX
      DDMAG=DDMAG+DDMAG
      IF (DDMAG.GE.1.0E+60)   DDMAG = 1.0E+60
      IF(DDMAX-DMAX)8,8,27
   27 DDMAX=DMAX
      GO TO 8
   13 IF(F-FA)28,23,23
   28 FC=FB
      DC=DB
   29 FB=F
      DB=D
      GO TO 30
   12 IF(F-FB)28,28,31
   31 FA=F
      DA=D
      GO TO 30
   11 IF(F-FB)32,10,10
   32 FA=FB
      DA=DB
      GO TO 29
```

```
   71 DL=1.
      DDMAX=5.
      FA=FP
      DA=-1.
      FB=FHOLD
      DB=0.
      D=1.
   10 FC=F
      DC=D
   30 A=(DB-DC)*(FA-FC)
      B=(DC-DA)*(FB-FC)
      IF((A+B)*(DA-DC))33,33,34
   33 FA=FB
      DA=DB
      FB=FC
      DB=DC
      GO TO 26
   34 D=0.5*(A*(DB+DC)+B*(DA+DC))/(A+B)
      DI=DB
      FI=FB
      IF(FB-FC)44,44,43
   43 DI=DC
      FI=FC
   44 GO TO (86,86,85),ITONE
   85 ITONE=2
      GO TO 45

   86 IF (ABS (D-DI)-DACC) 41,41,93
   93 IF (ABS (D-DI)-0.03*ABS (D)) 41,41,45
   45 IF ((DA-DC)*(DC-D)) 47,46,46
   46 FA=FB
      DA=DB
      FB=FC
      DB=DC
      GO TO 25
   47 IS=2
      IF ((DB-D)*(D-DC)) 48,8,8
   48 IS=3
      GO TO 8
   41 F=FI
      D=DI-DL
      DD=SQRT ((DC-DB)*(DC-DA)*(DA-DB)/(A+B))
      DO 49 I=1,N
      X(I)=X(I)+D*W(IDIRN)
      W(IDIRN)=DD*W(IDIRN)
   49 IDIRN=IDIRN+1
      W(ILINE)=W(ILINE)/DD
      ILINE=ILINE+1
      IF(IPRINT-1)51,50,51
C
   50 WRITE(NO,52)ITERC,NFCC,F,(X(I),I=1,N)
   52 FORMAT(/10H ITERATION,I5,I15,16H FUNCTION VALUES,10X,3HF =,E15.8/
     +2(E16.8,2X))
      GO TO(51,53),IPRINT
   51 GO TO (55,38),ITONE
   55 IF (FPREV-F-SUM) 94,95,95
```

```
   95 SUM=FPREV-F
      JIL=ILINE
   94 IF (IDIRN-JJ) 7,7,84
   84 GO TO (92,72),IND
   92 FHOLD=F
      IS=6
      IXP=JJ
      DO 59 I=1,N
      IXP=IXP+1
   59 W(IXP)=X(I)-W(IXP)
      DD=1.
      GO TO 58
   96 GO TO (112,87),IND
  112 IF(FP-F) 37,37,91
   91 D=2.*(FP+F-2.*FHOLD)/(FP-F)**2
      IF (D*(FP-FHOLD-SUM)**2-SUM) 87,37,37
   87 J=JIL*N+1
      IF (J-JJ) 60,60,61
   60 DO 62 I=J,JJ
      K=I-N
   62 W(K)=W(I)
      DO 97 I=JIL,N
   97 W(I-1)=W(I)
   61 IDIRN=IDIRN-N
      ITONE=3
      K=IDIRN

      IXP=JJ
      AAA=0.
      DO 67 I=1,N
      IXP=IXP+1
      W(K)=W(IXP)
      IF (AAA-ABS (W(K)/E(I))) 66,67,67
   66 AAA=ABS (W(K)/E(I))
   67 K=K+1
      DDMAG=1.
      W(N)=ESCALE/AAA
      ILINE=N
      GO TO 7
   37 IXP=JJ
      AAA=0.
      F=FHOLD
      DO 99 I=1,N
      IXP=IXP+1
      X(I)=X(I)-W(IXP)
      IF(AAA*ABS (E(I))-ABS (W(IXP))) 98,99,99
   98 AAA=ABS (W(IXP)/E(I))
   99 CONTINUE
      GO TO 72
   38 AAA=AAA*(1.+DI)
      GO TO (72,106),IND
   72 IF(IPRINT-2)53,50,50
   53 GO TO (109,88),IND
  109 IF (AAA - 0.1) 20,20,76
C
   76    IF(F-FP)35,78,78
   78 WRITE (NO,80)
```

```
   80 FORMAT(5X,31HACCURACY LIMITED BY ERRORS IN F)
      GO TO 20
C
   88 IND=1
   35 DDMAG=0.4*SQRT(ABS(FP-F))
      IF (DDMAG.GE.1.0E+60)   DDMAG = 1.0E+60
      ISGRAD=1
C
  108 ITERC=ITERC+1
      IF(ITERC-MAXIT)5,5,81
   81 WRITE (NO,82) MAXIT
   82 FORMAT(I5,29H ITERATIONS COMPLETED BY BOTM)
      IF(F-FKEEP)20,20,110
  110 F=FKEEP
      DO 111 I=1,N
      JJJ=JJJ+1
  111 X(I)=W(JJJ)
      GO TO 20
C
  101 JIL=1
      FP=FKEEP
      IF(F-FKEEP)105,78,104
  104 JIL=2
      FP=F
      F=FKEEP
C
  105 IXP=JJ
      DO 113 I=1,N
      IXP=IXP+1
      K=IXP+N
      GO TO (114,115),JIL
  114 W(IXP)=W(K)
      GO TO 113
  115 W(IXP)=X(I)
      X(I)=W(K)
  113 CONTINUE
      JIL=2
      GO TO 92
  106 IF(AAA-0.1) 20,20,107
C
   20 EF=F
      RETURN
C
  107 INN=1
      GO TO 35
C
      END
```

```
           POWELL-BOTM OPTIMIZATION ROUTINE

PARAMETERS

N =  2     MAXIT =  999    ESCALE =   .50

INITIAL GUESSES

X( 1) =   1.00000000E+00  X( 2) =   5.00000000E-01

ACCURACY REQUIRED FOR VARIABLES

E( 1) =   1.00000000E-01  E( 2) =   1.00000000E-01

ITERATION    1              4 FUNCTION VALUES       F = -.38738683E+04
   .19074992E+00     .50000000E+00

ITERATION    1              7 FUNCTION VALUES       F = -.38739053E+04
   .19074992E+00     .48653464E+00

ITERATION    2             10 FUNCTION VALUES       F = -.38739175E+04
   .20071930E+00     .48653464E+00

ITERATION    2             12 FUNCTION VALUES       F = -.38739215E+04
   .20071930E+00     .48207353E+00

ITERATION    2             14 FUNCTION VALUES       F = -.38739235E+04
   .20565857E+00     .47986330E+00

ITERATION    3             17 FUNCTION VALUES       F = -.38739235E+04
   .20565857E+00     .47986330E+00

ITERATION    3             19 FUNCTION VALUES       F = -.38739235E+04
   .20565857E+00     .47986330E+00

     VALUES OF THE VARIABLES

     X( 1) =    .20565857E+00

     X( 2) =    .47986330E+00

     OPTIMUM VALUE OF F =   -.38739235E+04
```

V. FLETCHER AND REEVES (FMCG ALGORITHM)*

A. Purpose

This program finds the unconstrained minimum of a multivariable, nonlinear function:

Minimize $\quad\quad F(X_1, X_2, \ldots, X_N)$.

B. Method

The basic procedure is described by Fletcher and Reeves (18). Derivatives of the objective function with respect to the independent variables are required. Unimodality is assumed and thus multiple starting points are recommended if a multimodal function is suspected. The algorithm proceeds as follows:

1) A starting point is selected.

2) The direction of steepest descent is determined by determining the following direction vector components (normalized form) at the starting point,

$$M_i^{(k)} = \left\{ \frac{-\partial F/\partial X_i}{\left[\sum_{j=1}^{N} \left(\frac{\partial F}{\partial X_j}\right)^2\right]^{1/2}} \right\}^{(k)}, \quad\quad i = 1, 2, \ldots, N$$

where k=0 for the starting point.

3) A one dimensional search is then conducted (see Chapter 7) along the direction of steepest descent utilizing the relation,

$$X_{i(new)} = X_{i(old)} + S M_i, \quad\quad i = 1, 2, \ldots, N$$

where S is the distance moved in the __M__ direction. When a minimum is obtained along the direction of steepest descent, a new "conjugate direction" (57) search direction is evaluated at the new point with the following normalized components:

---

*The FORTRAN program contained in this section is based on __Program FMCG (Fletcher and Reeves Unconstrained Search Technique)__, described on pages 225-229 of __System/360 Scientific Subroutine Package__, H20-0205-3. Used by permission of International Business Machines Corporation.

$$M_i^{(k)} = \frac{-\left(\frac{\partial F}{\partial X_i}\right)^{(k)} + \beta^{(k-1)} M_i^{(k-1)}}{\left[\sum_{j=1}^{N}\left(-\left(\frac{\partial F}{\partial X_j}\right)^{(k)} + \beta^{(k-1)} M_j^{(k-1)}\right)^2\right]^{1/2}} \quad , \quad i = 1, 2, \ldots, N$$

where

$$\beta^{(k-1)} = \frac{\sum_{i=1}^{N}\left[\left(\frac{\partial F}{\partial X_i}\right)^{(k)}\right]^2}{\sum_{i=1}^{N}\left[\left(\frac{\partial F}{\partial X_i}\right)^{(k-1)}\right]^2} \quad .$$

4) A one dimensional search is then conducted in this direction. When a minimum is found, an overall convergence check is made (see Description of Parameters). If convergence is achieved, the procedure terminates. If convergence is not achieved, new "conjugate direction" vector components are evaluated per step 3) at the minimum point from the current one dimensional search. This process is continued until convergence is achieved or N+1 directions have been searched. If a cycle of N+1 directions have been completed, a new cycle is started consisting of a steepest descent direction (step 2)) and N "conjugate directions" (step 3)).

A logic diagram illustrating the above procedure is given in Figure 9.V.

C. Program Description

1) Usage:

The program consists of a main program, the subroutine FMCG, and the user supplied subroutine FUNCT. All input and output is from the main program. Format changes may be required, depending on the problem under consideration.

2) Subroutines Required:

SUBROUTINE FMCG(N, M, X, F, G, EST, EPS, LIMIT, IER, H, KOUNT) called from main program, performs search.

SUBROUTINE FUNCT(N, ARG, VAL, GRAD) specifies objective function and derivatives of objective function with respect to the independent variables (user supplied).

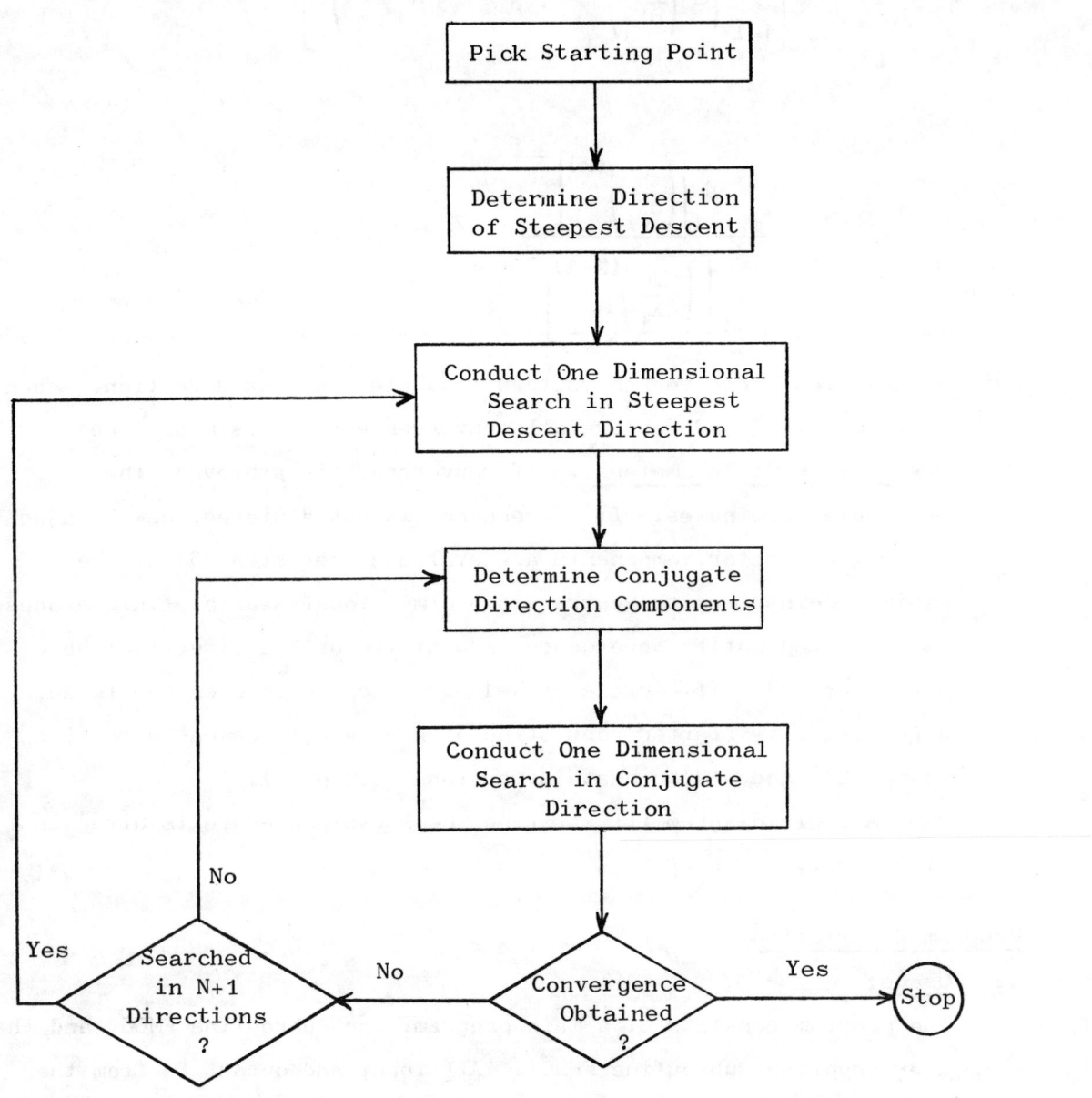

Figure 9.V. Fletcher and Reeves (FMCG ALGORITHM) Logic Diagram

3) Description of Parameters:

N	Number of independent variables
X	Independent variable vector (initial guesses on input, optimum values on output)
F	Final minimum value of the objective function
G	Final gradient vector at the minimum
VAL	Current value of objective function
ARG	Current vector of independent variable values
GRAD	Current gradient vector values
H	Storage vector
M	Storage vector dimension (2 * N)
NI	Card reader unit number
NO	Printer unit number
EST	Estimate of the minimum value of the objective function
EPS	Test value representing the expected absolute error in movement
LIMIT	Maximum number of iterations
IER	Error parameter IER = 0 means convergence was obtained IER = 1 means no convergence in LIMIT iterations IER = -1 means errors in gradient calculation IER = 2 means it is likely that a minimum does not exist
KOUNT	Iteration counter

4) DIMENSION Requirements:

The DIMENSION statement in the main program should be modified according to the requirements of each particular problem. The parameters included in the following DIMENSION statement conform to the parameter definitions.

DIMENSION X(N), G(N), H(M)

5) Input Formats:

CARD TYPE	FORMAT	CONTENTS
1	(5I10)	N, LIMIT
2	(8E10.4)	EST, EPS
3	(8E10.5)	(X(I), I=1, N)

(If N exceeds 8, additional cards of Type 3 are required.)

6) Output:

All output is from the main program. The method name is printed first followed by values of the input parameters. The error message IER, the final value of the objective function and independent variables and the number of required iterations (directions) are then listed.

7) <u>Summary of User Requirements</u>:
   a) Determine values for N, LIMIT, EST, EPS, (X(I), I=1, N).
   b) Specify NI and NO in the main program.
   c) Adjust the DIMENSION statement in the main program.
   d) Specify objective function and derivatives in FUNCT.
   e) Adjust input and output format statements as required by the problem under consideration.

Further details on this program are available from IBM (57).

D. <u>Test Problem</u>

The following test problem was taken from the literature (13). The calculations were performed on a CDC 6400 computer.

Function:
$$F = -3803.84 - 138.08X_1 - 232.92X_2 + 123.08X_1^2 + 203.64X_2^2 + 182.25X_1X_2$$

Starting Point: $X_1 = 1.0$, $X_2 = 0.5$

Parameters: LIMIT = 100, EST = 3000., EPS = 0.00001

Algorithm Answers:
F = -3873.9
X_1 =   0.20566
X_2 =   0.47986

Number of Iterations: 3

Central Processor Time: 3 seconds

The listing and output for this problem are contained in the following section.

E. Program Listings and Example Output

```
C
C     MAIN LINE PROGRAM FOR SUBROUTINE FMCG
C
C
C
      DIMENSION X(2), G(2), H(4)
C
      NI = 50
      NO = 66
C
      READ (NI,001) N, LIMIT
  001 FORMAT (5I10)
      READ (NI,002) EST, EPS
  002 FORMAT (8E10.4)
      READ (NI,002) (X(I), I=1,N)
      M = 2*N
C
      WRITE (NO,010)
  010 FORMAT (1H1,10X,36HFLETCHER-REEVES OPTIMIZATION ROUTINE )
      WRITE (NO,011) N, LIMIT, EST, EPS
  011 FORMAT (//,2X,10HPARAMETERS,//,2X,4HN = ,I2,4X,8HLIMIT = ,I4,4X,
     1 6HEST = ,E16.8,4X,6HEPS = ,E16.8 )
      WRITE (NO,012)
  012 FORMAT (//,2X,19HINITIAL VALUES OF X )
      WRITE (NO,014) (J, X(J), J=1,N)
  014 FORMAT (/,3(2X,2HX(,I2,4H) = ,E16.8 ))
C
      CALL FMCG (N,M,X,F,G,EST,EPS,LIMIT,IER,H,KOUNT)
C
      WRITE (NO,003) IER, F
    3 FORMAT (///,2X,6HIER = ,I3,6X,4HF = ,E15.8)
      WRITE (NO,004)
    4 FORMAT (//,2X,22HVALUES OF X AT MINIMUM)
      DO 100 J=1,N
      WRITE (NO,005) J, X(J)
    5 FORMAT (//,2X,2HX(,I2,4H) = ,E15.8)
  100 CONTINUE
      WRITE (NO,006) KOUNT
    6 FORMAT (///,2X,23HNUMBER OF ITERATIONS = ,I5)
C
      END
```

```
      SUBROUTINE FMCG (N,M,X,F,G,EST,EPS,LIMIT,IER,H,KOUNT)
C
      DIMENSION X(N), G(N), H(M)
C
C        COMPUTE FUNCTION VALUE AND GRADIENT VECTOR FOR INITIAL ARGUMENT
      CALL FUNCT(N,X,F,G)
C
C        RESET ITERATION COUNTER
      KOUNT=0
      IER=0
      N1=N+1
C
C        START ITERATION CYCLE FOR EVERY N+1 ITERATIONS
    1 DO 43 II=1,N1
C
C        STEP ITERATION COUNTER AND SAVE FUNCTION VALUE
      KOUNT=KOUNT+1
      OLDF=F
C
C        COMPUTE SQUARE OF GRADIENT AND TERMINATE IF ZERO
      GNRM=0.
      DO 2 J=1,N
    2 GNRM=GNRM+G(J)*G(J)
      IF(GNRM)46,46,3
C
C        EACH TIME THE ITERATION LOOP IS EXECUTED, THE FIRST STEP WILL
C        BE IN DIRECTION OF STEEPEST DESCENT
    3 IF(II-1)4,4,6
    4 DO 5 J=1,N
    5 H(J)=-G(J)
      GO TO 8
C
C        FURTHER DIRECTION VECTORS H WILL BE CHOOSEN CORRESPONDING
C        TO THE CONJUGATE GRADIENT METHOD
    6 AMBDA=GNRM/OLDG
      DO 7 J=1,N
    7 H(J)=AMBDA*H(J)-G(J)
C
C        CONPUTE TEST VALUE FOR DIRECTIONAL VECTOR AND DIRECTIONAL
C        DERIVATIVE
    8 DY=0.
      HNRM=0.
      DO 9 J=1,N
      K=J+N
C
C        SAVE ARGUMENT VECTOR
      H(K)=X(J)
      HNRM=HNRM+ABS(H(J))
    9 DY=DY+H(J)*G(J)
C
C        CHECK WHETHER FUNCTION WILL DECREASE STEPPING ALONG H AND
C        SKIP LINEAR SEARCH ROUTINE IF NOT
      IF(DY)10,42,42
C
C        COMPUTE SCALE FACTOR USED IN LINEAR SEARCH SUBROUTINE
   10 SNRM=1./HNRM
```

```
C
C             SEARCH MINIMUM ALONG DIRECTION H
C
C             SEARCH ALONG H FOR POSITIVE DIRECTIONAL DERIVATIVE
      FY=F
      ALFA=2.*(EST-F)/DY
      AMBDA=SNRM
C
C          USE ESTIMATE FOR STEPSIZE ONLY IF IT IS POSITIVE AND LESS THAN
C             SNRM. OTHERWISE TAKE SNRM AS STEPSIZE.
      IF(ALFA)13,13,11
   11 IF(ALFA-AMBDA)12,13,13
   12 AMBDA=ALFA
   13 ALFA=0.
C
C             SAVE FUNCTION AND DERIVATIVE VALUES FOR OLD ARGUMENT
   14 FX=FY
      DX=DY
C
C             STEP ARGUMENT ALONG H
      DO 15 I=1,N
   15 X(I)=X(I)+AMBDA*H(I)
C
C             COMPUTE FUNCTION VALUE AND GRADIENT FOR NEW ARGUMENT
      CALL FUNCT(N,X,F,G)
      FY=F
C
C             COMPUTE DIRECTIONAL DERIVATIVE DY FOR NEW ARGUMENT. TERMINATE
C             SEARCH, IF DY POSITIVE. IF DY IS ZERO THE MINIMUM IS FOUND
      DY=0.
      DO 16 I=1,N
   16 DY=DY+G(I)*H(I)
      IF(DY)17,38,20
C
C             TERMINATE SEARCH ALSO IF THE FUNCTION VALUE INDICATES THAT
C             A MINIMUM HAS BEEN PASSED
   17 IF(FY-FX)18,20,20

C
C             REPEAT SEARCH AND DOUBLE STEPSIZE FOR FURTHER SEARCHES
   18 AMBDA=AMBDA+ALFA
      ALFA=AMBDA
C
C             TERMINATE IF THE CHANGE IN ARGUMENT GETS VERY LARGE
      IF(HNRM*AMBDA-1.E10)14,14,19
C
C             LINEAR SEARCH TECHNIQUE INDICATES THAT NO MINIMUM EXISTS
   19 IER=2
      RETURN
C             END OF SEARCH LOOP
C
C             INTERPOLATE CUBICALLY IN THE INTERVAL DEFINED BY THE SEARCH
C             ABOVE AND COMPUTE THE ARGUMENT X FOR WHICH THE INTERPOLATION
C             POLYNOMIAL IS MINIMIZED
C
   20 T=0.
   21 IF(AMBDA)22,38,22
   22 Z=3.*(FX-FY)/AMBDA+DX+DY
```

```
      ALFA=AMAX1(ABS(Z),ABS(DX),ABS(DY))
      DALFA=Z/ALFA
      DALFA=DALFA*DALFA-DX/ALFA*DY/ALFA
      IF(DALFA)23,27,27
C
C        RESTORE OLD VALUES OF FUNCTION AND ARGUMENTS
   23 DO 24 J=1,N
      K=N+J
   24 X(J)=H(K)
      CALL FUNCT(N,X,F,G)
C
C        TEST FOR REPEATED FAILURE OF ITERATION
   25 IF(IER)47,26,47
   26 IER=-1
      GO TO 1
   27 W=ALFA*SQRT(DALFA)
      ALFA=(DY+W-Z)*AMBDA/(DY+2.*W-DX)
      DO 28 I=1,N
   28 X(I)=X(I)+(T-ALFA)*H(I)
C
C        TERMINATE, IF THE VALUE OF THE ACTUAL FUNCTION AT X IS LESS
C        THAN THE FUNCTION VALUES AT THE INTERVAL ENDS. OTHERWISE REDUCE
C        THE INTERVAL BY CHOOSING ONE END-POINT EQUAL TO X AND REPEAT
C        THE INTERPOLATION. WHICH END-POINT IS CHOOSEN DEPENDS ON THE
C        VALUE OF THE FUNCTION AND ITS GRADIENT AT X
C
      CALL FUNCT(N,X,F,G)
      IF(F-FX)29,29,30
   29 IF(F-FY)38,38,30
C
C        COMPUTE DIRECTIONAL DERVATIVE
   30 DALFA=0.
      DO 31 I=1,N
   31 DALFA=DALFA+G(I)*H(I)
      IF(DALFA)32,35,35
   32 IF(F-FX)34,33,35
   33 IF(DX-DALFA)34,38,34
   34 FX=F
      DX=DALFA
      T=ALFA
      AMBDA=ALFA
      GO TO 21
   35 IF(FY-F)37,36,37
   36 IF(DY-DALFA)37,38,37
   37 FY=F
      DY=DALFA
      AMBDA=AMBDA-ALFA
      GO TO 20
C
C        COMPUTE DIFFERENCE OF NEW AND OLD ARGUMENT VECTOR
   38 T=0.
      DO 39 J=1,N
      K=J+N
      H(K)=X(J)-H(K)
   39 T=T+ABS(H(K))
```

```
C
C           TEST LENGTH OF DIFFERENCE VECTOR IF AT LEAST N+1 ITERATIONS
C           HAVE BEEN EXECUTED. TERMINATE, IF LENGTH IS LESS THAN EPS
      IF(KOUNT-N1)41,40,40
   40 IF(T-EPS)45,45,41
C
C           TERMINATE, IF FUNCTION HAS NOT DECREASED DURING LAST
C           ITERATION, OTHERWISE SAVE GRADIENT NORM
   41 IF(OLDF-F+EPS)19,25,42
   42 OLDG=GNRM
C
C           TERMINATE, IF NUMBER OF ITERATIONS WOULD EXCEED LIMIT
      IF(KOUNT-LIMIT)43,44,44
   43 IER=0
C           END OF ITERATION CYCLE
C
C           START NEXT ITERATION CYCLE
      GO TO 1
C
C           NO CONVERGENCE AFTER LIMIT ITERATIONS
   44 IER=1
      IF(GNRM-EPS)46,46,47
C
C           TEST FOR SUFFICIENTLY SMALL GRADIENT
   45 IF(GNRM-EPS)46,46,25
   46 IER=C
   47 RETURN
      END




      SUBROUTINE FUNCT (N,ARG,VAL,GRAD)
C
      DIMENSION ARG(N), GRAD(N)
C
      X1 = ARG(1)
      X2 = ARG(2)
      X12 = (X1**2)
      X22 = (X2**2)
C
      SUMN = 3803.84 + 138.08*X1 + 232.92*X2 - 123.08*X12
     1  - 203.64*X22 - 182.25*X1*X2
C
      DERY1 = 138.08 - (2*123.08*X1) - 182.25*X2
      DERY2 = 232.92 - (2*203.64*X2) - 182.25*X1
C
      VAL = -SUMN
      GRAD(1) = - DERY1
      GRAD(2) = - DERY2
C
      RETURN
      END
```

FLETCHER-REEVES OPTIMIZATION ROUTINE

PARAMETERS

N = 2   LIMIT = 100   EST = .30000000E+04   EPS = .10000000E-04

INITIAL VALUES OF X

X(1) = .10000000E+01   X(2) = .50000000E+00

IER = 0   F = -.38739235E+04

VALUES OF X AT MINIMUM

X(1) = .20565857E+00

X(2) = .47986330E+00

NUMBER OF ITERATIONS = 3

VI. FLETCHER AND POWELL (FMFP ALGORITHM)*

A. Purpose

This program finds the minimum of a multivariable, unconstrained nonlinear function:

Minimize: $F(X_1, X_2, \ldots, X_N)$

B. Method

The method is based on a procedure proposed by Fletcher and Powell (17). Derivatives are required. Several alternate starting points are recommended for suspected multimodal functions. The algorithm proceeds as follows:

1) Select a starting point.

2) Compute a direction of search. In normalized form, this is as follows:

$$M_i^{(k)} = \left\{ \frac{-\sum_{j=1}^{N} H_{i,j}\left(\frac{\partial F}{\partial X_j}\right)}{\left[\sum_{\ell=1}^{N}\left(\sum_{j=1}^{N} H_{\ell,j}\left(\frac{\partial F}{\partial X_j}\right)\right)^2\right]^{1/2}} \right\}^{(k)}, \quad i = 1, 2, \ldots, N$$

where k is the iteration index (=0 at the starting point), M_i are the direction vector components and $\frac{\partial F}{\partial X_j}$ are the gradient vector components. $H_{i,j}$ are the elements of a symmetric positive definite matrix (N×N) which is initially chosen to be the identity matrix. Thus the initial direction of search is the path of steepest descent.

3) A one dimensional search is conducted in the direction chosen by the previous step (see Chapter 7) until a minimum is located utilizing the relation,

$$X_{i(new)} = X_{i(old)} + SM_i, \quad i = 1, 2, \ldots, N$$

where S is the step size in the direction of search.

---

* The FORTRAN program contained in this section is based on <u>Program FMFP</u> (Fletcher and Powell Unconstrained Search Technique), described on pages 221-225 of <u>System/360 Scientific Subroutine Package</u>, H20-0205-3. Used by permission of International Business Machines Corporation.

4) A convergence check is then made (see Description of Parameters). If convergence is achieved, the procedure is terminated. If convergence is not achieved, a new search direction is chosen per Step 2) except $\underline{H}^{(k+1)}$ is calculated as follows:

$$\underline{H}^{(k+1)} = \underline{H}^{(k)} + \underline{A}^{(k)} - \underline{B}^{(k)}$$

where

$$\underline{A}^{(k)} = \frac{\underline{\Delta X}^{(k)} (\underline{\Delta X}^{(k)})^t}{(\underline{\Delta X}^{(k)})^t (\underline{\Delta G}^{(k)})}$$

$$\underline{B}^{(k)} = \frac{\underline{H}^{(k)} \underline{\Delta G}^{(k)} (\underline{\Delta G}^{(k)})^t (\underline{H}^{(k)})}{(\underline{\Delta G}^{(k)})^t \underline{H}^{(k)} \underline{\Delta G}^{(k)}}$$

$$\underline{\Delta X}^{(k)} = \underline{X}^{(k+1)} - \underline{X}^{(k)} \quad \text{(difference in location between iterations)}$$

$$\underline{\Delta G}^{(k)} = \frac{\partial F}{\partial \underline{X}}^{(k+1)} - \frac{\partial F}{\partial \underline{X}}^{(k)} \quad \text{(difference in gradients between iterations)}$$

A new one dimensional search is performed in the new direction. The process is repeated until convergence is obtained.

A logic diagram illustrating the above procedure is given in Figure 9.VI.

C. Program Description

1) Usage:

The program consists of a main program, the primary subroutine FMFP, and a user supplied subroutine FUNCT. All input and output is from the main program. Format changes may be required depending on the problem under consideration.

2) Subroutines Required:

SUBROUTINE FMFP(FUNCT, N, M, X, F, G, EST, EPS, LIMIT, IER, H, KOUNT) called from main program, performs search.

SUBROUTINE FUNCT(N, ARG, VAL, GRAD) specifies objective function and derivatives of objective function with respect to the independent variables (user supplied).

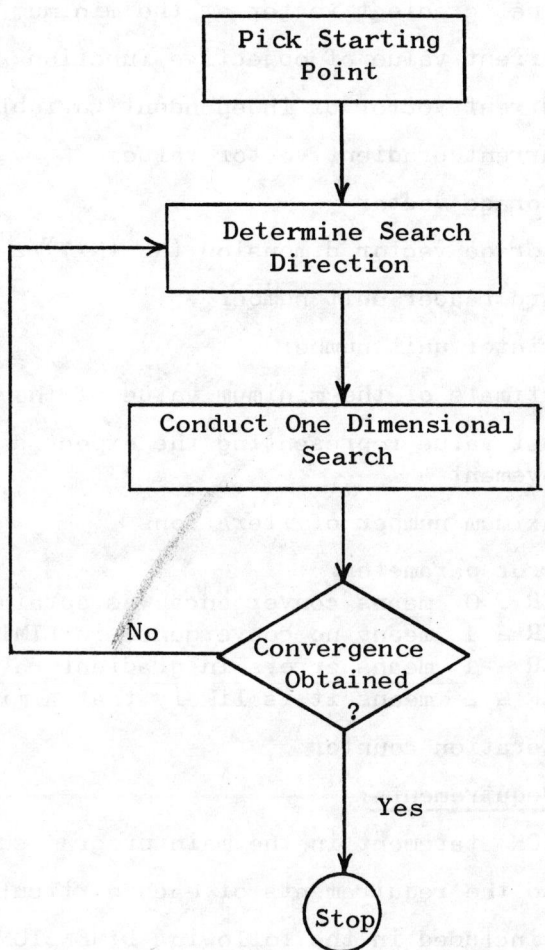

Figure 9.VI. Fletcher and Powell (FMFP ALGORITHM) Logic Diagram

3) **Description of Parameters:**

   N   Number of independent variables

   X   Independent variable vector (initial values on input, optimum values on output)

   F   Final minimum value of the objective function

   G   Final gradient vector at the minimum

   VAL   Current value of objective function

   ARG   Current vector of independent variable values

   GRAD   Current gradient vector values

   H   Storage vector

   M   Storage vector dimension $(N*(N+7))/2$

   NI   Card reader unit number

   NO   Printer unit number

   EST   Estimate of the minimum value of the objective function

   EPS   Test value representing the expected absolute error in movement

   LIMIT   Maximum number of iterations

   IER   Error parameter
   - IER = 0 means convergence was obtained
   - IER = 1 means no convergence in LIMIT iterations
   - IER = -1 means errors in gradient calculation
   - IER = 2 means it is likely that a minimum does not exist

   KOUNT   Iteration counter

4) **DIMENSION Requirements:**

   The DIMENSION statement in the main program should be modified according to the requirements of each particular problem. The parameters included in the following DIMENSION statement conform to the parameter definitions.

   DIMENSION X(N), G(N), H(M)

5) **Input Formats:**

CARD TYPE	FORMAT	CONTENTS
1	(8I10)	N, LIMIT
2	(8F10.0)	EST, EPS
3	(8F10.0)	(X(I), I=1, N)

   (If N exceeds 8, additional cards of Type 3 are required.)

6) **Output:**

   All output is from the main program. The method name is printed first followed by values of the input parameters. The error message IER, the final value of the objective function and independent variables and the number of required iterations (directions) are then listed.

7) <u>Summary of User Requirements</u>:
   a) Determine values for N, LIMIT, EST, EPS, (X(I), I=1, N).
   b) Specify NI and NO in the main program.
   c) Adjust the DIMENSION statement in the main program.
   d) Specify the objective function and derivatives in FUNCT.
   e) Adjust input and output format statements as required by the problem under consideration.

Further details on this program are available from IBM (56).

D. <u>Test Problem</u>

The following test problem was taken from the literature (13). The calculations were performed on a CDC 6400 computer.

Function:
$$F = -3803.84 - 138.08X_1 - 232.92X_2 + 123.08X_1^2 + 203.64X_2^2 + 182.25X_1X_2$$

Starting Point: $X_1 = 1.0$, $X_2 = 0.5$

Parameters: LIMIT = 100, EST = -2000., EPS = 0.00001

Algorithm Answers:
$F = -3873.9$
$X_1 = 0.20566$
$X_2 = 0.47986$

Number of Iterations: 3

Central Processor Time: 3 seconds

The listing and output for this problem are contained in the following section.

E. Program Listings and Example Output

```
C
C      MAIN LINE PROGRAM FOR SUBROUTINE FMFP
C
       DIMENSION X(2), G(2), H(9)
       EXTERNAL FUNCT
C
       NI = 50
       NO = 66
C
       READ (NI,001) N, LIMIT
   001 FORMAT (8I10)
       READ (NI,002) EST, EPS
   002 FORMAT (8F10.0)
       READ (NI,002) (X(I), I=1,N)
C
       WRITE (NO,003)
   003 FORMAT (1H1,10X,25HFLETCHER POWELL ALGORITHM)
       WRITE (NO,004) N, LIMIT, EST, EPS
   004 FORMAT (//,2X,10HPARAMETERS,//,2X,4HN = ,I2,4X,8HLIMIT = ,I4,4X,
      1 6HEST = ,E16.8,4X,6HEPS = ,E16.8 )
       WRITE (NO,005)
   005 FORMAT (//,2X,19HINITIAL VALUES OF X )
       DO 100  J=1,N
       WRITE (NO,006) J, X(J)
   006 FORMAT (/,2X,2HX(,I2,4H) = ,1PE16.8)
   100 CONTINUE
C
       M = N*(N+7)/2
C
       CALL FMFP (FUNCT,N,M,X,F,G,EST,EPS,LIMIT,IER,H,KOUNT)
C
       WRITE (NO,007)
   007 FORMAT (///,8X,32HMINIMIZATION PROCEDURE COMPLETED )
       WRITE (NO,008) IER, KOUNT
   008 FORMAT (//,2X,6HIER = ,I2,8X,23HNUMBER OF ITERATIONS = ,I4)
       WRITE (NO,009) F
   009 FORMAT (//,2X,21HMINIMUM VALUE OF F = ,1PE16.8)
       WRITE (NO,010)
   010 FORMAT (//,2X,17HFINAL VALUES OF X )
       DO 200  J=1,N
       WRITE (NO,006) J, X(J)
   200 CONTINUE
C
       END
```

```
      SUBROUTINE FMFP(FUNCT,N,M,X,F,G,EST,EPS,LIMIT,IER,H,KOUNT)
      DIMENSION H(M),X(N),G(N)
C
C        COMPUTE FUNCTION VALUE AND GRADIENT VECTOR FOR INITIAL ARGUMENT
      CALL FUNCT(N,X,F,G)
C
C        RESET ITERATION COUNTER AND GENERATE IDENTITY MATRIX
      IER=0
      KOUNT=0
      N2=N+N
      N3=N2+N
      N31=N3+1
    1 K=N31
      DO 4 J=1,N
      H(K)=1.
      NJ=N-J
      IF(NJ)5,5,2
    2 DO 3 L=1,NJ
      KL=K+L
    3 H(KL)=0.
    4 K=KL+1
C
C        START ITERATION LOOP
    5 KOUNT=KOUNT+1
C
C        SAVE FUNCTION VALUE, ARGUMENT VECTOR AND GRADIENT VECTOR
      OLDF=F
      DO 9 J=1,N
      K=N+J
      H(K)=G(J)
      K=K+N
      H(K)=X(J)
C
C        DETERMINE DIRECTION VECTOR H.

      K=J+N3
      T=0.
      DO 8 L=1,N
      T=T-G(L)*H(K)
      IF(L-J)6,7,7
    6 K=K+N-L
      GO TO 8
    7 K=K+1
    8 CONTINUE
    9 H(J)=T
C
C        CHECK WHETHER FUNCTION WILL DECREASE STEPPING ALONG H.
      DY=0.
      HNRM=0.
      GNRM=0.
C
C        CALCULATE DIRECTIONAL DERIVATIVE AND TEST VALUES FOR DIRECTION
C        VECTOR H AND GRADIENT VECTOR G.
      DO 10 J=1,N
      HNRM=HNRM+ABS(H(J))
      GNRM=GNRM+ABS(G(J))
```

```
   10 DY=DY+H(J)*G(J)
C
C        REPEAT SEARCH IN DIRECTION OF STEEPEST DESCENT IF DIRECTIONAL
C        DERIVATIVE APPEARS TO BE POSITIVE OR ZERO.
      IF(DY)11,51,51
C
C        REPEAT SEARCH IN DIRECTION OF STEEPEST DESCENT IF DIRECTION
C        VECTOR H IS SMALL COMPARED TO GRADIENT VECTOR G.
   11 IF(HNRM/GNRM-EPS)51,51,12
C
C        SEARCH MINIMUM ALONG DIRECTION H
C
C        SEARCH ALONG H FOR POSITIVE DIRECTIONAL DERIVATIVE
   12 FY=F
      ALFA=2.*(EST-F)/DY
      AMBDA=1.
C
C        USE ESTIMATE FOR STEPSIZE ONLY IF IT IS POSITIVE AND LESS THAN
C        1. OTHERWISE TAKE 1. AS STEPSIZE
      IF(ALFA)15,15,13
   13 IF(ALFA-AMBDA)14,15,15
   14 AMBDA=ALFA
   15 ALFA=0.
C
C        SAVE FUNCTION AND DERIVATIVE VALUES FOR OLD ARGUMENT
   16 FX=FY
      DX=DY
C
C        STEP ARGUMENT ALONG H
      DO 17 I=1,N
   17 X(I)=X(I)+AMBDA*H(I)
C
C        COMPUTE FUNCTION VALUE AND GRADIENT FOR NEW ARGUMENT
      CALL FUNCT(N,X,F,G)
      FY=F
C
C        COMPUTE DIRECTIONAL DERIVATIVE DY FOR NEW ARGUMENT. TERMINATE
C        SEARCH, IF DY IS POSITIVE. IF DY IS ZERO THE MINIMUM IS FOUND
      DY=0.
      DO 18 I=1,N
   18 DY=DY+G(I)*H(I)
      IF(DY)19,36,22
C
C        TERMINATE SEARCH ALSO IF THE FUNCTION VALUE INDICATES THAT
C        A MINIMUM HAS BEEN PASSED
   19 IF(FY-FX)20,22,22
C
C        REPEAT SEARCH AND DOUBLE STEPSIZE FOR FURTHER SEARCHES
   20 AMBDA=AMBDA+ALFA
      ALFA=AMBDA
C        END OF SEARCH LOOP
C
C        TERMINATE IF THE CHANGE IN ARGUMENT GETS VERY LARGE
      IF(HNRM*AMBDA-1.E10)16,16,21
C
C        LINEAR SEARCH TECHNIQUE INDICATES THAT NO MINIMUM EXISTS
   21 IER=2
      RETURN
```

```
C
C           INTERPOLATE CUBICALLY IN THE INTERVAL DEFINED BY THE SEARCH
C           ABOVE AND COMPUTE THE ARGUMENT X FOR WHICH THE INTERPOLATION
C           POLYNOMIAL IS MINIMIZED
   22 T=0.
   23 IF(AMBDA)24,36,24
   24 Z=3.*(FX-FY)/AMBDA+DX+DY
      ALFA=AMAX1(ABS(Z),ABS(DX),ABS(DY))
      DALFA=Z/ALFA
      DALFA=DALFA*DALFA-DX/ALFA*DY/ALFA
      IF(DALFA)51,25,25
   25 W=ALFA*SQRT(DALFA)
      ALFA=(DY+W-Z)*AMBDA/(DY+2.*W-DX)
      DO 26 I=1,N
   26 X(I)=X(I)+(T-ALFA)*H(I)
C
C           TERMINATE, IF THE VALUE OF THE ACTUAL FUNCTION AT X IS LESS
C           THAN THE FUNCTION VALUES AT THE INTERVAL ENDS. OTHERWISE REDUCE
C           THE INTERVAL BY CHOOSING ONE END-POINT EQUAL TO X AND REPEAT
C           THE INTERPOLATION. WHICH END-POINT IS CHOOSEN DEPENDS ON THE
C           VALUE OF THE FUNCTION AND ITS GRADIENT AT X
C
      CALL FUNCT(N,X,F,G)
      IF(F-FX)27,27,28
   27 IF(F-FY)36,36,28
   28 DALFA=0.
      DO 29 I=1,N

   29 DALFA=DALFA+G(I)*H(I)
      IF(DALFA)30,33,33
   30 IF(F-FX)32,31,33
   31 IF(DX-DALFA)32,36,32
   32 FX=F
      DX=DALFA
      T=ALFA
      AMBDA=ALFA
      GO TO 23
   33 IF(FY-F)35,34,35
   34 IF(DY-DALFA)35,36,35
   35 FY=F
      DY=DALFA
      AMBDA=AMBDA-ALFA
      GO TO 22
C
C           COMPUTE DIFFERENCE VECTORS OF ARGUMENT AND GRADIENT FROM
C           TWO CONSECUTIVE ITERATIONS
   36 DO 37 J=1,N
      K=N+J
      H(K)=G(J)-H(K)
      K=N+K
   37 H(K)=X(J)-H(K)
C
C           TERMINATE, IF FUNCTION HAS NOT DECREASED DURING LAST ITERATION
      IF(OLDF-F+EPS)51,38,38
C
C           TEST LENGTH OF ARGUMENT DIFFERENCE VECTOR AND DIRECTION VECTOR
C           IF AT LEAST N ITERATIONS HAVE BEEN EXECUTED. TERMINATE, IF
C           BOTH ARE LESS THAN EPS
```

```
   38 IER=0
      IF(KOUNT-N)42,39,39
   39 T=0.
      Z=0.
      DO 40 J=1,N
      K=N+J
      W=H(K)
      K=K+N
      T=T+ABS(H(K))
   40 Z=Z+W*H(K)
      IF(HNRM-EPS)41,41,42
   41 IF(T-EPS)56,56,42
C
C        TERMINATE, IF NUMBER OF ITERATIONS WOULD EXCEED LIMIT
   42 IF(KOUNT-LIMIT)43,50,50
C
C        PREPARE UPDATING OF MATRIX H
   43 ALFA=0.
      DO 47 J=1,N
      K=J+N3
      W=0.
      DO 46 L=1,N
      KL=N+L

      W=W+H(KL)*H(K)
      IF(L-J)44,45,45
   44 K=K+N-L
      GO TO 46
   45 K=K+1
   46 CONTINUE
      K=N+J
      ALFA=ALFA+W*H(K)
   47 H(J)=W
C
C        REPEAT SEARCH IN DIRECTION OF STEEPEST DESCENT IF RESULTS
C        ARE NOT SATISFACTORY
      IF(Z*ALFA)48,1,48
C
C        UPDATE MATRIX H
   48 K=N31
      DO 49 L=1,N
      KL=N2+L
      DO 49 J=L,N
      NJ=N2+J
      H(K)=H(K)+H(KL)*H(NJ)/Z-H(L)*H(J)/ALFA
   49 K=K+1
      GO TO 5
C        END OF ITERATION LOOP
C
C        NO CONVERGENCE AFTER LIMIT ITERATIONS
   50 IER=1
      RETURN
C
C        RESTORE OLD VALUES OF FUNCTION AND ARGUMENTS
   51 DO 52 J=1,N
      K=N2+J
   52 X(J)=H(K)
      CALL FUNCT(N,X,F,G)
```

```
C
C         REPEAT IN DIRECTION OF STEEPEST DESCENT IF DERIVATIVE
C         FAILS TO BE SUFFICIENTLY SMALL
      IF(GNRM-EPS)55,55,53
C
C         TEST FOR REPEATED FAILURE OF ITERATION
   53 IF(IER)56,54,54
   54 IER=-1
      GO TO 1
   55 IER=0
   56 RETURN
      END




      SUBROUTINE FUNCT (N,ARG,VAL,GRAD)
C
C     ARGUMENT LIST
C
C     ARG      = VECTOR OF X VALUES.
C     VAL      = OBJECTIVE FUNCTION EQUATION.
C     GRAD     = VECTOR OF OBJECTIVE FUNCTION DERIVATIVES,(N LONG).
C
      DIMENSION ARG(N), GRAD(N)
C
      X1  = ARG(1)
      X2  = ARG(2)
      X12 = X1**2
      X22 = X2**2
C
      VAL = 3803.84 + 138.08*X1 + 232.92*X2 - 123.08*X12 - 203.64*X22
     1 - 182.25*X1*X2
C
      DERY1 = 138.08 - (2*123.08*X1) - 182.25*X2
      DERY2 = 232.92 - (2*203.64*X2) - 182.25*X1
C
      VAL = -VAL
      GRAD(1) = -DERY1
      GRAD(2) = -DERY2
C
      RETURN
      END
```

FLETCHER POWELL ALGORITHM

PARAMETERS

N = 2   LIMIT = 100   EST = -.20000000E+04   EPS = 0.10000000E-04

INITIAL VALUES OF X

X(1) =   1.0000000E+00
X(2) =   5.0000000E-01

MINIMIZATION PROCEDURE COMPLETED

IER = 0   NUMBER OF ITERATIONS = 3

MINIMUM VALUE OF F = -3.8739235E+03

FINAL VALUES OF X

X(1) =   2.0565857E-01
X(2) =   4.7986330E-01

Chapter 10

MULTIVARIABLE CONSTRAINED METHODS

The multivariable constrained search methods are the practical classification of search methods. Multivariable, unconstrained building blocks are normally utilized (see Chapter 9) which in turn incorporate single variable search methods (see Chapters 7, 8). The field is not as advanced as the unconstrained case. The techniques for handling the constraints can be divided into two broad categories,

1) feasibility check
2) modified objective function.

The feasibility check methods are similar to the unconstrained methods except that a check section is added to see if a constraint is violated. If this occurs, the current point is relocated inside the feasible region in a prescribed manner. The modified objective function technique incorporates the constraints into the objective function thus producing an unconstrained problem. Penalty functions are used which apply a penalty to the objective functions at non feasible points thus forcing the search back into the feasible region.

The techniques may be subcategorized with regard to linearities, type of constraints and derivative requirements (see Chapter 1). In this chapter, the following routines will be presented:

   I.   Box (COMPLEX ALGORITHM)
  II.   Constrained Rosenbrock (HILL ALGORITHM)
 III.   Rosen (PROJG ALGORITHM)
  IV.   Fiacco and McCormick (SUMT ALGORITHM)
   V.   Constrained Fletcher-Powell (CONMIN ALGORITHM)

I. BOX (COMPLEX ALGORITHM)*

A. Purpose

This program finds the maximum of a multivariable, nonlinear function subject to nonlinear inequality constraints:

Maximize $\quad F(X_1, X_2, \ldots, X_N)$

Subject to $\quad G_k \leq X_k \leq H_k, \quad k = 1, 2, \ldots, M$

The implicit variables $X_{N+1}, \ldots, X_M$ are dependent functions of the explicit independent variables $X_1, X_2, \ldots, X_N$. The upper and lower constraints H_k and G_k are either constants or functions of the independent variables.

B. Method

The procedure is based on the "complex" method of M. J. Box (7). This method is a sequential search technique which has proven effective in solving problems with nonlinear objective functions subject to nonlinear inequality constraints. No derivatives are required. The procedure should tend to find the global maximum due to the fact that the initial set of points are randomly scattered throughout the feasible region. If linear constraints are present or equality constraints are involved, other methods should prove to be more efficient (5). The algorithm proceeds as follows:

1) An original "complex" of $K \geq N + 1$ points is generated consisting of a feasible starting point and $K - 1$ additional points generated from random numbers and constraints for each of the independent variables:

$$X_{i,j} = G_i + r_{i,j}(H_i - G_i),$$

$$i = 1, 2, \ldots, N$$

and

$$j = 1, 2, \ldots, K-1$$

where $r_{i,j}$ are random numbers between 0 and 1.

---

*Program code developed by Joel A. Richardson, Arizona State University, Tempe, Arizona.

2) The selected points must satisfy both the explicit and implicit constraints. If at any time the explicit constraints are violated, the point is moved a small distance δ inside the violated limit. If an implicit constraint is violated, the point is moved one half of the distance to the centroid of the remaining points

$$X_{i,j}(new) = (X_{i,j}(old) + \overline{X}_{i,c})/2$$

$$i = 1, 2, \ldots, N$$

where the coordinates of the centroid of the remaining points, $\overline{X}_{i,c}$, are defined by

$$\overline{X}_{i,c} = \frac{1}{K-1}\left[\sum_{j=1}^{K} X_{i,j} - X_{i,j}(old)\right], \quad i = 1, 2, \ldots, N.$$

This process is repeated as necessary until all the implicit constraints are satisfied.

3) The objective function is evaluated at each point. The point having the lowest function value is replaced by a point which is located at a distance α times as far from the centroid of the remaining points as the distance of the rejected point on the line joining the rejected point and the centroid:

$$X_{i,j}(new) = \alpha\,(\overline{X}_{i,c} - X_{i,j}(old)) + \overline{X}_{i,c}$$

$$i = 1, 2, \ldots, N$$

Box (7) recommends a value of $\alpha = 1.3$.

4) If a point repeats in giving the lowest function value on consecutive trials, it is moved one half the distance to the centroid of the remaining points.

5) The new point is checked against the constraints and is adjusted as before if the constraints are violated.

6) Convergence is assumed when the objective function values at each point are within β units for γ consecutive iterations. An iteration is defined as the calculations required to select a new point which satisfies the constraints and does not repeat in yielding the lowest function value.

A flow sheet illustrating the above procedure is given in Figure 10.I.

C. Program Description

1) Usage:

The program consists of a main program, three general subroutines (CONSX, CHECK, CENTR) and two user supplied subroutines (FUNC, CONST). Initial guesses of the independent variables, random numbers, solution parameters, dimension limits and printer code designation are passed to the subroutines from the main program. Final function and independent variable values are transferred to the main program for printout. Subroutine CONSX is the primary subroutine and coordinates the special purpose subroutines (CHECK, CENTR, FUNC and CONST). Intermediate printouts are provided in this subroutine, if the user desires. Format changes may be required depending on the problem under consideration.

2) Subroutines Required:

SUBROUTINE CONSX (N, M, K, ITMAX, ALPHA, BETA, GAMMA, DELTA, X, R, F, IT, IEV2, NO, G, H, XC, IPRINT) called from main program and coordinates all special purpose subroutines (CHECK, CENTR, FUNC, CONST).

SUBROUTINE CHECK (N,M,K,X,G,H,I,KODE,XC,DELTA,K1) checks all points against explicit and implicit constraints and applies correction if violations are found.

SUBROUTINE CENTR (N,M,K,IEV1,I,XC,X,K1) calculates the centroid of points.

SUBROUTINE FUNC (N,M,K,X,F,I) specifies objective function (user supplied).

SUBROUTINE CONST (N,M,K,X,G,H,I) specifies explicit and implicit constraint limits (user supplied), order explicit constraints first.

3) Description of Parameters:

N	Number of explicit independent variables - Define in main program
M	Number of sets of constraints - Define in main program
K	Number of points in the complex - Define in main program
ITMAX	Maximum number of iterations - Define in main program
IC	Number of implicit variables - Define in main program

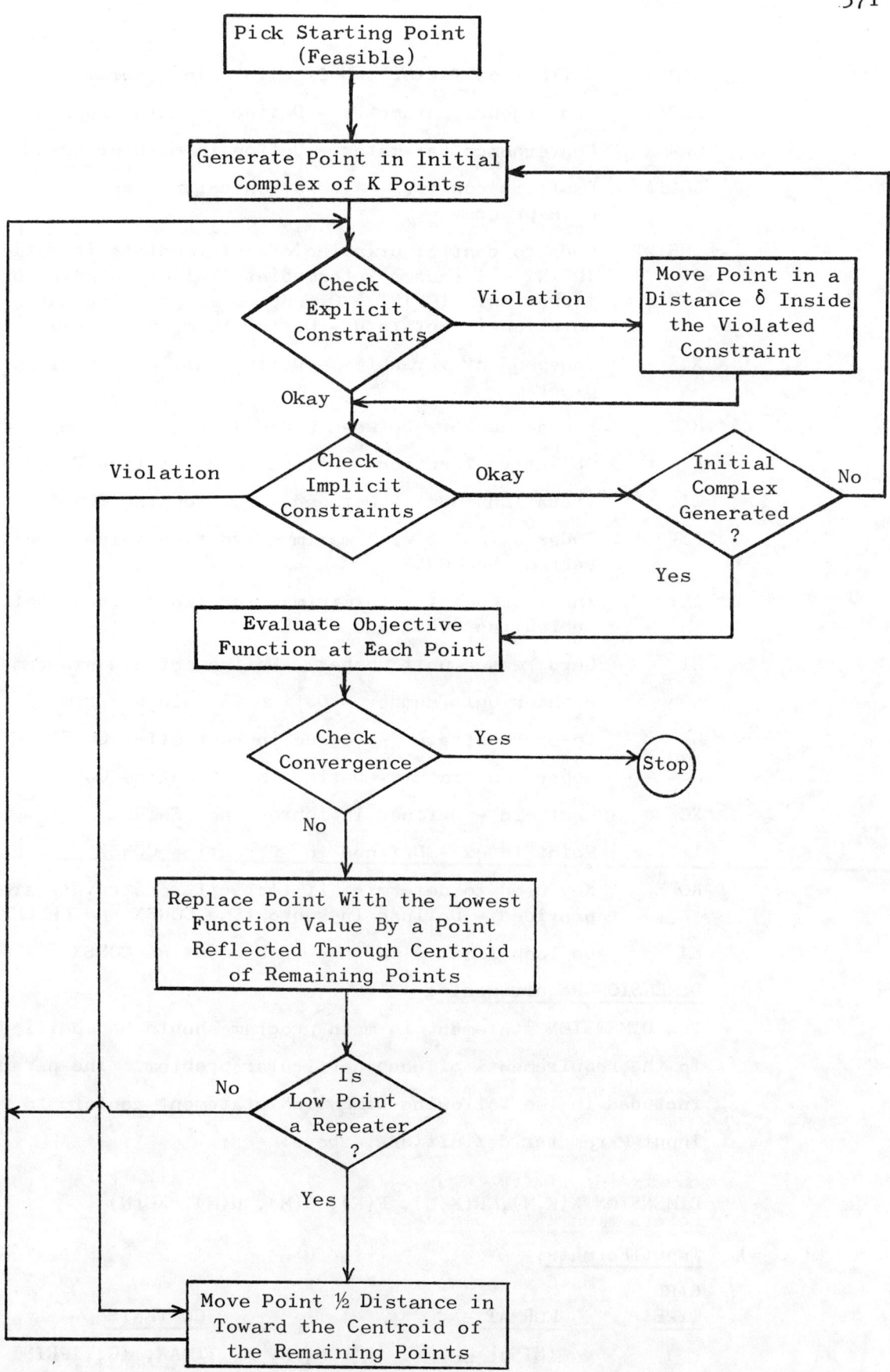

Figure 10.I. Box (COMPLEX ALGORITHM) Logic Diagram

ALPHA	Reflection factor - Define in main program
BETA	Convergence parameter - Define in main program
GAMMA	Convergence parameter - Define in main program
DELTA	Explicit constraint violation correction - Define in main program
IPRINT	Code to control printing of intermediate iterations. IPRINT = 1 causes intermediate values to print on each iteration. IPRINT = 0 suppresses printing until final solution is obtained - Define in main program
X	Independent variables - Define initial values in main program
R	Random numbers between 0 and 1 - Define in main program
F	Objective function - Define in subroutine FUNC
IT	Iteration index - Defined in subroutine CONSX
IEV2	Index of point with maximum function value - Defined in subroutine CONSX
IEV1	Index of point with minimum function value - Defined in subroutine CONSX and CHECK
NI	Card reader unit number - Define in main program
NO	Printer unit number - Define in main program
G	Lower constraint - Define in subroutine CONST
H	Upper constraint - Define in subroutine CONST
XC	Centroid - Defined in subroutine CENTR
I	Point index - Defined in subroutine CONSX
KODE	Key used to determine if implicit constraints are provided - Defined in subroutine CONSX and CHECK
K1	Do loop limit - Defined in subroutine CONSX

4) <u>DIMENSION Requirements:</u>

The DIMENSION statement in main program should be modified according to the requirements of each particular problem. The parameters included in the following DIMENSION statement conform to the Input Parameter definitions, above:

DIMENSION X(K,M), R(K,N), F(K), G(M), H(M), XC(N)

5) <u>Input Formats:</u>

CARD TYPE	FORMAT	CONTENTS
1	(8I5)	N, M, K, ITMAX, IC, IPRINT
2	(2E10.4,I5)	ALPHA, BETA, GAMMA

3	(8E10.4)	(X(1,J), J=1,N)

(If N exceeds 8, an additional CARD TYPE 3 is required.)

4	(16F5.4)	(R(II,JJ), JJ = 1,N)

(A total of (K-1) cards of CARD TYPE 4 are required.)

6) Output:

The main program first prints out values of all variables described in previous section, Input Formats.

Subroutine CONSX provides intermediate output on each iteration, provided the user specifies IPRINT = 1 on Card Type 1. If IPRINT = 0, only the final solution is printed.

When the solution has converged to within the allowable range, or when the maximum number of iterations has been executed, the main program prints the final value of the function, the X vector, and the total number of iterations.

7) Summary of User Requirements:

a) Determine values for N, M, K, ITMAX, IC, ALPHA, BETA, GAMMA, DELTA, IPRINT. Guidelines for specifying the parameters are as follows:

$K = N+1$

$ALPHA = 1.3$

$BETA$ = Some small number, say magnitude of function times 10^{-4}

$GAMMA = 5$

$DELTA$ = Some small number, say magnitude of X vector times 10^{-4}

b) Determine initial estimates for optimum values of independent variables: enter as (X(1,J), J=1,N). Initial point must satisfy explicit and implicit constraints.

c) Obtain random numbers between 0 and 1: enter as ((R(II,JJ), JJ = 1,N; II = 2,K).

d) Adjust DIMENSION and FORMAT statements as necessary.

e) Specify objective function by writing SUBROUTINE FUNC.

f) Define H (upper constraints) and G (lower constraints) in SUBROUTINE CONST. Explicit constraints <u>must</u> precede implicit constraints.

g) Specify NI and NO.

D. <u>Test Problem</u>

The following test problem was formulated by Rosenbrock (41). The calculations were performed on a CDC 6400 computer.

Function: $F = X_1 X_2 X_3$

Constraints: $0 \leq X_i \leq 42 \qquad i = 1, 2, 3$

$0 \leq (X_4 = X_1 + 2X_2 + 2X_3) \leq 72$

Starting Point: $X_1 = 1.0, \; X_2 = 1.0, \; X_3 = 1.0$

Parameters: K = 6, ALPHA = 1.3, BETA = .1, GAMMA = 5, DELTA = .0001

Algorithm Answers: F = 3455.986
X_1 = 23.958
X_2 = 12.007
X_3 = 12.014

Number of Iterations: 90

Central Processor Time: 7 seconds

The listing and output for this problem are contained in the following section.

```
C
C       MAIN LINE PROGRAM FOR COMPLEX ALGORITHM OF BOX
C
        DIMENSION X(6,4), R(6,3), F(6), G(4), H(4), XC(3)
        INTEGER GAMMA
C

        NI = 50
        NO = 66
C
        READ (NI,001) N, M, K, ITMAX, IC, IPRINT
    001 FORMAT (8I5)
        READ (NI,002) ALPHA, BETA, GAMMA
    002 FORMAT (2E10.4,I5)
        DELTA = 0.0001
        READ (NI,004) (X(1,J), J=1,N)
    004 FORMAT (8E10.4)
        DO 100 II=2,K
        READ (NI,003) (R(II,JJ), JJ=1,N)
    003 FORMAT (16F5.4)
    100 CONTINUE
C
        WRITE (NO,010)
    010 FORMAT (1H1,//,18X,24HCOMPLEX PROCEDURE OF BOX)
        WRITE (NO,018)
    018 FORMAT (//,2X,10HPARAMETERS )
        WRITE (NO,011) N, M, K, ITMAX, IC, ALPHA, BETA, GAMMA, DELTA
    011 FORMAT (//,2X,4HN = ,I2,3X,4HM = ,I2,3X,4HK = ,I2,2X,8HITMAX = ,
       1I4,2X,5HIC = ,I2,//,2X,8HALPHA = ,F5.2,5X,7HBETA = ,F10.5,3X,
       28HGAMMA = ,I2,3X,8HDELTA = ,F6.5)
        IF (IPRINT) 40, 50, 40
     40 WRITE (NO,012)
    012 FORMAT (//,2X,14HRANDOM NUMBERS)
        DO 200 J=2,K
        WRITE (NO,013) (J, I, R(J,I), I=1,N)
    013 FORMAT (/,3(2X,2HR(,I2,1H,,I2,4H) = ,F6.4,2X))
    200 CONTINUE
C
     50 CALL CONSX (N,M,K,ITMAX,ALPHA,BETA,GAMMA,DELTA,X,R,F,IT,
       1IEV2,NO,G,H,XC,IPRINT)
C
        IF (IT-ITMAX) 20,20,30
     20 WRITE (NO,014) F(IEV2)
    014 FORMAT (///,2X,30HFINAL VALUE OF THE FUNCTION = ,E20.8)
        WRITE (NO,015)
    015 FORMAT (//,2X,14HFINAL X VALUES)
        DO 300 J=1,N
        WRITE (NO,016) J, X(IEV2,J)
    016 FORMAT (/,2X,2HX(,I2,4H) = ,E20.8)
    300 CONTINUE
        GO TO 999
```

```
C
   30 WRITE (NO,017) ITMAX
  017 FORMAT (///,2X,38HTHE NUMBER OF ITERATIONS HAS EXCEEDED ,I4,10X,
     118HPROGRAM TERMINATED)
  999 STOP
      END
      SUBROUTINE CONSX (N,M,K,ITMAX,ALPHA,BETA,GAMMA,DELTA,X,R,F,
     1IT,IEV2,NO,G,H,XC,IPRINT)
C     COORDINATES SPECIAL PURPOSE SUBROUTINES
C
C
C     ARGUMENT LIST
C
C
C     IT    = ITERATION INDEX.
C     IEV1  = INDEX OF POINT WITH MINIMUM FUNCTION VALUE.
C     IEV2  = INDEX OF POINT WITH MAXIMUM FUNCTION VALUE.
C     I     = POINT INDEX.
C     KODE  = CONTROL KEY USED TO DETERMINE IF IMPLICIT CONSTRAINTS
C               ARE PROVIDED.
C     K1    = DO LOOP LIMIT
C
C
C     ALL OTHERS PREVIOUSLY DEFINED IN MAIN LINE.
C
      DIMENSION X(K,M), R(K,N), F(K), G(M), H(M), XC(N)
      INTEGER GAMMA
C
      IT = 1
      KODE = 0
      IF (M-N) 20,20,10
   10 KODE = 1
   20 CONTINUE
      DO 40 II=2,K
      DO 30 J=1,N
   30 X(II,J) = 0.0
   40 CONTINUE
C
C     CALCULATE COMPLEX POINTS AND CHECK AGAINST CONSTRAINTS
C
C
      DO 65 II=2,K
      DO 50 J=1,N
      I = II
      CALL CONST (N,M,K,X,G,H,I)
      X(II,J) = G(J) + R(II,J)*(H(J)-G(J))
   50 CONTINUE
      K1 = II
      CALL CHECK (N,M,K,X,G,H,I,KODE,XC,DELTA,K1)
      IF (II-2) 51, 51, 55
   51 IF (IPRINT) 52, 65, 52
   52 WRITE (NO,018)
  018 FORMAT (//,2X,30HCOORDINATES OF INITIAL COMPLEX)
      IO = 1
      WRITE (NO,019) (IO, J, X(IO,J), J=1,N)
  019 FORMAT (/,3(2X,2HX(,I2,1H,,I2,4H) = ,1PE13.6))
   55 IF (IPRINT) 56, 65, 56
   56 WRITE (NO,019) (II, J, X(II,J), J=1,N)
   65 CONTINUE
      K1 = K
      DO 70 I=1,K
      CALL FUNC (N,M,K,X,F,I)
```

```
      70 CONTINUE
         KOUNT = 1
         IA = 0
C
C        FIND POINT WITH LOWEST FUNCTION VALUE
C
         IF (IPRINT) 72, 80, 72
      72 WRITE (NO,021)
     021 FORMAT (/,2X,22HVALUES OF THE FUNCTION )
         WRITE (NO,022) (J, F(J), J=1,K)
     022 FORMAT (/,3(2X,2HF(,I2,4H) = ,1PE13.6))
      80 IEV1 = 1
         DO 100 ICM=2,K
         IF (F(IEV1)-F(ICM)) 100,100,90
      90 IEV1 = ICM
     100 CONTINUE
C
C        FIND POINT WITH HIGHEST FUNCTION VALUE
C
         IEV2 = 1
         DO 120 ICM=2,K
         IF (F(IEV2)-F(ICM)) 110,110,120
     110 IEV2 = ICM
     120 CONTINUE
C
C        CHECK CONVERGENCE CRITERIA
C
         IF (F(IEV2)-(F(IEV1)+BETA)) 140,130,130
     130 KOUNT = 1
         GO TO 150
     140 KOUNT = KOUNT + 1
         IF (KOUNT-GAMMA) 150,240,240
C
C        REPLACE POINT WITH LOWEST FUNCTION VALUE
C
     150 CALL CENTR (N,M,K,IEV1,I,XC,X,K1)
         DO 160  JJ=1,N
     160 X(IEV1,JJ) = (1.0+ALPHA)*(XC(JJ))-ALPHA*(X(IEV1,JJ))
         I = IEV1
         CALL CHECK (N,M,K,X,G,H,I,KODE,XC,DELTA,K1)
         CALL FUNC (N,M,K,X,F,I)
C
C        REPLACE NEW POINT IF IT REPEATS AS LOWEST FUNCTION VALUE
C
     170 IEV2 = 1
         DO 190 ICM=2,K
         IF (F(IEV2)-F(ICM)) 190,190,180
     180 IEV2 = ICM
     190 CONTINUE
         IF (IEV2-IEV1) 220,200,220
     200 DO 210 JJ=1,N
         X(IEV1,JJ)=(X(IEV1,JJ) + XC(JJ))/2.0
     210 CONTINUE
         I = IEV1
         CALL CHECK   (N,M,K,X,G,H,I,KODE,XC,DELTA,K1)
         CALL FUNC (N,M,K,X,F,I)
```

```
      GO TO 170
  220 CONTINUE

      IF (IPRINT) 230, 228, 230
  230 WRITE (NO,023) IT
  023 FORMAT (//,2X,17HITERATION NUMBER ,I5)
      WRITE (NO,024)
  024 FORMAT (/,2X,30HCOORDINATES OF CORRECTED POINT)
      WRITE (NO,019) (IEV1, JC, X(IEV1,JC), JC=1,N)
      WRITE (NO,021)
      WRITE (NO,022) (I, F(I), I=1,K)
      WRITE (NO,025)
  025 FORMAT (/,2X,27HCOORDINATES OF THE CENTROID)
      WRITE (NO,026) (JC, XC(JC), JC=1,N)
  026 FORMAT (/,3(2X,2HX(,I2,6H,C) = ,1PE14.6,4X))
  228 IT = IT + 1
      IF (IT-ITMAX) 80,80,240
  240 RETURN
      END




      SUBROUTINE CHECK (N,M,K,X,G,H,I,KODE,XC,DELTA,K1)
C
C     ARGUMENT LIST
C
C     ALL ARGUMENTS DEFINED IN MAIN LINE AND CONSX
C
      DIMENSION X(K,M), G(M), H(M), XC(N)
C
   10 KT = 0
      CALL CONST (N,M,K,X,G,H,I)
C
C     CHECK AGAINST EXPLICIT CONSTRAINTS
C
      DO 50 J=1,N
      IF (X(I,J)-G(J)) 20,20,30
   20 X(I,J) = G(J) + DELTA
      GO TO 50
   30 IF (H(J)-X(I,J)) 40,40,50
   40 X(I,J) = H(J) - DELTA
   50 CONTINUE
C
      IF (KODE) 110,110,60
C
C     CHECK AGAINST THE IMPLICIT CONSTRAINTS
C
   60 NN = N + 1
      DO 100 J=NN,M
      CALL CONST (N,M,K,X,G,H,I)
      IF (X(I,J)-G(J)) 80,70,70
   70 IF (H(J)-X(I,J)) 80,100,100
   80 IEV1 = I
      KT = 1
```

```
      CALL CENTR (N,M,K,IEV1,I,XC,X,K1)
      DO 90 JJ=1,N
      X(I,JJ) = (X(I,JJ) + XC(JJ))/2.0
   90 CONTINUE
  100 CONTINUE
      IF (KT) 110, 110, 10
  110 RETURN
      END




      SUBROUTINE CENTR (N,M,K,IEV1,I,XC,X,K1)
C
      DIMENSION X(K,M), XC(N)
C
      DO 20  J=1,N
      XC(J) = 0.0
      DO 10   IL=1,K1
   10 XC(J) = XC(J) + X(IL,J)
      RK = K1
   20 XC(J) = (XC(J)-X(IEV1,J))/(RK-1.0)
      RETURN
      END




      SUBROUTINE FUNC (N,M,K,X,F,I)
C
      DIMENSION X(K,M), F(K)
C
C     POST OFFICE TEST PROBLEM
C
C
      F(I) = X(I,1) * X(I,2) * X(I,3)
C
      RETURN
      END
```

```
      SUBROUTINE CONST (N,M,K,X,G,H,I)
C
      DIMENSION X(K,M), G(M), H(M)
C
C     POST OFFICE TEST PROBLEM
C
      G(1) = 0.0
      H(1) = 42.0
      G(2) = 0.0
      H(2) = 42.0
      G(3) = 0.0
      H(3) = 42.0
      G(4) = 0.0
      H(4) = 72.0
      X(I,4) = X(I,1) + 2.0*X(I,2) + 2.0*X(I,3)
C
      RETURN
      END
```

COMPLEX PROCEDURE OF BOX

PARAMETERS

N = 3   M = 4   K = 6   ITMAX = 500   IC = 1

ALPHA = 1.30   BETA = .10000   GAMMA = 5   DELTA = .00010

RANDOM NUMBERS

R(2, 1) = .3987   R(2, 2) = .5165   R(2, 3) = .8829

R(3, 1) = .9306   R(3, 2) = .9059   R(3, 3) = .4409

R(4, 1) = .7880   R(4, 2) = .9408   R(4, 3) = .1780

R(5, 1) = .7943   R(5, 2) = .0091   R(5, 3) = .4468

R(6, 1) = .4037   R(6, 2) = .8764   R(6, 3) = .7506

COORDINATES OF INITIAL COMPLEX

X(1, 1) = 1.000000E+00   X(1, 2) = 1.000000E+00   X(1, 3) = 1.000000E+00

X(2, 1) = 8.872700E+00   X(2, 2) = 1.134650E+01   X(2, 3) = 1.904090E+01

X(3, 1) = 1.347356E+01   X(3, 2) = 1.414189E+01   X(3, 3) = 1.214479E+01

X(4, 1) = 1.411057E+01   X(4, 2) = 1.650050E+01   X(4, 3) = 9.915422E+00

X(5, 1) = 3.336060E+01   X(5, 2) = 3.822000E-01   X(5, 3) = 1.876560E+01

X(6, 1) = 1.451247E+01   X(6, 2) = 1.219104E+01   X(6, 3) = 1.459232E+01

VALUES OF THE FUNCTION

F(1) = 1.000000E+00   F(2) = 1.916925E+03   F(3) = 2.314087E+03
F(4) = 2.308621E+03   F(5) = 2.392693E+02   F(6) = 2.581705E+03

ITERATION NUMBER   1

COORDINATES OF CORRECTED POINT

X(1, 1) = 1.751054E+01   X(1, 2) = 1.131512E+01   X(1, 3) = 1.545616E+01

VALUES OF THE FUNCTION

F(1) = 3.062387E+03   F(2) = 1.916925E+03   F(3) = 2.314087E+03
F(4) = 2.308621E+03   F(5) = 2.392693E+02   F(6) = 2.581705E+03

COORDINATES OF THE CENTROID

X(1,C) = 1.686598E+01   X(2,C) = 1.091242E+01   X(3,C) = 1.489181E+01

(Every 10th iteration is included.)

ITERATION NUMBER   10

COORDINATES OF CORRECTED POINT

X(4, 1) = 2.167078E+01   X(4, 2) = 9.151034E+00   X(4, 3) = 1.598269E+01

VALUES OF THE FUNCTION

F(1) = 3.062387E+03   F(2) = 2.644029E+03   F(3) = 2.974589E+03
F(4) = 3.169529E+03   F(5) = 2.931249E+03   F(6) = 2.741044E+03

COORDINATES OF THE CENTROID

X(1,C) = 2.127606E+01   X(2,C) = 8.653194E+00   X(3,C) = 1.598193E+01

ITERATION NUMBER   20

COORDINATES OF CORRECTED POINT

X(6, 1) = 1.922043E+01   X(6, 2) = 1.342782E+01   X(6, 3) = 1.279129E+01

VALUES OF THE FUNCTION

F(1) = 3.284752E+03   F(2) = 3.266554E+03   F(3) = 3.373840E+03
F(4) = 3.293414E+03   F(5) = 3.307825E+03   F(6) = 3.301286E+03

COORDINATES OF THE CENTROID

X(1,C) = 2.044508E+01   X(2,C) = 1.137108E+01   X(3,C) = 1.431966E+01

ITERATION NUMBER   30

COORDINATES OF CORRECTED POINT

X(6, 1) = 2.400270E+01   X(6, 2) = 1.064443E+01   X(6, 3) = 1.320940E+01

VALUES OF THE FUNCTION

F(1) = 3.387708E+03   F(2) = 3.380427E+03   F(3) = 3.373840E+03
F(4) = 3.377641E+03   F(5) = 3.386615E+03   F(6) = 3.374937E+03

COORDINATES OF THE CENTROID

X(1,C) = 2.395942E+01   X(2,C) = 1.112380E+01   X(3,C) = 1.272432E+01

ITERATION NUMBER   40

COORDINATES OF CORRECTED POINT

X(4, 1) =   2.397161E+01   X(4, 2) =   1.086374E+01   X(4, 3) =   1.313392E+01

VALUES OF THE FUNCTION

F(1) =   3.432884E+03   F(2) =   3.408083E+03   F(3) =   3.412829E+03
F(4) =   3.420352E+03   F(5) =   3.417478E+03   F(6) =   3.418797E+03

COORDINATES OF THE CENTROID

X(1,C) =   2.413074E+01   X(2,C) =   1.102158E+01   X(3,C) =   1.286221E+01

ITERATION NUMBER   50

COORDINATES OF CORRECTED POINT

X(1, 1) =   2.328930E+01   X(1, 2) =   1.164534E+01   X(1, 3) =   1.269517E+01

VALUES OF THE FUNCTION

F(1) =   3.443081E+03   F(2) =   3.437832E+03   F(3) =   3.436757E+03
F(4) =   3.439093E+03   F(5) =   3.445292E+03   F(6) =   3.442985E+03

COORDINATES OF THE CENTROID

X(1,C) =   2.337682E+01   X(2,C) =   1.196570E+01   X(3,C) =   1.231243E+01

ITERATION NUMBER   60

COORDINATES OF CORRECTED POINT

X(4, 1) =   2.453972E+01   X(4, 2) =   1.157250E+01   X(4, 3) =   1.214872E+01

VALUES OF THE FUNCTION

F(1) =   3.448947E+03   F(2) =   3.451475E+03   F(3) =   3.452101E+03
F(4) =   3.450066E+03   F(5) =   3.451915E+03   F(6) =   3.452102E+03

COORDINATES OF THE CENTROID

X(1,C) =   2.372175E+01   X(2,C) =   1.172522E+01   X(3,C) =   1.240903E+01

```
ITERATION NUMBER    70

COORDINATES OF CORRECTED POINT

X( 5, 1) =  2.427262E+01   X( 5, 2) =  1.197253E+01   X( 5, 3) =  1.188740E+01

VALUES OF THE FUNCTION

F( 1) =  3.455329E+03   F( 2) =  3.453899E+03   F( 3) =  3.454482E+03
F( 4) =  3.454371E+03   F( 5) =  3.454534E+03   F( 6) =  3.454816E+03

COORDINATES OF THE CENTROID

X( 1,C) =    2.407121E+01       X( 2,C) =    1.189983E+01       X( 3,C) =    1.206190E+01




ITERATION NUMBER    80

COORDINATES OF CORRECTED POINT

X( 6, 1) =  2.420284E+01   X( 6, 2) =  1.198064E+01   X( 6, 3) =  1.191776E+01

VALUES OF THE FUNCTION

F( 1) =  3.455763E+03   F( 2) =  3.455858E+03   F( 3) =  3.455550E+03
F( 4) =  3.455914E+03   F( 5) =  3.455810E+03   F( 6) =  3.455739E+03

COORDINATES OF THE CENTROID

X( 1,C) =    2.409085E+01       X( 2,C) =    1.198448E+01       X( 3,C) =    1.196985E+01
```

ITERATION NUMBER   90

COORDINATES OF CORRECTED POINT

X(5, 1) =   2.395813E+01   X(5, 2) =   1.200706E+01   X(5, 3) =   1.201386E+01

VALUES OF THE FUNCTION

F(1) =   3.455979E+03   F(2) =   3.455957E+03   F(3) =   3.455981E+03
F(4) =   3.455982E+03   F(5) =   3.455986E+03   F(6) =   3.455969E+03

COORDINATES OF THE CENTROID

X(1,C) =   2.401633E+01   X(2,C) =   1.199316E+01   X(3,C) =   1.199865E+01

FINAL VALUE OF THE FUNCTION =   .34559857E+04

FINAL X VALUES

X(1) =   .23958128E+02

X(2) =   .12007056E+02

X(3) =   .12013858E+02

II. CONSTRAINED ROSENBROCK (HILL ALGORITHM)*

A. Purpose

This program finds the maximum or minimum of a multivariable, nonlinear function subject to nonlinear inequality constraints:

Optimize $\quad F(X_1, X_2, \ldots, X_N)$

Subject to $\quad G_k \leq X_k \leq H_k, \quad k = 1, 2, \ldots, M$

The implicit variables $X_{N+1}, \ldots, X_M$ are dependent functions of the explicit independent variables, $X_1, X_2, \ldots, X_N$. The upper and lower constraints H_k and G_k are either constants or functions of the independent variables.

B. Method

The procedure is based on the "automatic" method proposed by H. H. Rosenbrock (41, 42). This method is a sequential search technique which has proven effective in solving some problems where the variables are constrained. The algorithm proceeds per the unconstrained Rosenbrock procedure (see Section III, Chapter 9) until convergence is reached or a boundary zone in the vicinity of the constraints is entered. The boundary zones are defined as follows:

Lower Zone: $\quad G_k \leq X_k \leq (G_k + (H_k - G_k) \cdot 10^{-4})$

Upper Zone: $\quad H_k \geq X_k \geq (H_k - (H_k - G_k) \cdot 10^{-4})$

$$k = 1, 2, \ldots, M.$$

The procedure requires a starting point that satisfies the constraints and does not lie in the boundary zones. The search computations are then the same as those for the unconstrained case except that after each function evaluation, the following steps are carried out:

1) Define by F^o the current best objective function value for a point where the constraints are satisfied, and F^* the current best objective function value for a point where the constraints are

---

*Computer code developed by C. B. Yancey and R. C. Spear, Naval Weapons Center, China Lake, California. Used by permission.

satisfied and in addition the boundary zones are not violated. F^o and F^* are initially set equal to the objective function value at the starting point.

2) If the current point objective function evaluation, F, is worse than F^o or if the constraints are violated, the trial is a failure and the unconstrained procedure is continued.

3) If the current point lies within a boundary zone, the objective function is modified as follows:

$$F(new) = F(old) - (F(old) - F^*)(3\lambda - 4\lambda^2 + 2\lambda^3)$$

where

$$\lambda = \frac{\text{distance into boundary zone}}{\text{width of boundary zone}}$$

$$= \frac{G_k + (H_k - G_k) \cdot 10^{-4} - X_k}{(H_k - G_k) \cdot 10^{-4}} \quad \text{(lower zone)}$$

$$= \frac{X_k - (H_k - (H_k - G_k) \cdot 10^{-4})}{(H_k - G_k) \cdot 10^{-4}} \quad \text{(upper zone)}$$

At the inner edge of the boundary zone, $\lambda = 0$, i.e., the function is unaltered ($F(new) = F(old)$). At the constraint, $\lambda = 1$, and thus $F(new) = F^*$. Thus the function value is replaced by the best current function value in the feasible region and not in a boundary zone. For a function which improves as the constraint is approached, the modified function has an optimum in the boundary zone.

4) If an improvement in the objective function has been obtained without violating the boundary zones or constraints, F^* is set equal to F^o and the procedure continued.

5) The search procedure is terminated when the convergence criteria is satisfied (see <u>Description of Parameters</u>).

A flow sheet showing the modifications in the Rosenbrock procedure to handle constraints is given in Figure 10.II.

C. <u>Program Description</u>

1) <u>Usage</u>:

The program consists of a main program and four user supplied function subprograms (Function F, Function CX, Function CG,

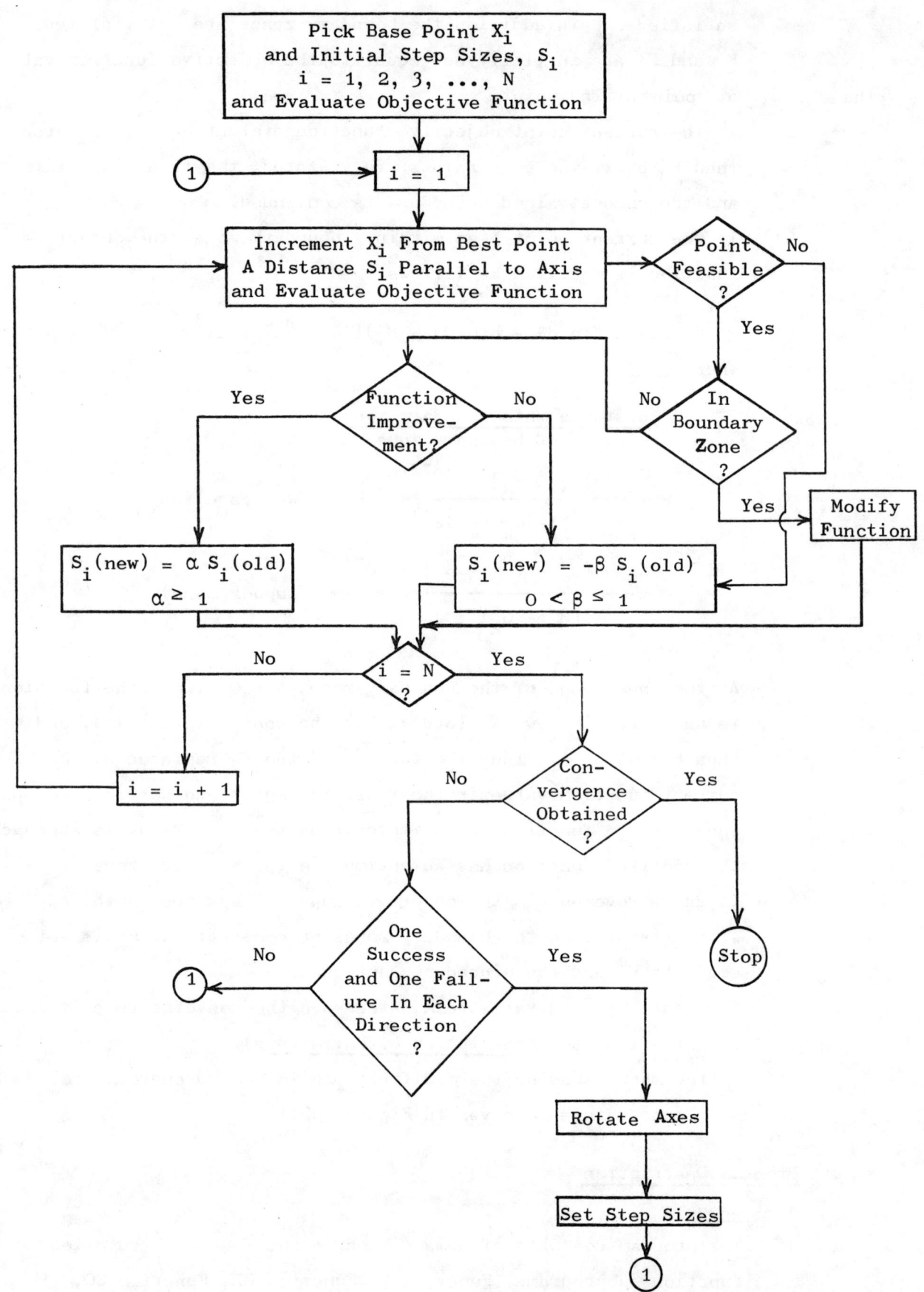

Figure 10.II. Constrained Rosenbrock (HILL ALGORITHM) Logic Diagram

Function CH). All input and output is executed in the main program. Format changes may be required depending on the problem under consideration.

2) <u>Subprograms Required</u>:

FUNCTION F (X,DA,N,NPAR) specifies objective function (user supplied).

FUNCTION CX (X,DA,N,NPAR,K) specifies function to be constrained (user supplied).

FUNCTION CG (X,DA,N,NPAR,K) specifies lower bound of constraints (user supplied).

FUNCTION CH (X,DA,N,NPAR,K) specifies upper bound of constraints (user supplied).

3) Description of Parameters:

M	Problem controller; +1 for maximization, -1 for minimization
P	Number of variables
K	Point index
L	Number of variables plus implicit constraints
LOOPY	Maximum number of stages to be calculated
PR	Printing controller (number of stages between outputs)
ND	Storage controller; 1 for storage in DA, 0 for no storage
DA	General storage vector
NDATA	Number of data points to be read into DA
NPAR	Dimension limit in subroutines
E	Vector of initial step sizes
X	Vector of initial guesses for independent variables
N	Dimension limit in subroutines
NSTEP	Control on step sizes for each rotation; 0 for original step sizes, 1 for step sizes from previous rotation
DELY	Error in objective function (difference between current value and previous stage value)
NI	Card Reader unit number - Define in main program
NO	Printer unit number - Define in main program
V	Direction vector

4) <u>DIMENSION Requirements</u>:

The DIMENSION statement in the main program should be modified according to the requirements of each particular problem. The parameters included in the following DIMENSION statement conform to the Input Parameter definitions above:

DIMENSION X(P), E(P), V(P,P) SA(P), D(P), G(L), H(L), AL(L), PH(L), A(L,L), B(L,L), BX(L), DA(NDATA), VV(L,L), EINT(P), VM(L)

5) Input Formats:

CARD TYPE	FORMAT	CONTENTS
1	(8I5)	M,P,L,LOOPY,PR,ND,NDATA,NSTEP
2	(1E10.4)	(X(K), K = 1,P) (one card for each X value)
3	(1E10.4)	(E(J), J = 1,P) (one card for each E value)

6) Output:

The main program furnishes all printouts of STAGE, PROGRESS, LATERAL PROGRESS, function value and variable values. PROGRESS represents the total progress in the direction of maximum change in the objective function (equals $A_{i,j}^{(k)}$ - see unconstrained Rosenbrock procedure in Chapter 9). LATERAL PROGRESS denotes the change in the second most sensitive direction. When the problem has converged or the maximum number of iterations has been reached, the final values of the independent variables and objective function are printed out, plus the final step sizes, direction vector components and number of function evaluations.

7) Summary of User Requirements:

a) Determine values for M,P,L,LOOPY,PR,ND,NDATA,NSTEP. If PR is set equal to LOOPY no intermediate printout will be given.

b) Determine the initial estimates of X(K), K = 1, ..., P. Initial point must satisfy both the explicit and implicit constraints and not lie within any boundary zone.

c) Determine the initial step sizes to be used E(J), J = 1, ..., P.

d) Adjust DIMENSION and FORMAT statements as required.

e) Specify the objective function F, implicit variables, lower bounds CG, and upper bounds CH in the function subroutines.

f) Specify NI and NO.

D. Test Problem

The following test problem was taken from the literature (13). The caluclations were performed on a CDC 6400 computer.

Function:

Maximize $\quad F = 3803.84 + 138.08X_1 + 232.92X_2 - 123.08X_1^2$
$\quad\quad\quad\quad - 203.64X_2^2 - 182.25X_1X_2$

Constraints: $\quad 0.0 \leq X_1 \leq 2.0$

$\quad\quad\quad\quad\quad 0.0 \leq X_2 \leq 2.5$

Starting Point: $\quad X_1 = 1.0, \quad X_2 = 0.5$

Parameters: $\quad E(1) = E(2) = 0.1, M = +1, P = 2, L = 2,$
$\quad\quad\quad\quad LOOPY = 100, PR = 1, ND = 0, NDATA = 0, NSTEP = 0$

Algorithm Answers: $\quad F = 3873.9$
$\quad\quad\quad\quad X_1 = 0.20566$
$\quad\quad\quad\quad X_2 = 0.47986$

Number of Stages: 8
Number of Function Evaluations: 212
Central Processor Time: 4 seconds

A listing of the program and test problem plus output is contained in the following section.

E. Program Listings and Example Output

```
C
C       MAIN LINE PROGRAM FOR ROSENBROCK HILLCLIMB
C
        DIMENSION X(2), E(2), V(2,2), SA(2), D(2), G(2), H(2), AL(2),
       1 PH(2), A(2,2), B(2,2), BX(2), DA(1), VV(2,2), EINT(2), VM(2)
        COMMON KOUNT
        INTEGER P
        INTEGER PR
        INTEGER R
        INTEGER C
        REAL LC
C
        NI = 50
        NO = 66
C
        READ (NI,001) M, P, L, LOOPY, PR, ND, NDATA, NSTEP
    001 FORMAT (8I5)
     10 DO 100  K=1,P
        READ (NI,002) X(K)
    002 FORMAT (1E10.4)
    100 CONTINUE
        DO 200  J=1,P
        READ (NI,002) E(J)
    200 CONTINUE
        WRITE (NO,013)
    013 FORMAT (1H1,10X,30HROSENBROCK HILLCLIMB PROCEDURE)
C
        IF (ND-1) 30, 20, 30
     20 DO 300 KA=1,NDATA
        READ (NI,002) DA(KA)
    300 CONTINUE
C
     30 LAP=PR-1
        LOOP=0
        ISW=0
        INIT=0
        KOUNT = 0
        TERM = 0.0
        DELY = 1.0E-10
        F1 = 0.0
        NPAR = NDATA
        N=L
        DO 40   K=1,L
     40 AL(K) = (CH(X,DA,N,NPAR,K)-CG(X,DA,N,NPAR,K))*.0001
        DO 60   I=1,P
        DO 60   J=1,P
        V(I,J) = 0.0
        IF (I-J) 60,61,60
     61 V(I,J) = 1.0
     60 CONTINUE
        DO 65 KK = 1,P
        EINT(KK) = E(KK)
     65 CONTINUE
```

```
C
C
 1000 DO 70   J=1,P
      IF (NSTEP.EQ.0)   E(J) = EINT(J)
      SA(J) = 2.0
   70 D(J) = 0.0
      FBEST = F1
   80 I = 1
      IF (INIT.EQ.0) GO TO 120
   90 DO 110  K=1,P
  110 X(K) = X(K) + E(I)*V(I,K)
      DO 50   K=1,L
   50 H(K) = F0
C
C
  120 F1 = F(X,DA,N,NPAR)
      F1 = M*F1
      IF (ISW.EQ.0) F0 = F1
      ISW = 1
      IF (ABS(FBEST-F1)-DELY) 122,122,125
  122 TERM = 1.0
      GO TO 450
  125 CONTINUE
C
C
      J = 1
C
  130 XC = CX(X,DA,N,NPAR,J)
      LC = CG(X,DA,N,NPAR,J)
      UC = CH(X,DA,N,NPAR,J)
      IF (XC.LE.LC) GO TO 420
      IF (XC.GE.UC) GO TO 420
      IF (F1.LT.F0) GO TO 420
      IF (XC.LT.LC+AL(J))  GO TO 140
      IF (XC.GT.UC-AL(J))  GO TO 140
      H(J) = F0
      GO TO 210
C
C
  140 CONTINUE
C
      BW = AL(J)
C
      IF (XC.LE.LC.OR.UC.LE.XC) GO TO 150
      IF (LC.LT.XC.AND.XC.LT.LC+BW) GO TO 160
      IF (UC-BW.LT.XC.AND.XC.LT.UC) GO TO 170
      PH(J) = 1.0
      GO TO 210
C
C
  150 PH(J) = 0.0
      GO TO 190
  160 PW = (LC+BW-XC)/BW
      GO TO 180
  170 PW = (XC-UC+BW)/BW
  180 PH(J) = 1.0-3.0*PW+4.0*PW*PW-2.0+PW*PW*PW
```

```
C
  190 F1 = H(J)+(F1-H(J))*PH(J)
C
  210 CONTINUE
      IF (J.EQ.L) GO TO 220
      J = J+1
      GO TO 130
C
  220 INIT = 1
      IF (F1.LT.F0) GO TO 420
      D(I) = D(I) + E(I)
      E(I) = 3.0 * E(I)
      F0 = F1
      IF (SA(I).GE.1.5)  SA(I) = 1.0
C
  230 DO 240 JJ=1,P
      IF (SA(JJ).GE.0.5) GO TO 440
  240 CONTINUE
C
C     AXES ROTATION
C
      DO 250 R=1,P
      DO 250 C=1,P
  250 VV(C,R) = 0.0
      DO 260 R=1,P
      KR = R
      DO 260 C=1,P
      DO 265 K=KR,P
  265 VV(R,C) = D(K) * V(K,C) + VV(R,C)
  260 B(R,C) = VV(R,C)
      BMAG = 0.0
      DO 280 C=1,P
      BMAG = BMAG + B(1,C)*B(1,C)
  280 CONTINUE
      BMAG = SQRT(BMAG)
      BX(1) = BMAG
      DO 310 C=1,P
  310 V(1,C) = B(1,C)/BMAG
C
      DO 390 R=2,P
C
      IR = R-1
      DO 390 C=1,P
      SUMVM = 0.0
      DO 320 KK=1,IR
      SUMAV= 0.0
      DO 330 KJ=1,P
  330 SUMAV = SUMAV + VV(R,KJ)*V(KK,KJ)
  320 SUMVM = SUMAV*V(KK,C) + SUMVM
  390 B(R,C) = VV(R,C) - SUMVM
      DO 340 R=2,P
      BBMAG = 0.0
      DO 350 K=1,P
  350 BBMAG = BBMAG + B(R,K)*B(R,K)
      BBMAG = SQRT(BBMAG)
      DO 340 C=1,P
  340 V(R,C) = B(R,C)/BBMAG
```

```
              LOOP = LOOP+1
              LAP = LAP+1
              IF (LAP.EQ.PR) GO TO 450
              GO TO 1000
C
       420 IF (INIT.EQ.0)  GO TO  450
              DO 430 IX=1,P
       430 X(IX) = X(IX)-E(I)*V(I,IX)
              E(I) = -0.5*E(I)
              IF (SA(I).LT.1.5) SA(I)=0.0
              GO TO 230
C
       440 CONTINUE
              IF (I.EQ.P) GO TO 80
              I=I+1
              GO TO 90
C
       450 WRITE (NO,003)
       003 FORMAT (//,2X,5HSTAGE,8X,8HFUNCTION,12X,8HPROGRESS,9X,
             116HLATERAL PROGRESS )
              WRITE (NO,004) LOOP, F0, BMAG, BBMAG
       004 FORMAT (1H ,I5,3E20.8)
              WRITE (NO,014) KOUNT
       014 FORMAT (/,2X,33HNUMBER OF FUNCTION EVALUATIONS = ,I8)
              WRITE (NO,005)
       005 FORMAT (/,2X,25HVALUES OF X AT THIS STAGE )
C      PRINT CURRENT VALUES OF X
              WRITE (NO,006) (JM, X(JM), JM=1,P)
       006 FORMAT (/,2X,3(2HX(,I2,4H) = ,1PE14.6,4X))
C
              LAP = 0
              IF (INIT.EQ.0)  GO TO 470
              IF (TERM.EQ.1.0) GO TO 480
              IF (LOOP.GE.LOOPY) GO TO 480
              GO TO 1000
C
       470 WRITE (NO,007)
       007 FORMAT (///,2X,81HTHE STARTING POINT MUST NOT VIOLATE THE CONSTRAI
             1NTS.  IT APPEARS TO HAVE DONE SO.)
       480 CONTINUE
       490 WRITE (NO,008)
       008 FORMAT (///,2X,29HFINAL DIRECTION VECTOR MATRIX)
              DO 500 J=1,P
       500 WRITE (NO,009) (J, I, V(J,I), I=1,P)
       009 FORMAT (/,2X,3(2HV(,I2,1H,,I2,4H) = ,F10.8,4X))
              WRITE (NO,011)
       011 FORMAT (//,2X,16HFINAL STEP SIZES )
              WRITE (NO,012) (J, E(J), J=1,P)
       012 FORMAT (/,2X,3(2HS(,I2,4H) = ,1PE14.6,4X))
C
              END
```

```fortran
      FUNCTION F (X,DA,N,NPAR)
C
      DIMENSION X(N), DA(NPAR)
      COMMON KOUNT
C
      X1 = X(1)
      X2 = X(2)
      X12 = X1**2
      X22 = X2**2
C
      F = 3803.84 + 138.08*X1 + 232.92*X2 - 123.08*X12 - 203.64*X22
     1  - 182.25*X1*X2
C
      KOUNT = KOUNT + 1
C
      RETURN
      END




      FUNCTION CX (X,DA,N,NPAR,K)
C
      DIMENSION X(N), DA(NPAR)
C
      CX = X(K)
C
      RETURN
      END




      FUNCTION CG (X,DA,N,NPAR,K)
C
      DIMENSION X(N), DA(NPAR)
C
      CG = 0.0
C
      RETURN
      END




      FUNCTION CH (X,DA,N,NPAR,K)
C
      DIMENSION X(N), DA(NPAR)
C
      GO TO (1,2), K
C
    1 CH = 2.0
      GO TO 3
    2 CH = 2.5
C
    3 RETURN
      END
```

ROSENBROCK HILLCLIMB PROCEDURE

STAGE	FUNCTION	PROGRESS	LATERAL PROGRESS
1	.38690576E+04	.53851648E+00	.18569534E+00

NUMBER OF FUNCTION EVALUATIONS =   8

VALUES OF X AT THIS STAGE

X(1) =   4.000000E-01   X(2) =   3.000000E-01   X(

STAGE	FUNCTION	PROGRESS	LATERAL PROGRESS
2	.38737881E+04	.22360680E+00	.89442719E-01

NUMBER OF FUNCTION EVALUATIONS =   17

VALUES OF X AT THIS STAGE

X(1) =   2.328742E-01   X(2) =   4.485563E-01   X(

STAGE	FUNCTION	PROGRESS	LATERAL PROGRESS
3	.38739227E+04	.44194174E-01	.61871843E-02

NUMBER OF FUNCTION EVALUATIONS =   29

VALUES OF X AT THIS STAGE

X(1) =   2.043273E-01   X(2) =   4.822935E-01   X(

STAGE	FUNCTION	PROGRESS	LATERAL PROGRESS
4	.38739235E+04	.32211763E-02	.75792383E-03

NUMBER OF FUNCTION EVALUATIONS =   47

VALUES OF X AT THIS STAGE

X(1) =   2.057495E-01   X(2) =   4.794033E-01   X(

STAGE	FUNCTION	PROGRESS	LATERAL PROGRESS
5	.38739235E+04	.49071658E-03	.48585800E-04

NUMBER OF FUNCTION EVALUATIONS =   73

VALUES OF X AT THIS STAGE

X(1) =   2.055777E-01   X(2) =   4.798629E-01   X(

STAGE	FUNCTION	PROGRESS	LATERAL PROGRESS
6	.38739235E+04	.10918301E-03	.43673203E-04

NUMBER OF FUNCTION EVALUATIONS =   98

VALUES OF X AT THIS STAGE

X(1) =   2.056863E-01   X(2) =   4.798514E-01   X(

STAGE   FUNCTION   PROGRESS   LATERAL PROGRESS
  7   .38739235E+04   .22217132E-04   .58686764E-05

NUMBER OF FUNCTION EVALUATIONS =   136

VALUES OF X AT THIS STAGE

X(1) =   2.056657E-01   X(2) =   4.798597E-01   X(

STAGE   FUNCTION   PROGRESS   LATERAL PROGRESS
  8   .38739235E+04   .72579565E-05   .38094247E-06

NUMBER OF FUNCTION EVALUATIONS =   176

VALUES OF X AT THIS STAGE

X(1) =   2.056591E-01   X(2) =   4.798628E-01   X(

STAGE   FUNCTION   PROGRESS   LATERAL PROGRESS
  8   .38739235E+04   .72579565E-05   .38094247E-06

NUMBER OF FUNCTION EVALUATIONS =   212

VALUES OF X AT THIS STAGE

X(1) =   2.056588E-01   X(2) =   4.798621E-01   X(

FINAL DIRECTION VECTOR MATRIX

V(1, 1) = -.90605896   V(1, 2) = .42315146   V(

V(2, 1) = .42315146   V(2, 2) = .90605896   V(

FINAL STEP SIZES

S(1) =   3.814697E-07   S(2) =   -7.629395E-07   S(

III. ROSEN (PROJG ALGORITHM)*

A. <u>Purpose</u>

This program finds the maximum of a multivariable, nonlinear function subject to linear constraints:

Maximize $\quad F(X_1, X_2, \ldots, X_N)$

Subject to $\quad G_k \leq 0, \quad k = 1, 2, \ldots, M$

The constraints, G_k, are linear functions of the independent variables.

B. <u>Method</u>

The procedure is based on the gradient projection method first developed by J. B. Rosen in 1960 (40). The procedure is limited here to linear constraints. The algorithm proceeds as follows:

1) A feasible starting point and initial step size S are selected.

2) The derivatives of the objective function with respect to the independent variables, $\frac{\partial F}{\partial X_i}$, $i = 1, 2, \ldots, N$, are evaluated at this base point and the normalized direction vector components, M_i, are determined:

$$M_i = \frac{\pm \frac{\partial F}{\partial X_i}}{\sqrt{\sum_{j=1}^{N} \frac{\partial F}{\partial X_j}^2}}$$

If $\frac{\partial F}{\partial X_i} \leq$ LIMIT for all $i = 1, 2, \ldots, N$, then the procedure is stopped. The required point has been found.

3) A new point is then located as follows:

$$\text{new } X_i = \text{old } X_i + S M_i$$

$$i = 1, 2, \ldots, N$$

---

*Program code developed by Matthew A. Nichols, Arizona State University, Tempe, Arizona.

The objective function is then evaluated. The following possibilities are then considered:

a) If an improvement in the objective function occurs without violating the constraints, the step size is doubled for the next move and new direction vector components are evaluated at the improved point. The process is then continued.

b) If the function is not improved at the new point, then the step size is halved for the next move from the last successful point.

c) If an improvement in the objective function is obtained but one or more constraints are violated, a return is made to the last feasible successful point and the step size determined that places the next point on the violated constraints. New direction vector components are then determined along the constraints at the point on the constraints as follows:

$$M_i = \frac{\left[\frac{\partial F}{\partial X_i} + \sum_{k=1}^{\ell} \lambda_k \frac{\partial G_k}{\partial X_i}\right]}{\left[\sum_{j=1}^{N} \left(\frac{\partial F}{\partial X_j} + \sum_{k=1}^{\ell} \left(\lambda_k \frac{\partial G_k}{\partial X_j}\right)\right)^2\right]^{1/2}}$$

$i = 1, 2, \ldots, N$

ℓ = number of violated constraints.

λ_k, $k = 1, 2, \ldots, \ell$ is determined from the following ℓ equations:

$$\sum_{i=1}^{N} \sum_{j=1}^{\ell} \left(\lambda_j \frac{\partial G_j}{\partial X_i} \cdot \frac{\partial G_k}{\partial X_i}\right) = -\sum_{i=1}^{N} \left(\frac{\partial G_k}{\partial X_i} \frac{\partial F}{\partial X_i}\right)$$

If $\frac{\partial F}{\partial X_j} + \sum_{k=1}^{\ell} \left(\lambda_k \frac{\partial G_k}{\partial X_j}\right) \leq$ LIMIT for all $j = 1, 2, \ldots, N$, then convergence is assumed. Otherwise the procedure is resumed using the new search directions.

A flow sheet illustrating the above procedure is given in Figure 10.III.

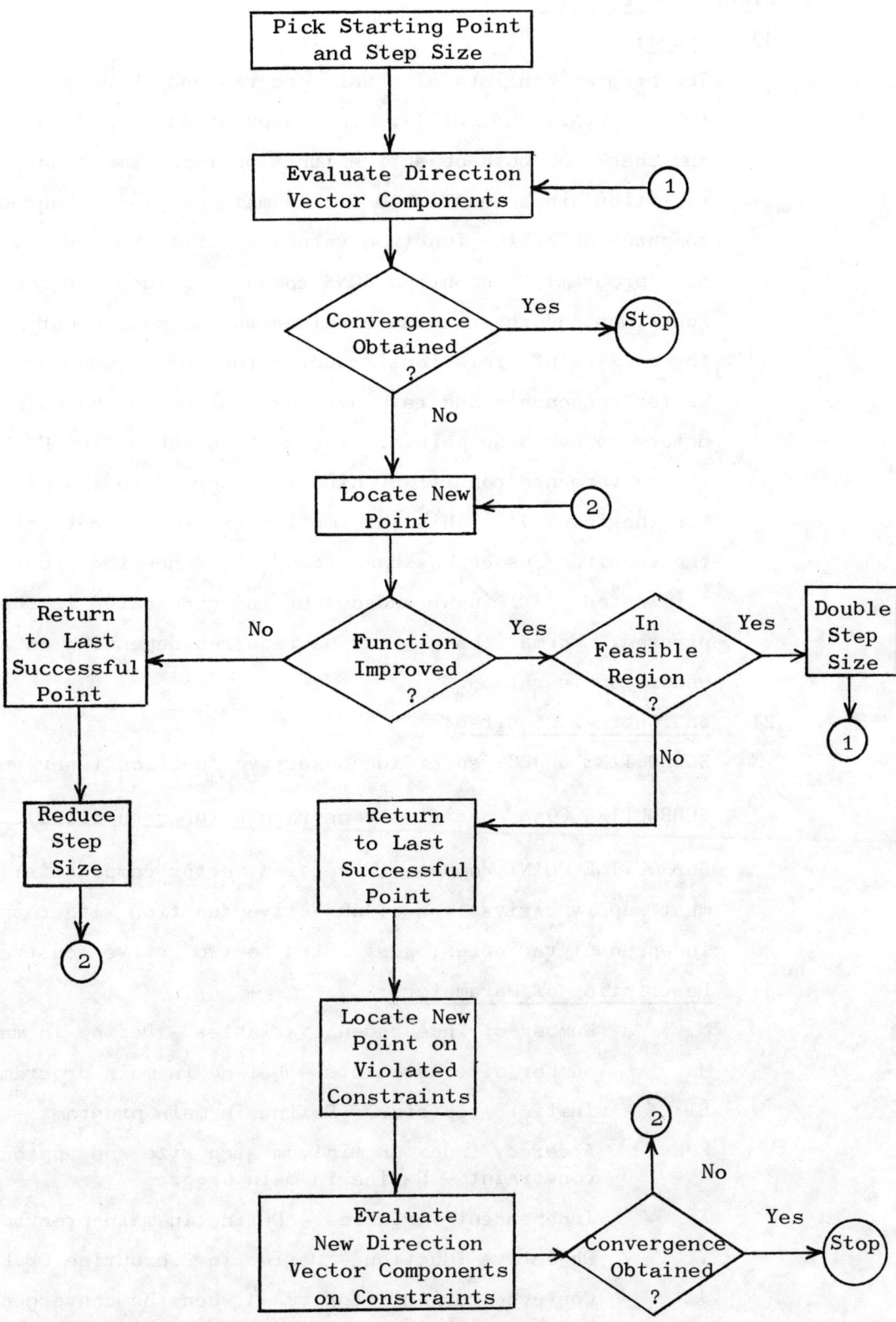

Figure 10.III. Rosen (PROJG ALGORITHM) Logic Diagram

C. Program Description

1) Usage:

The program consists of a main program and three subroutines (OBJECT, CONS, and POINT). The computation of new base points, and checks on both objective function improvement and constraint violation are accomplished in the main program. Subroutine OBJECT computes objective function values and returns these values to the main program. Subroutine CONS computes values for the constraint functions and returns these values to the main program to be checked for constraint violations. Subroutine POINT computes direction vector components and returns these values to the main program to determine new base points. In addition subroutine POINT specifies the convergence condition which is returned to the main program via the index JA. This information is used to determine when the required answer has been found, and when the program is terminated. All input and output is coordinated in the main program. Format changes may be required depending on the problem under consideration.

2) Subroutines Required:

SUBROUTINE OBJECT specifies objective function (user supplied).

SUBROUTINE CONS specifies constraints (user supplied).

SUBROUTINE POINT computes direction vector components; user must supply derivatives of objective function with respect to independent variables; restricted to two active constraints maximum.

3) Description of Parameters:

N	Number of independent variables - Define in main program
M	Number of constraints - Define in main program
SI	Initial step size - Define in main program
EPS	Accuracy index on minimum step size and approach to constraint - Define in main program
X	Independent variables - Define in main program
Y	Objective function - Define in subroutine OBJECT
JA	Convergence indicator (JA=1 when the convergence limit has been satisfied (see Method); the convergence limit is defined in Subroutine POINT
K	Constraint violation indicator (still inside feasible region when K=0) - Define in main program

C	Lagrangian multiplier - Define in subroutine POINT	
D	Lagrangian multiplier - Define in subroutine POINT	
IC	First violated constraint number - Define in main program	
ID	Second violated constraint number - Define in main program	
PROB	Problem name - Define in main program	
NAME	User information - Define in main program	
G	Constraint function - Define in subroutine CONS	
GX	Constraint independent variable coefficients - Define in main program	
YX	Objective function derivatives - Define in subroutine POINT	
NI	Card reader unit number - Define in main program	
NO	Printer unit number - Define in main program	
IPRINT	Printing controller; =0 for final answers only, =1 for complete printout	

4) <u>DIMENSION Requirements</u>:

The DIMENSION and COMMON statements in the main program and subroutines should be modified according to the requirements of each particular problem. The parameters included in the following DIMENSION and COMMON statements conform to the <u>Input Parameter</u> definitions, above:

DIMENSION RID(N)

COMMON X(N), G(M), YX(N), GX(M,N), DIR(N)

The dimensions for PROB and NAME depend on the number of characters desired.

5) Input Formats:

CARD TYPE	FORMAT	CONTENTS
1	(10A8)	(PROB(I), I=1,10)
2	(10A8)	(NAME(J), J=1,10)
3	(8I10)	N, M, IPRINT
4	(8E10.4)	SI, EPS
5	(8E10.4)	(X(I), I=1,N)

(If N exceeds 8, an additional CARD TYPE 5 is required.)

6	(8E10.4)	(GX(J,I), I=1,N)

(M cards of CARD TYPE 6 are required.)

6) **Output:**

The main program first prints out the search procedure name, problem name, and user information. Input parameter information is then printed. If IPRINT = 1, the independent variables and objective function values are printed out for each point until convergence is obtained. If IPRINT = 0, only the final answers are printed out. A message is then printed stating the nature of the answer -- interior stationary point, maximum on the constraints or minimum on the constraints corresponding to the values of λ_i obtained ($=0$, ≤ 0, ≥ 0). Normally a maximum on the constraints or in the interior is to be expected.

7) **Summary of User Requirements:**

a) Determine problem name and user information, if desired.

b) Determine values for N, M, IPRINT, SI, EPS, NI and NO.

c) Determine initial estimates for the independent variables (X(I), I=1,N). The initial point must satisfy the constraints.

d) Specify the constraint coefficients (GX(J,I), I=1,N; J=1,M).

e) Adjust DIMENSION and COMMON statements in main program and subroutines.

f) Specify objective function in Subroutine OBJECT.

g) Specify constraints in Subroutine CONS.

h) Specify objective function derivatives in Subroutine POINT. Also adjust convergence limit in this subroutine if desired (Cards 18, 39, and 69--currently set at 0.0001).

i) Change input and output FORMAT statements as required by problem.

D. **Test Problem**

The following test problem was obtained from the literature (13). The calculations were performed on a CDC 6400 computer.

Function:

Maximize $F = 3803.84 + 138.08X_1 + 232.92X_2 - 123.08X_1^2 - 203.64X_2^2 - 182.25X_1X_2$

Constraints: $0.1 \leq X_1 \leq 2.5$
$0.1 \leq X_2 \leq 2.0$

Starting Point: $X_1 = 1.0$, $X_2 = 0.5$

Parameters: $N = 2$, $M = 4$, $S = 0.1$, $EPS = 10^{-5}$,
Convergence limit = 0.0001, IPRINT = 1

Algorithm Answers: $F = 3873.9$
$X_1 = 0.20565$
$X_2 = 0.47987$

Number of Points: 22

Central Processor Time: 4 seconds

A listing of the program and test problem plus output is contained in the following section.

E. Program Listings and Example Output

```
C
C      MAIN LINE PROGRAM FOR ROSEN ALGORITHM
C
       DIMENSION PROB(10), NAME(10), RID(2)
       COMMON J, N, M, X(2), G(4), YX(2), GX(4,2), DIR(2), Y, IC, JA,
      1 C, ID
C
       NI = 50
       NO = 66
C
       READ (NI,001) (PROB(I), I=1,10)
  001 FORMAT (10A8)
       READ (NI,001) (NAME(J), J=1,10)
       READ (NI,002) N, M, IPRINT
  002 FORMAT (8I10)
       READ (NI,003) SI, EPS
  003 FORMAT (8E10.4)
       READ (NI,003) (X(I), I=1,N)
       DO 100 J=1,M
  100 READ (NI,003) (GX(J,I), I=1,N)
C
       S=SI
       JA=0
       J=0
       IC=0
       C=0.
       CALL OBJECT
       CALL POINT
       YOLD=Y
C
       WRITE (NO,004)
  004 FORMAT (1H1,10X,31HROSEN GRADIENT SEARCH PROCEDURE )
       WRITE (NO,005) (PROB(I), I=1,10)
  005 FORMAT (/,2X,10A8)
       WRITE (NO,005) (NAME(J), J=1,10)
       WRITE (NO,006) N, M, SI, EPS
  006 FORMAT (//,2X,10HPARAMETERS,/,2X,4HN = ,I2,4X,4HM = ,I2,/,2X,
      1 12HSTEP SIZE = ,E16.8,4X,11HACCURACY = ,E16.8 )
       WRITE (NO,007)
  007 FORMAT (//,2X,11HVALUES OF X )
       WRITE (NO,008) (I, X(I), I=1,N)
  008 FORMAT (/,3(2X,2HX(,I2,4H) = ,E16.8))
       WRITE (NO,009) Y
  009 FORMAT (//,2X,24HVALUE OF THE FUNCTION = ,E16.8)
C
  500 DO 10 I=1,N
       X(I)=X(I)+S*DIR(I)
       RID(I)=-DIR(I)
   10 CONTINUE
       CALL CONS
       K=0
       DO 70 I=1,M
       IF(G(I)-EPS) 70,70,11
```

```
   11 K=K+1
      IF(K-1)31,31,32
   31 IC=I
      GO TO 70
   32 ID=I
   70 CONTINUE
      IF(K)12,12,15
   12 CALL OBJECT
      CALL POINT
      YNEW=Y
      FD=YNEW-YOLD
      IF(FD)15,15,20
C
   15 DO 16 I=1,N
      X(I)=X(I)+S*RID(I)
   16 CONTINUE
      S=.5*S
      SMIN = ABS(S)
      IF (SMIN - EPS) 60, 500, 500
C
   20 IF (IPRINT) 40, 40, 30
   30 WRITE (NO,021)
  021 FORMAT (//,2X,14HNEW BASE POINT)
      WRITE (NO,007)
      WRITE (NO,008) (JJ, X(JJ), JJ=1,N)
      WRITE (NO,009) Y
   40 CONTINUE
      S=2.*S
      YOLD=Y
      GO TO 500
C
   60 IF (IPRINT)  80, 80, 140
  140 WRITE (NO,022) IC
  022 FORMAT (//,2X,36HNEW BASE POINT ON CONSTRAINT NUMBER ,I2 )
      WRITE (NO,007)
      WRITE (NO,008) (J, X(J), J=1,N)
      WRITE (NO,009) Y
   80 J = K
      CALL POINT
      IF (JA) 500, 500, 999
C
  999 IF (C)   90, 110, 120
   90 WRITE (NO,023)
  023 FORMAT (///,2X,20HOPTIMUM IS A MAXIMUM )
      GO TO 130
  110 WRITE (NO,024)
  024 FORMAT (///,2X,39HOPTIMUM IS AN INTERIOR STATIONARY POINT )
      GO TO 130
  120 WRITE (NO,025)
  025 FORMAT (///,2X,20HOPTIMUM IS A MINIMUM )
  130 WRITE (NO,007)
      WRITE (NO,008) (II, X(II), II=1,N)
      WRITE (NO,009) Y
C
      END
```

```
      SUBROUTINE OBJECT
C
      COMMON J, N, M, X(2), G(4), YX(2), GX(4,2), DIR(2), Y, IC, JA,
     1 C, ID
C
      X1 = X(1)
      X2 = X(2)
      X12 = X1**2
      X22 = X2**2
      Y = 3803.84 + 138.08*X1 + 232.92*X2 - 123.08*X12 - 203.64*X22
     1 - 182.25*X1*X2
C
      RETURN
      END




      SUBROUTINE CONS
C
      COMMON J, N, M, X(2), G(4), YX(2), GX(4,2), DIR(2), Y, IC, JA,
     1 C, ID
C
      G(1) =   X(1) - 2.5
      G(2) =   X(2) - 2.0
      G(3) = - X(1) + 0.1
      G(4) = - X(2) + 0.1
C
      RETURN
      END




      SUBROUTINE POINT
C
      COMMON J, N, M, X(2), G(4), YX(2), GX(4,2), DIR(2), Y, IC, JA,
     1 C, ID
C
C     PLACE GRADIENTS AS STATEMENT FUNCTIONS IN THIS SECTION
C
      YX(1) = 138.08 - (2*123.08*X(1)) - 182.25*X(2)
      YX(2) = 232.92 - (2*203.64*X(2)) - 182.25*X(1)
C
C     END GRADIENT SECTION
C
      IF(J-1)5,50,70
    5 SSD=0.
      DO 7 I=1,N
      SSD=SSD + (YX(I)*YX(I))
    7 CONTINUE
      IF(SSD-.0001)9,9,11
    9 JA=1
      RETURN
```

```
C
   11 SD=SQRT(SSD)
      DO 13 I=1,N
      DIR(I)=YX(I)/SD
   13 CONTINUE
      RETURN
C
   50 A=0.
      B=0.
      DO 55 I=1,N
      A=A-GX(IC,I)*YX(I)
      B=B+GX(IC,I)*GX(IC,I)
   55 CONTINUE
      C=A/B
      SSD=0.
      DO 57 I=1,N
      SSD=SSD+(YX(I)+C*GX(IC,I))**2
   57 CONTINUE
      IF(SSD-.0001)59,59,61
   59 JA=1
      RETURN
C
   61 SD=SQRT(SSD)
      DO 63 I=1,N
      DIR(I)=(YX(I)+C*GX(IC,I))/SD
   63 CONTINUE
      RETURN
   70 BA=0.
      BB=0.
      AA=0.
      AB=0.
      AC=0.
      AD=0.
      DO 75 I=1,N
      AA=AA+GX(IC,I)*GX(IC,I)
      AB=AB+GX(ID,I)*GX(IC,I)
      AC=AC+GX(IC,I)*GX(ID,I)
      AD=AD+GX(ID,I)*GX(ID,I)
      BA=BA-GX(IC,I)*YX(I)
      BB=BB-GX(ID,I)*YX(I)
   75 CONTINUE
      DET=(AA*AD)-(AB*AC)
      C=((BA*AD)-(AB*BB))/DET
      D=((AA*BB)-(BA*AC))/DET
      SSD=0.
      DO 77 I=1,N
      SSD=SSD+(YX(I)+C*GX(IC,I)+D*GX(ID,I))**2
   77 CONTINUE
      IF(SSD-.0001)79,79,81
   79 JA=1
      RETURN
C
   81 SD = SQRT(SSD)
      DO 83 I=1,N
      DIR(I)=(YX(I)+C*GX(IC,I)+D*GX(ID,I))/SD
   83 CONTINUE
C
      RETURN
      END
```

ROSEN GRADIENT SEARCH PROCEDURE

LITERATURE SAMPLE PROBLEM -- QUADRATIC FUNCTION

TESTED BY PAUL S. INGLISH -- 1 APRIL 1972

PARAMETERS
N = 2   M = 4
STEP SIZE =   .10000000E+00   ACCURACY =   .10000000E-04

VALUES OF X

X(1) =   .10000000E+01   X(2) =   .50000000E+00   X(

VALUE OF THE FUNCTION =   .37932650E+04

NEW BASE POINT

VALUES OF X

X(1) =   .92068671E+00   X(2) =   .43909514E+00   X(

VALUE OF THE FUNCTION =   .38159712E+04

NEW BASE POINT

VALUES OF X

X(1) =   .75487933E+00   X(2) =   .32725637E+00   X(

VALUE OF THE FUNCTION =   .38473300E+04

(Intervening iterations omitted.)

NEW BASE POINT

VALUES OF X

X(1) =   .20566038E+00   X(2) =   .47989386E+00   X(

VALUE OF THE FUNCTION =   .38739235E+04

NEW BASE POINT

VALUES OF X

X(1) =   .20564792E+00   X(2) =   .47987287E+00   X(

VALUE OF THE FUNCTION =   .38739235E+04

NEW BASE POINT ON CONSTRAINT NUMBER 3

VALUES OF X

X(1) =   .20564792E+00   X(2) =   .47987287E+00   X(

VALUE OF THE FUNCTION =   .38739235E+04

OPTIMUM IS AN INTERIOR STATIONARY POINT

VALUES OF X

X(1) =   .20564792E+00   X(2) =   .47987287E+00   X(

VALUE OF THE FUNCTION =   .38739235E+04

IV. FIACCO AND McCORMICK (SUMT ALGORITHM)*

A. Purpose

This program finds the minimum of a multivariable, nonlinear function subject to nonlinear inequality and equality constraints:

$$\text{Minimize} \quad F(X_1, X_2, \ldots, X_N)$$

$$\text{Subject to} \quad G_k(X_1, X_2, \ldots, X_N) \geq 0, \quad k = 1, 2, \ldots, M$$

$$H_k(X_1, X_2, \ldots, X_N) = 0, \quad k = M+1, M+2, \ldots, M+P.$$

B. Method

The procedure was developed by Fiacco and McCormick (16). The technique uses the problem constraints and the original objective function to form an unconstrained objective function which is minimized by any appropriate unconstrained, multivariable technique (see Chapter 9). The algorithm proceeds as follows:

1) A modified objective function is formulated consisting of the original function and penalty functions with the form

$$P = F - r \sum_{k=1}^{M} \ln G_k + \sum_{k=M+1}^{M+P} H_k^2 / r$$

where r is a positive constant. As the algorithm progresses, r is reevaluated to form a monotonically decreasing sequence $r_1 > r_2 > \ldots > 0$. As r becomes small, under suitable conditions P approaches F and the problem is solved.

2) Select a starting point (feasible or nonfeasible) and an initial value for r.

3) Determine the minimum of the modified objective function for the current value of r using an appropriate technique (several options available).

---

*Computer code developed by W. Charles Mylander, R. L. Holmes and G. P. McCormick, Research Analysis Corporation, McLean, Va. Used by permission.

4) Estimate the optimal solution using extrapolation formulas (16).

5) Select a new value for r (16) and repeat the procedure until the convergence criteria is satisfied.

A logic diagram for this method is given in Figure 10.IV.

C. Program Description

1) Usage:

The program consists of a main program, two control subroutines (BODY, FEAS), twenty one special purpose subroutines (CONVRG, ESTIM, EVALU, FINAL, GRAD, INVERS, OPT, OUTPUT, PEVALU, PUNCH, REJECT, RHOCOM, SECORD, STORE, TCHECK, TIMEC, SET, XMOVE, DIFF1, DIFF2, CHCKER) and four user supplied subroutines (READIN, RESTNT, GRAD1, MATRIX). Input is coordinated by the main program and READIN. Output is from the main program and subroutines BODY, CHCKER, CONVRG, ESTIM, FEAS, INVERS, OPT, OUTPUT, PUNCH, TCHECK, TIMEC. Format changes may be required for different problems.

2) Subroutines Required:

SUBROUTINE BODY coordinates all subroutines.

SUBROUTINE CHCKER computes first derivatives of objective function using GRAD1 and DIFF1.

SUBROUTINE CONVRG(N1) checks for convergence.

SUBROUTINE DIFF1(IN) computes first derivatives by central difference.

SUBROUTINE DIFF2(IN) computes second derivatives by central difference.

SUBROUTINE ESTIM estimates Lagrange multiplier (λ_i) values and final solution extrapolation.

SUBROUTINE EVALU evaluates objective function and constraints.

SUBROUTINE FEAS determines feasibility of starting point; if not feasible, a feasible point is sought; if no feasible point possible, error message printed.

SUBROUTINE FINAL(N2) checks for convergence.

SUBROUTINE GRAD(IS) computes first derivatives of penalty function.

SUBROUTINE INVERS(NSME) solves linear system of equations.

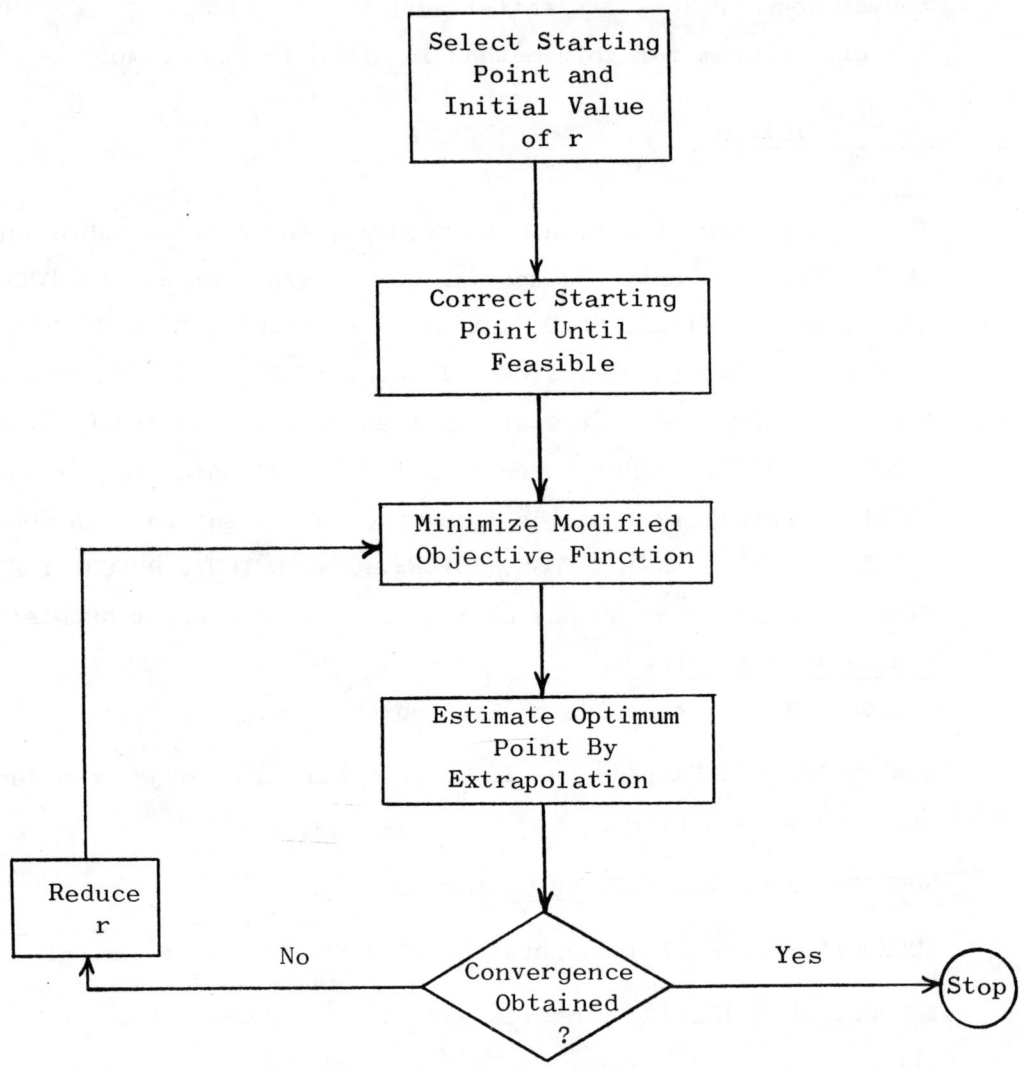

Figure 10.IV. Fiacco and McCormick (SUMT ALGORITHM) Logic Diagram

SUBROUTINE OPT performs one dimensional search by Golden Section method (modified Fibonacci).

SUBROUTINE OUTPUT (K) prints out results of each iteration.

SUBROUTINE PEVALU computes value of penalty function and dual.

SUBROUTINE PUNCH punches stopping point and parameters.

SUBROUTINE REJECT returns stored values to original locations.

SUBROUTINE RHOCOM computes initial value of r.

SUBROUTINE SECORD(IS) computes second derivatives of penalty function.

SUBROUTINE SET(TMMAX) stores the time at start of run.

SUBROUTINE STORE stores values of current point.

SUBROUTINE TCHECK checks run time for 90% of maximum specified.

SUBROUTINE TIMEC checks elapsed time of run for possible termination.

SUBROUTINE XMOVE determines search directions by one of several options (modified Newton-Raphson, Steepest Descent, Fletcher-Powell).

SUBROUTINE READIN reads data (user supplied).

SUBROUTINE RESTNT(IN,VAL) specifies objective function and constraints (user supplied).

SUBROUTINE GRAD1(IN) evaluates analytical first derivatives for objective function and constraints (user supplied).

SUBROUTINE MATRIX(J,L) evaluates analytical second derivatives for objective function and constraints (user supplied).

3) <u>Description of Parameters:</u>

NI	Card reader unit number
NO	Printer unit number
NP	Card punch unit number
EPSI	Tolerence, ϵ, used to decide if an unconstrained minimum has been achieved for each subproblem (see NT9)
RHOIN	Possible initial value of r, r_1, (often set at 1.0) (see NT1)
THETAO	Tolerance, θ_o, used to decide if the solution to the problem has been approximated (see NT5)
RATIO	Parameter, C, (>1) used to compute consecutive values of r; $r_{i+1} = r_i/C$ (often set at 16.0)

TMMAX	Maximum amount of time for solving problem (sec.)
M	Number of inequality constraints
N	Number of independent variables
MZ	Number of equality constraints
X	Independent variables (starting point on input)
NEWOLD	READIN key: if NEWOLD = 1, read in data for new problem if NEWOLD = 2, only read in new B vector if NEWOLD = 3, use same constants as in previous problem
NR	READIN data index limit
MR	READIN data index limit
D	READIN data vector
E	READIN data vector
A	READIN data matrix
C	READIN data matrix
B	READIN data vector
NT1	Option key for r values, as follows: =1 The value for r is made by finding an approximation solution $\min\{[\nabla P(X^o, r)][\nabla^2 P(X^o, r)]^{-1} \nabla P(X^o, r)\}$ which is a good approximation only when X^o is close to the boundary (i.e., for some i, $G_i(X)$ is close to zero) or when $\nabla^2 F(X^o)=0$ and when MZ=0 =2 r_1 is given by formula 8.65 [Ref. (16) p. 191] (Only can be used when MZ=0) =3 r_1 = RHOIN (normally set to 3)
NT2	Option key for constraints as follows: =1 The requirements (trivial constraints) that $X_i \geq 0$ for i = 1, ..., N are to be automatically included in the problem =2 The only constraints on the problem are those inputed by the user
NT3	Option key for printout as follows: =1 Standard printout (this includes a call to OUTPUT after the solution of every subproblem. Also the estimates of the "Lagrange multipliers" and first- and second-order solution estimates are printed) =2 For additional printout (includes standard printout and every intermediate point, gradient of P and the vector S) (normally set to 1)
NT4	Option key on final convergence as follows: =1 Final convergence is determined on the basis of current solution to the subproblem

=2 Final convergence is determined on the basis of the first order estimates. The first order estimate of the solution vector must be close to feasible. See below.

=3 Final convergence is determined on the basis of the second order estimates. The second order estimate of the solution vector must be close to feasible before the convergence check is made. If $\underline{X}$ is a solution estimate it is considered close to being feasible if $G_i(\underline{X}) + \theta_o \geq 0$, $i = 1, 2, \ldots, M$.
(normally set to 1)

NT5   Option key on final convergence as follows:
The convergence criterion determining the problem has been solved (Only use = 1, when NT4 $\neq$ 1)

=1 Quit when $\dfrac{G - F[X(r_k)]}{G[X(r_k), \mu(r_k), \lambda(r_k)]} < \theta_o$

=2 Quit when $\left| r \sum_{j=1}^{m} \ln G_j[X(r_k)] \right| < \theta_o$

=3 Quit when $\dfrac{\text{first order estimate of } v_o}{G[X(r_k), \mu(r_k), \lambda(r_k)]} - 1 < \theta_o$

(normally set to 1)

NT6   Option key for next problem as follows:
=1 After final convergence the program reads in new data and solves the next problem

=2 After final convergence has been determined a call to PUNCH is made before proceeding on to the next problem
(normally set to 1)

NT7   Option key for extrapolation as follows:

=1 No extrapolation

=2 Extrapolate through last two minima

=3 Extrapolate through last three minima
(normally set to 1)

NT8   Not used

NT9   Option key for subproblem convergence (fixed value of r) as follows:
=1 Quit when $\left| \nabla_X P^t(X^i, r) \left[\dfrac{\partial^2 P(X, r)}{\partial X_i \partial X_j} \right]^{-1} \nabla_X P(X^i, r) \right| < \epsilon$

=2 Quit when $\left| \nabla_X P^t(X^i, r) \left[\dfrac{\partial^2 P(X, r)}{\partial X_i \partial X_j} \right]^{-1} \nabla_X P(X^i, r) \right|$
$< \dfrac{P(X^{i-1}) - P(X^i)}{5}$

=3 Quit when $\left| \nabla_X P(X^i, r) \right| < \epsilon$

NT10   Option key on problem linearity as follows:
=1   At least one nonlinear constraint

=2   Linear constraints

=3   Linear constraints and linear objective function (i.e., a linear programming problem)

When Option 10 = 3 MATRIX (the user subroutine supplying the second partial derivatives) will not be called, and when Option 10 = 2 it will be called only to get the second partials of F(X).

XEP1   Finite difference parameter used in DIFF1 if numerical derivatives used. Usually XEP1 = 0.0001 is satisfactory.

XEP2   Iteration improvement limit. When minimizing the P-function for a given value of r (RHO) the value of P must decrease by an amount exceeding XEP2 for each iteration after the first. If it does not, then the code prints out the message "apparently roundoff errors prevent a more accurate determination of the minimum of this subproblem," and it is assumed a minimum has been found. (Normally, set XEP2 equal to 0.)

NEXOP1   Key for checking derivatives as follows:
=1   Solve problems without checking derivatives

=2   Solve problem after checking only first derivatives

=3   Do not solve problem after checking only first derivatives

=4   Solve problem after checking first and second derivatives

=5   Do not solve problem after checking first and second derivatives

If the user wishes to check his first or second partial derivatives by having them printed along with experimental SUMT's numerical approximations, he can accomplish this by appropriately setting NEXOP1.

In all cases the first values printed out are those gotten from the user's code, and the second values are those gotten by numerical differencing.

NEXOP2   Key for choosing unconstrained minimization technique as follows:
=1   The method for minimizing the unconstrained penalty function is to be the generalized Newton-Raphson method as modified to handle indefinite Hessian matrices. This method requires function values, first and second derivatives.

=2   Same as 1, except that when an "orthogonal move" is made because of an indefinite Hessian matrix $-\nabla P$ is added to the orthogonal move vector.

=3   Steepest descent is used to minimize P-function.

=4 The method for minimizing the unconstrained penalty function is McCormick's modification of the Fletcher-Powell method as reported in the Fiacco-McCormick book ( ). This requires function values and first derivatives.

VAL   Objective function value in RESTNT

A,B,C,D,E User supplied data in READIN for evaluating objective function and constraints.

4) <u>DIMENSION Requirements</u>:

The general subroutines in the code are currently dimensioned to handle a problem with up to 100 variables (N) and 200 constraints (M+MZ). This should be sufficient for most problems attempted with this method. If the limits need to be changed, a detailed description of the various parameters are given elsewhere (34).

5) <u>Input Formats</u>:

CARD TYPE	FORMAT	CONTENTS
1 (parameters)	(8E10.4)	EPSI, RHOIN, THETAO, RATIO, TMMAX
2 (parameters)	(8I10)	M, N, MZ
3 (initial X)	(8E10.4)	(X(I), I=1, N)
4 (user info.)	(8I10)	NEWOLD, NR, MR
5 (user info.)	(8E10.4)	(D(I), I=1, NR)
6 (user info.)	(8E10.4)	(E(I), I=1, NR)
7 (user info.)	(8E10.4)	(A(I,K), K=1, NR), I=1, MR
8 (user info.)	(8E10.4)	(C(J,K), K=1, NR), J=1, NR
9 (user info.)	(8E10.4)	(B(I), I=1, MR)
10 (options)	(16I5)	NT1, NT2, NT3, NT4, NT5, NT6, NT7, NT8, NT9, NT10
11 (tolerances)	(8E10.4)	XEP1, XEP2
12 (options)	(8I10)	NEXOP1, NEXOP2

Card types 3, 4, 5, 6, 7, 8 and 9 may have to be repeated if the limits are exceeded. Card types 1, 2, 3, 10, 11, 12 are read in by the main program. Card types 4, 5, 6, 7, 8, 9 are under control of READIN--this subroutine allows the user to read in the data necessary, if any, to evaluate the objective function and constraints. For multiple problems, the entire data card set is repeated.

6) <u>Output</u>:

The program title is printed first followed by a listing of the parameters, option keys, and starting point information. This is followed by intermediate result printouts until convergence is achieved. At convergence, the final answer information is printed. Execution time is listed with the intermediate results. Format statement changes may be required for different problems.

7) <u>Summary of User Requirements</u>:

   a) Determine values for the card input data (see <u>Description of Parameters</u> and <u>Input Formats</u>).

   b) Specify NI, NO and NP in the main program.

   c) Adjust the DIMENSION, COMMON and FORMAT statements as required by the problem under study (see <u>DIMENSION Requirements</u>).

   d) Specify objective function and constraints in subroutine RESTNT (using data from READIN if desired).

   e) Specify analytical derivatives, if used, for first and second derivatives of objective function and constraints in GRAD1 and MATRIX.

   Further details on this program are available elsewhere (34).

D. Test Problem

The following test problem was formulated by Rosenbrock (41). The calculations were performed on a CDC 6400 computer.

Function: $F = -X_1 X_2 X_3$

Constraints: $0 \leq X_i \leq 42$, $\quad i = 1, 2, 3$
$0 \leq (X_4 = X_1 + 2X_2 + 2X_3) \leq 72$

Starting Point: $X_1 = 10.0$, $X_2 = 10.0$, $X_3 = 10.0$

Parameters:
- N = 3
- M = 4
- MZ = 0
- TMMAX = 60.0
- r(start) = 1.0
- RATIO = 4.0
- EPSI = 10^{-9}
- THETAO = 10^{-6}

Options:
- NT1 = 1
- NT2 = 1
- NT3 = 1
- NT4 = 1
- NT5 = 2
- NEXOP1 = 4
- NT6: 2
- NT7: 1
- NT8: 1
- NT9: 1
- NT10: 2
- NEXOP2 = 4

Algorithm Answers: F = -3456.000
X_1 = 24.000
X_2 = 12.000
X_3 = 12.000

Central Processor Time: 25 seconds

The listing and output for this problem are contained in the following section.

E. **Program Listings and Example Output**

```
C
C        MAIN LINE PROGRAM FOR SUMT ALGORITHM.
C
         COMMON /SHARE/ X(100),DEL(100),A(100,100),N,M,MN,NP1,NM1
         COMMON /EQAL/ H, H1, MZ
         COMMON /OPTNS/ NT1,NT2,NT3,NT4,NT5,NT6,NT7,NT8,NT9,NT10
         COMMON /VALUE/ F,G,P0,RSIGMA,RJ(100),RHO
         COMMON /CRST/ DELX(100),DELX0(100),RHOIN,RATIO,EPSI,THETA0,
        1 RSIG1, G1, X1(100), X2(100), X3(100), XR2(100), XR1(100), PR1,
        2 PR2, P1, F1, RJ1(200), DOTT, PGRAD(100), DIAG(100),
        3 PREV3, ADELX, NTCTR, NUMINI, NPHASE, NSATIS
         COMMON /EXPOPT/ NEXOP1, NEXOP2, XEP1, XEP2
         COMMON /DEVC/ NI, NO, NP
C
         NI = 50
         NO = 66
         NP = 58
C
C        PARAMETER CARD
     10  READ (NI,001) EPSI, RHOIN, THETA0, RATIO, TMMAX
    001  FORMAT (8E10.4)
         READ (NI,002) M, N, MZ
    002  FORMAT (3I10)
         IF (N) 40, 40, 20
     20  CALL SET (TMMAX)
C        INITIAL X VECTOR
         READ (NI,001) (X(I), I=1,N)
C        SUBROUTINE READIN IS UNDER PROGRAMMER CONTROL CALL READIN
         CALL READIN
C        OPTION CARD FOLLOWS PROGRAMMERS DATA
         READ (NI,003) NT1,NT2,NT3,NT4,NT5,NT6,NT7,NT8,NT9,NT10
    003  FORMAT (16I5)
C        READ TOLERANCES
         READ (NI,001) XEP1, XEP2
C        READ SECOND OPTION CARD
         READ (NI,002) NEXOP1, NEXOP2
C
         WRITE (NO,004)
    004  FORMAT (1H1,10X,49HNONLINEAR PROGRAMMING ROUTINE  --   SUMT VERSION
        1 4 )
         WRITE (NO,005) N,M,MZ,TMMAX,RHOIN,RATIO,EPSI,THETA0
    005  FORMAT (//,2X,10HPARAMETERS,//,2X,4HN = ,I2,4X,4HM = ,I2,4X,5HMZ =
        1 ,I2,4X,8HTMMAX = ,F8.3,/,2X,6HRHO = ,E10.4,4X,8HRATIO = ,E10.4,
        2 4X,9HEPSILON = ,E11.4,4X,8HTHETA = ,E10.4 )
         WRITE (NO,006) NT1,NT2,NT3,NT4,NT5,NT6,NT7,NT8,NT9,NT10,NEXOP1,
        1 NEXOP2
    006  FORMAT (//,2X,16HOPTIONS SELECTED,//,2X,6HNT1 = ,I1,3X,6HNT2 = ,
        1 I1,3X,6HNT3 = ,I1,3X,6HNT4 = ,I1,3X,6HNT5 = ,I1,/,2X,6HNT6 = ,
        2 I1,3X,6HNT7 = ,I1,3X,6HNT8 = ,I1,3X,6HNT9 = ,I1,3X,7HNT10 = ,I1,
        3 /,2X,9HNEXOP1 = ,I1,3X,9HNEXOP2 = ,I1 )
         WRITE (NO,007) XEP1, XEP2
    007  FORMAT (//,2X,37HTOLERANCES FOR DIFFERENCING AND MOVES ,//,2X,
        1 7HXEP1 = ,E10.4,4X,7HXEP2 = ,E10.4 )
```

```
C
      NTCTR=0
      NP1=N+1
      NM1=N-1
      CALL TIMEC
      NPHASE=4
C     JUST TO GET AN INITIAL PRINTOUT
      CALL EVALU
      P0=0.0
      G=0.0
      H=0.0
      RSIGMA=0.0
      CALL OUTPUT (2)
      CALL STORE
      IF (NEXOP1.GT.1) CALL CHCKER
      IF (NEXOP1.EQ.3) STOP 01072
      IF (NEXOP1.EQ.5) STOP 01104
      CALL FEAS
C     NPHASE=5 INDICATES EITHER TIME RAN OUT OR NO FEASIBLE POINT WAS
C     FOUND.
      GO TO (30,30,30,30,10), NPHASE
   30 NPHASE=2
      NTCTR=0
      CALL BODY
C
      WRITE (NO,008) F
  008 FORMAT (///,2X,19HFINAL VALUE OF F = ,1PE16.8 )
      WRITE (NO,009)
  009 FORMAT (//,2X,14HFINAL X VALUES  )
      WRITE (NO,011) (I, X(I), I=1,N)
  011 FORMAT (/,3(2X,2HX(,I2,4H) = ,1PE16.8 ))
C
      GO TO 10
C
   40 STOP
      END




      SUBROUTINE BODY
C
C     BODY COORDINATES THE FLOW AMONG THE SUBROUTINES THAT ACTUALLY DO
C     THE CALCULATIONS REQUIRED BY THE VARIOUS PARTS OF THE ALGORITHM.
C
      COMMON /SHARE/ X(100),DEL(100),A(100,100),N,M,MN,NP1,NM1
      COMMON /OPTNS/ NT1,NT2,NT3,NT4,NT5,NT6,NT7,NT8,NT9,NT10
      COMMON /VALUE/ F,G,P0,RSIGMA,RJ(100),RHO
      COMMON /CRST/ DELX(100),DELX0(100),RHOIN,RATIO,EPSI,THETA0,
     1 RSIG1, G1, X1(100), X2(100), X3(100), XR2(100), XR1(100), PR1,
     2 PR2, P1, F1, RJ1(200), DOTT, PGRAD(100), DIAG(100),
     3 PREV3, ADELX, NTCTR, NUMINI, NPHASE, NSATIS
      COMMON /CONPAR/ NF1,NF2,NF3
      COMMON /DEVC/ NI, NO, NP
C
      NF2=2
```

```
              NF3=2
              MN=0
              NUMINI=0
C             OPTION OF GETTING INITIAL RHO
              CALL RHOCOM
              CALL EVALU
10            CALL XMOVE
              GO TO (30,20), NT3
20            CALL TIMEC
              CALL OUTPUT (1)
              GO TO 40
30            CALL TCHECK
C             IN FEASIBILITY PHASE 4 MEANS FEASIBILITY ACHIEVED
40            GO TO (50,50,50,200), NSATIS
50            CALL CONVRG (N1)
              GO TO (60,10,125), N1
C             MINIMUM ACHIEVED IF   N1=1
60            GO TO (70,80), NT3
70            CALL TIMEC
              CALL OUTPUT (1)
C             NUMBER OF MINIMA ACHIEVED INCREASED BY 1
80            NUMINI=NUMINI+1
              MN=0
              GO TO (190,90,90), NPHASE
90            CALL ESTIM
C             FINAL MIGHT HAVE BEEN CALLED BY ESTIM-CONVERGED IF N2=1
              GO TO (100,110,120), NT4
C             NT4=1 FINAL CONVERGENCE ON 0 ORDER ESTIMATES, NT4=2 CONVERGE ON
C             FIRST ORDER ESTIMATES, NT4=3 CONVERGE ON SECOND ORDER ESTIMATES.
100           CALL FINAL (NF1)
              GO TO (130,140), NF1
110           GO TO (130,140), NF2
120           GO TO (130,140), NF3
125           NPHASE=5
130           RETURN
140           RHO=RHO/RATIO
C             USING PREVIOSLY COMPUTED VALUES FOR F, AND RJ P IS RECOMPUTED WITH
C             THE NEW VALUE OF RHO.
              CALL PEVALU
C             A VECTOR IS LEFT IN DELX(I) BY ESTIM
              IF (NUMINI-2) 10,150,150
150           GO TO (10,160,160), NT7
160           CALL GRAD (2)
              CALL OPT
              GO TO (180,170), NT3
170           WRITE (NO,210)
210           FORMAT (//,2X,30HMOVED ON EXTRAPOLATION VECTOR  )
              CALL OUTPUT (1)
180           GO TO 50
C             DUAL VALUE GREATER THAN 0 MEANS NO FEASIBLE POINT EXISTS
190           IF (G) 90,90,200
C
200           RETURN
              END
```

```
      SUBROUTINE CHCKER
C
C     CHCKER COMPUTES AND LIST THE FIRST PARTIAL DERIVATIVES USING GRAD1
C     AND THEN USING NUMERICAL DIFFERENCING (DIFF1).  IF REQUESTED THE
C     SECOND PARTIAL DERIVATIVES ARE COMPUTED AND LISTED USING MATRIX
C     AND DIFF2.
C
      COMMON /SHARE/ X(100),DEL(100),A(100,100),N,M,MN,NP1,NM1
      COMMON /EQAL/ H, H1, MZ
      COMMON /EXPOPT/ NEXOP1, NEXOP2, XEP1, XEP2
      COMMON /DEVC/ NI, NO, NP
C
      MMZ=1+M+MZ
      DO 5 J=1,N
      DEL(J)=1.2345678
5     CONTINUE
      DO 10 I=1,MMZ
      IN=I-1
      IF (IN) 170, 170, 180
170   WRITE (NO,001)
001   FORMAT (//,2X,38HVALUES OF OBJECTIVE FUNCTION PARTIALS  )
      GO TO 190
180   WRITE (NO,002) IN
002   FORMAT (//,2X,29HVALUES FOR CONSTRAINT NUMBER ,I2 )
      CALL GRAD1 (IN)
190   WRITE (NO,003)
003   FORMAT (/,2X,25HANALYTICAL FIRST PARTIALS  )
      WRITE (NO,004) (J, DEL(J), J=1,N)
004   FORMAT (/,3(2X,4HDEL(,I2,4H) = ,E15.7))
      CALL DIFF1 (IN)
      WRITE (NO,006)
006   FORMAT (/,2X,24HNUMERICAL FIRST PARTIALS  )
      WRITE (NO,004) (J, DEL(J), J=1,N)
10    CONTINUE
C     SOMETIMES ONLY FIRST DERIVATIVES ARE TO BE CHECKED
      IF (NEXOP1.LT.4) GO TO 160
      DO 150 I=1,MMZ
      IN=I-1
      IF (IN) 200, 200, 210
200   WRITE (NO,001)
      GO TO 220
210   WRITE (NO,002) IN
220   IT=2
      DO 30 K=1,N
      DO 20 J=1,N
20    A(K,J)=0.
30    CONTINUE
      CALL MATRIX (IN,IT)
      IF (IT.EQ.1) GO TO 150
      DO 50 K=2,N
      KM1=K-1
      DO 40 J=1,KM1
      IF (A(K,J).EQ.0.0) GO TO 40
      NEXOP1=5
      WRITE (NO,007) K, J
```

```
007         FORMAT (/,2X,2HA(,I2,1H,,I2,10H) .NE. 0.0 )
            GO TO 60
40          CONTINUE
50          CONTINUE
60          WRITE (NO,009)
009         FORMAT (/,2X,26HANALYTICAL SECOND PARTIALS )
            DO 90 K=1,N
            DO 70 J=K,N
            IF (A(K,J).NE.0.0) GO TO 80
70          CONTINUE
80          WRITE (NO,008) (K, J, A(K,J), J=1,N)
008         FORMAT (/,3(2X,2HA(,I2,1H,,I2,4H) = ,E12.6))
90          CONTINUE
            DO 110 K=1,N
            DO 100 J=1,N
100         A(K,J)=0.
110         CONTINUE
            WRITE (NO,011)
011         FORMAT (/,2X,25HNUMERICAL SECOND PARTIALS )
            CALL DIFF2 (IN)
            DO 140 K=1,N
            DO 120 J=K,N
            IF (A(K,J).NE.0) GO TO 130
120         CONTINUE
            GO TO 140
130         WRITE (NO,008) (K, J, A(K,J), J=1,N)
140         CONTINUE
150         CONTINUE
160         CONTINUE
C
            RETURN
            END




            SUBROUTINE SET (TMMAX)
C
C           SET STORES THE TIME AT WHICH THE PROBLEM IS BEGUN
C
            COMMON /TSW/ NSWW
            COMMON /TMX/ TMO,EXT,EXT90
C
C           SECOND GIVES JOB CPU EXECUTION TIME IN 1/1000 OF A SECOND
C
            CALL SECOND (TMO)
            EXT=TMMAX+TMO
            EXT90= TMO + 0.90*TMMAX
            NSWW=1
C
            RETURN
            END
```

```
      SUBROUTINE CONVRG (N1)
C
C     AFTER EACH ITERATION OF THE ALGORITHM TO LOCATE THE MINIMUM OF THE
C     PENALTY FUNCTION, CONVRG DETERMINES IF THE CURRENT POINT IS CLOSE
C     ENOUGH TO THE POINT GIVING THE MINIMUM VALUE OF THE P FUNCTION.
C     N1 SET EQUAL TO 1 IF MINIMUM HAS BEEN FOUND.
C     N1 SET EQUAL TO 2 IF MINIMUM HAS NOT BEEN FOUND AND TIME IS NOT UP
C     N1 SET EQUAL TO 3 OTHERWISE
C
      COMMON /SHARE/ X(100),DEL(100),A(100,100),N,M,MN,NP1,NM1
      COMMON /OPTNS/ NT1,NT2,NT3,NT4,NT5,NT6,NT7,NT8,NT9,NT10
      COMMON /VALUE/ F,G,P0,RSIGMA,RJ(100),RHO
      COMMON /CRST/ DELX(100),DELX0(100),RHOIN,RATIO,EPSI,THETA0,
     1 RSIG1, G1, X1(100), X2(100), X3(100), XR2(100), XR1(100), PR1,
     2 PR2, P1, F1, RJ1(200), DOTT, PGRAD(100), DIAG(100),
     3 PREV3, ADELX, NTCTR, NUMINI, NPHASE, NSATIS
      COMMON /EXPOPT/ NEXOP1, NEXOP2, XEP1, XEP2
      COMMON /TSW/ NSWW
      COMMON /DEVC/ NI, NO, NP
C
      N1=2
      IF (NT8.LE.1) Q1=P0
      NT8=2
      IF (MN.LE.1) Q1=P0
      GO TO (10,20,30), NT9
10    IF (ABS(DOTT).LT.EPSI) GO TO 70
      GO TO 40
20    IF (ABS(DOTT).LT.(P1-P0)/5.0) GO TO 70
      GO TO 40
30    IF (ADELX.LT.EPSI) GO TO 70
40    GO TO (50,60), NSWW
50    IF (MN.LE.1) RETURN
      IF (P0+XEP2 .LT. Q1) GO TO 75
      WRITE (NO,80)
80    FORMAT (/,2X, 51HROUNDOFF ERRORS PREVENT MORE ACCURATE DETERMINATI
     1ON ,/,2X,34HOF THE MINIMUM OF THIS SUBPROBLEM.   )
      GO TO 70
60    CALL PUNCH
      WRITE (NO,90)
90    FORMAT (///,10X,50H***** TIME LIMIT.  CALLING EXIT FROM CONVRG ***
     1**   )
      N1=3
C
C     FOUND THE MINIMUM TO THE SUBPROBLEM.
      RETURN
70    N1=1
75    Q1 = P0
C
      RETURN
      END
```

```fortran
      SUBROUTINE DIFF2 (IN)
C
C     DIFF2 COMPUTES THE SECOND DERIVATIVES BY NUMERICAL DIFFERENCING.
C
      COMMON /SHARE/ X(100),DEL(100),A(100,100),N,M,MN,NP1,NM1
      COMMON /EXPOPT/ NEXOP1, NEXOP2, XEP1, XEP2
      COMMON /STIRX/ XSTR(100), XSSS(100), DDLL(100)
C
      DO 10 J=1,N
   10 XSSS(J)=X(J)
      DO 50 J=1,N
      IF (J.EQ.1) GO TO 20
      JM1=J-1
      X(JM1)=XSSS(JM1)
   20 X(J)=XSSS(J)+XEP1
      CALL GRAD1 (IN)
      DO 30 I=1,N
   30 DDLL(I)=DEL(I)
      X(J)=XSSS(J)-XEP1
      CALL GRAD1 (IN)
      DO 40 I=J,N
   40 A(J,I)=(DDLL(I)-DEL(I))/(2.*XEP1)
   50 CONTINUE
      X(N)=XSSS(N)
C
      RETURN
      END
```

```fortran
      SUBROUTINE DIFF1 (IN)
C
C     DIFF1 COMPUTES THE FIRST DERIVATIVES BY NUMERICAL DIFFERENCING.
C     USER CAN CALL FOR DIFFERENCING OF SELECTED FUNCTIONS
C
      COMMON /SHARE/ X(100),DEL(100),A(100,100),N,M,MN,NP1,NM1
      COMMON /EXPOPT/ NEXOP1, NEXOP2, XEP1, XEP2
      COMMON /STIRX/ XSTR(100), XSSS(100), DDLL(100)
C
      DO 10 J=1,N
   10 XSTR(J)=X(J)
      DO 30 J=1,N
      IF (J.EQ.1) GO TO 20
      JM1=J-1
      X(JM1)=XSTR(JM1)
   20 X(J)=XSTR(J)+XEP1
      CALL RESTNT (IN,ZZ2)
      X(J)=XSTR(J)-XEP1
      CALL RESTNT (IN,ZZ1)
   30 DEL(J)=(ZZ2-ZZ1)/(2.*XEP1)
      X(N)=XSTR(N)
C
      RETURN
      END
```

```
      SUBROUTINE ESTIM
C
C     ESTIM PERFORMS THE COMPUTATIONS TO ESTIMATE THE LAGRANGE MULTIPLI-
C     ERS AND MAKE THE FIRST- AND SECOND-ORDER ESTIMATES OF THE FINAL
C     SOLUTION OF THE PROBLEM.
C
      COMMON /SHARE/ X(100),DEL(100),A(100,100),N,M,MN,NP1,NM1
      COMMON /EQAL/ H, H1, MZ
      COMMON /VALUE/ F,G,P0,RSIGMA,RJ(100),RHO
      COMMON /OPTNS/ NT1,NT2,NT3,NT4,NT5,NT6,NT7,NT8,NT9,NT10
      COMMON /CRST/ DELX(100),DELX0(100),RHOIN,RATIO,EPSI,THETA0,
     1 RSIG1, G1, X1(100), X2(100), X3(100), XR2(100), XR1(100), PR1,
     2 PR2, P1, F1, RJ1(200), DOTT, PGRAD(100), DIAG(100),
     3 PREV3, ADELX, NTCTR, NUMINI, NPHASE, NSATIS
      COMMON /CONPAR/ NF1,NF2,NF3
      COMMON /DEVC/ NI, NO, NPP
C
      CALL STORE
      Z10=RATIO**2
      Z9=RATIO
      Z1=1.0/Z9+1.0/Z10
      Z2=Z1+1./Z9**3
      Z3=1./Z9**3
      Z4=Z10+Z9
      Z5=Z9**3
      Z6=1.0/((Z10-1.0)*(Z9-1.0))
      Z7=1./Z9
      Z8=1./(Z9-1.)
      RQ=1.0/RHO
      IF (NUMINI-2) 150,80,10
10    WRITE (NO,330)
330   FORMAT (//,2X,20H2ND ORDER ESTIMATES  )
      P0=(PR2-Z4*PR1+Z5*P1)*Z6
      G=(RATIO*G1-GR1)/(RATIO-1.)
      DO 20 I=1,N
20    X(I)=(XR2(I)-Z4*XR1(I)+Z5*X1(I))*Z6
      NP=NPHASE
      NPHASE=4
      CALL EVALU
      NPHASE=NP
      CALL OUTPUT (2)
C     CHECK TO SEE IF ESTIMATES HAVE CONVERGED
      GO TO (70,30,70), NPHASE
30    DO 50 J=1,M
      IF (RJ(J)) 40,50,50
40    IF (THETA0+RJ(J)) 70,50,50
50    CONTINUE
      GO TO (70,70,60), NT4
60    CALL FINAL (NF3)
70    CONTINUE
80    WRITE (NO,340)
340   FORMAT (//,2X,20H1ST ORDER ESTIMATES  )
      P0=(Z9*P1-PR1)*Z8
      G=(RATIO*G1-GR1)/(RATIO-1.)
      DO 90 I=1,N
```

```
90      X(I)=(Z9*X1(I)-XR1(I))*Z8
        NP=NPHASE
        NPHASE=4
        CALL EVALU
        NPHASE=NP
        CALL OUTPUT (2)
C       CHECK TO SEE IF ESTIMATES HAVE CONVERGED
        GO TO (140,100,140), NPHASE
100     DO 120 J=1,M
        IF (RJ(J)) 110,120,120
110     IF (RJ(J)+THETA0) 140,120,120
120     CONTINUE
        GO TO (140,130,140), NT4
130     CALL FINAL (NF2)
140     CONTINUE
150     WRITE (NO,350)
350     FORMAT (//,2X,27HSOLUTION OF THE SUBPROBLEM  )
        IF (M) 180,180,160
160     DO 170 J=1,M
170     RJ(J)=RHO/RJ1(J)
180     IF (MZ) 210,210,190
190     DO 200 J=1,MZ
        MNJ=M+J
200     RJ(MNJ)=2.*RJ1(MNJ)*RQ
210     GO TO (220,240), NT2
220     DO 230 I=1,N
230     X(I)=RHO/X1(I)
240     CALL OUTPUT (2)
        CALL REJECT
        IF (NUMINI-2) 280,300,250
250     GO TO (280,310,260), NT7
C       SECOND  ORDER MOVE  FOR NEXT MINIMUM
260     DO 270 I=1,N
270     DELX(I)=Z1*X1(I)-Z2*XR1(I)+Z3*XR2(I)
280     PR2=PR1
        GR2=GR1
        PR1=P1
        GR1=G1
        DO 290 I=1,N
        XR2(I)=XR1(I)
290     XR1(I)=X1(I)
        RETURN
300     GO TO (280,310,310), NT7
310     DO 320 I=1,N
320     DELX(I)=(X1(I)-XR1(I))*Z7
        GO TO 280
C
        END
```

```
      SUBROUTINE EVALU
C
C     IN THE NORMAL PHASE EVALU CALLS THE USER-SUPPLIED ROUTINES TO EVAL
C     UATE THE OBJECTIVE FUNCTION AND THE CONSTRAINT FUNCTIONS AT THE
C     CURRENT POINT. IN THE FEASIBILITY PHASE THIS ROUTINE PUTS THE
C     NEGATIVE SUM OF THE VIOLATED CONSTRAINTS IN LOCATION F.
C
      COMMON /SHARE/ X(100),DEL(100),A(100,100),N,M,MN,NP1,NM1
      COMMON /EQAL/ H, H1, MZ
      COMMON /OPTNS/ NT1,NT2,NT3,NT4,NT5,NT6,NT7,NT8,NT9,NT10
      COMMON /VALUE/ F,G,P0,RSIGMA,RJ(100),RHO
      COMMON /CRST/ DELX(100),DELX0(100),RHOIN,RATIO,EPSI,THETAC,
     1 RSIG1, G1, X1(100), X2(100), X3(100), XR2(100), XR1(100), PR1,
     2 PR2, P1, F1, RJ1(200), DOTT, PGRAD(100), DIAG(100),
     3 PREV3, ADELX, NTCTR, NUMINI, NPHASE, NSATIS
C
      H=0.0
      RSIGMA=0.0
      F=0.0
      NSATIS=2
C     NPASE DETERMINES PHASE OF PROGRAM
C     1  PROBLEM IN FEASIBILITY PHASE
C     2  PROBLEM IN REGULAR PHASE
C     3  PROBLEM IN GUESS PHASE
C     4  EVALUATE ALL FUNCTIONS REGARDLESS OF PHASE
      GO TO (10,100,190,200), NPHASE
C     FEASIBILITY
10    GO TO (20,40), NT2
C     NON-NEGATIVIES INCLUDED
20    DO 30 I=1,N
      IF (X(I)) 260,260,30
30    RSIGMA=RSIGMA-RHO*ALOG(X(I))
40    IF (M.EQ.0) GO TO 90
      DO 80 J=1,M
      CALL RESTNT (J,RJ(J))
      IF (RJ1(J).LE.0.0) GO TO 50
      IF (RJ(J).GT.0.0) GO TO 60
C     VIOLATION OF A PREVIOUSLY SATISFIED CONSTRAINT
      GO TO 260
50    IF (RJ(J).GT.0.0) GO TO 70
C     ALL VIOLATED CONSTRAINTS ADDED INTO OBJECTIVE FUNCTION
      F=F-RJ(J)
      GO TO 80
60    RSIGMA=RSIGMA-RHO*ALOG(RJ(J))
      GO TO 80
C     INDICATES SATISFACTION OF CONSTRAINT (1 OR MORE)
70    NSATIS=1
      RSIGMA=RSIGMA-RHO*ALOG(RJ(J))
80    CONTINUE
90    CONTINUE
C     EQUALITIES NOT COMPUTED IN FEASIBILITY PHASE
      P0=F+RSIGMA
      G=F-RHO*FLOAT(M)
      IF (NT2.EQ.1) G=G-RHO*FLOAT(N)
      RETURN
```

```
C       REGULAR PHASE
100     GO TO (110,130), NT2
C       NON NEGATIVITIES INCLUDED
110     DO 120 I=1,N
        IF (X(I)) 260,260,120
120     RSIGMA=RSIGMA-RHO*ALOG(X(I))
130     IF (M.EQ.0) GO TO 150
        DO 140 J=1,M
        CALL RESTNT (J,RJ(J))
        IF (RJ(J).LE.0.0) GO TO 260
        RSIGMA=RSIGMA-RHO*ALOG(RJ(J))
140     CONTINUE
C       EVALUATE AND ADD IN EQUALITY CONSTRAINTS
150     CONTINUE
        CALL RESTNT (0,F)
        IF (MZ) 180,180,160
160     DO 170 I=1,MZ
        J=I+M
        CALL RESTNT (J,RJ(J))
C       ADD INTO THIRD TERM OF P FUNCTION
        H=H+(RJ(J))**2
170     CONTINUE
        H=H/RHO
180     P0=RSIGMA+H
        P0=F+P0
        G=2.*H-RHO*FLOAT(M)
        G=G+F
        IF (NT2.EQ.1) G=G-RHO*FLOAT(N)
C       DUAL VALUE
        RETURN
C       GUESS PHASE NOT CODED
190     RETURN
C       STRAIGHT FUNCTION EVALUATION ( MAIN+FEASIBLE ONLY)
200     CONTINUE
        IF (M.EQ.0) GO TO 220
        DO 210 I=1,M
        CALL RESTNT (I,RJ(I))
210     CONTINUE
220     CALL RESTNT (0,F)
C       EQUALITY CONSTRAINTS
        IF (MZ) 250,250,230
230     DO 240 I=1,MZ
        KZ=M+I
240     CALL RESTNT (KZ,RJ(KZ))
250     RETURN
C       CONSTRAINTS VIOLATED NOT SO BEFORE
260     NSATIS=3
        P0=10.E35
C
        RETURN
        END
```

```
      SUBROUTINE FEAS
C
C     FEAS DETERMINES WHETHER THE STARTING POINT IS FEASABLE.  IF IT IS
C     NOT, FEAS LOOKS FOR A FEASABLE ONE.  IF NONE EXISTS, A MESSAGE IS
C     PRINTED AND CONTROL RETURNS TO MAIN.
C
      COMMON /SHARE/ X(100),DEL(100),A(100,100),N,M,MN,NP1,NM1
      COMMON /OPTNS/ NT1,NT2,NT3,NT4,NT5,NT6,NT7,NT8,NT9,NT10
      COMMON /VALUE/ F,G,P0,RSIGMA,RJ(100),RHO
      COMMON /CRST/ DELX(100),DELX0(100),RHOIN,RATIO,EPSI,THETA0,
     1 RSIG1, G1, X1(100), X2(100), X3(100), XR2(100), XR1(100), PR1,
     2 PR2, P1, F1, RJ1(200), DOTT, PGRAD(100), DIAG(100),
     3 FREV3, ADELX, NTCTR, NUMINI, NPHASE, NSATIS
      COMMON /DEVC/ NI, NO, NP
C
      NPHASE=1
      GO TO (10,50), NT2
10    NFIX=1
      DO 30 I=1,N
      IF (X(I)) 20,20,30
20    NFIX=2
      X(I)=1.E-05
30    CONTINUE
      GO TO (50,40), NFIX
40    NPHASE=4
      CALL EVALU
      NPHASE=1
      WRITE (NO,130)
130   FORMAT (//,2X,72HMADE VARIABLES WHICH VIOLATED NON-NEGATIVE CONSTR
     1AINTS SLIGHTLY POSITIVE    )
      CALL OUTPUT (2)
50    IF (M) 90,90,60
60    DO 70 I=1,M
      IF (RJ(I)) 100,100,70
70    CONTINUE
80    CALL TIMEC
      WRITE (NO,140)
140   FORMAT (//,2X,38HTHE FEASIBLE STARTING POINT AND VALUES   )
      G=0.0
      CALL RESTNT (0,F)
      CALL OUTPUT (2)
90    RETURN
100   CALL BODY
      IF(NPHASE .EQ. 5) RETURN
      DO 110 I=1,M
      IF (RJ(I)) 120,120,110
110   CONTINUE
      GO TO 80
120   WRITE (NO,150)
150   FORMAT (/////,2X,88HTHIS PROBLEM POSSESSES NO FEASIBLE STARTING PO
     1INT.  WILL LOOK FOR DATA TO NEXT PROBLEM.    )
C     TO INDICATE TO MAIN TO START ON NEXT PROBLEM.
      NPHASE=5
      GO TO 90
C
      END
```

```
      SUBROUTINE FINAL (N2)
C
C     FINAL CONTAINS THE TESTS USED TO DETERMINE WHETHER A POINT
C     SATISFIES THE FINAL CONVERGENCE CRITERION CHOOSEN TO DETERMINE IF
C     THE NLP PROBLEM HAS BEEN SOLVED.
C     N2 SET EQUAL TO 1 IF CONVERGENCE CRITERION IS SATISFIED.
C     N2 SET EQUAL TO 2 OTHERWISE.
C
      COMMON /SHARE/ X(100),DEL(100),A(100,100),N,M,MN,NP1,NM1
      COMMON /OPTNS/ NT1,NT2,NT3,NT4,NT5,NT6,NT7,NT8,NT9,NT10
      COMMON /VALUE/ F,G,P0,RSIGMA,RJ(100),RHO
      COMMON /CRST/ DELX(100),DELX0(100),RHOIN,RATIO,EPSI,THETA0,
     1 RSIG1, G1, X1(100), X2(100), X3(100), XR2(100), XR1(100), PR1,
     2 PR2, P1, F1, RJ1(200), DOTT, PGRAD(100), DIAG(100),
     3 PREV3, ADELX, NTCTR, NUMINI, NPHASE, NSATIS
C
      GO TO (10,20,30), NT5
10    EPSIL=ABS(F/G-1.)
      IF (EPSIL-THETA0) 50,50,70
20    IF (ABS(RSIGMA)-THETA0) 50,50,70
30    IF (NUMINI-1) 50,40,40
40    PEST=PR1-(PR1-P0)/(1.-1./SQRT(RATIO))
      EPSIL=ABS(PEST/G-1.)
      IF (EPSIL-THETA0) 50,70,70
50    N2=1
      GO TO (80,60), NT6
60    CALL PUNCH
      GO TO 80
70    N2=2
C
80    RETURN
      END




      SUBROUTINE   GRAD1(IN)
C
C     THIS ROUTINE EVALUATES THE GRADIENT FOR THE OBJECTIVE FUNCTION AND
C     CONSTRAINTS FOR A GIVEN VALUE OF THE X VECTOR
C
      COMMON /SHARE/ X(100),DEL(100),A(100,100),N,M,MN,NP1,NM1
      COMMON /CONST/ AM, B, C, D, E
      DIMENSION AM(50,25), B(25), C(25,25), D(25), E(25)
C
      CALL DIFF1 (IN)
C
      RETURN
      END
```

```
      SUBROUTINE GRAD (IS)
C
C     GRAD COMPUTES THE GRADIENT OF THE PENALTY FUNCTION AND THE OUTER
C     PRODUCT FACTORS OF THE MATRIX OF SECOND PARTIALS OF P.
C     IF (IS=1) ACCUM. MATRIX OF 2ND PARTIALS   IF(IS=2) DONT
C
      COMMON /SHARE/ X(100),DEL(100),A(100,100),N,M,MN,NP1,NM1
      COMMON /EQAL/ H, H1, MZ
      COMMON /OPTNS/ NT1,NT2,NT3,NT4,NT5,NT6,NT7,NT8,NT9,NT10
      COMMON /VALUE/ F,G,P0,RSIGMA,RJ(100),RHO
      COMMON /CRST/ DELX(100),DELX0(100),RHOIN,RATIO,EPSI,THETA0,
     1 RSIG1, G1, X1(100), X2(100), X3(100), XR2(100), XR1(100), PR1,
     2 PR2, P1, F1, RJ1(200), DOTT, PGRAD(100), DIAG(100),
     3 PREV3, ADELX, NTCTR, NUMINI, NPHASE, NSATIS
C
      GO TO (10,30), IS
10    DO 20 I=1,N
      DO 20 J=1,I
20    A(I,J)=0.
30    DO 40 I=1,N
40    DELX0(I)=0.
C     THIS SECTION WORKS CORRECTLY IN FEASIBILITY PHASE AS WELL AS
C     NORMAL PHASE
      GO TO (50,80), NT2
50    DO 70 I=1,N
      DELX0(I)=-RHO/X(I)
      GO TO (60,70), IS
60    A(I,I)=(-DELX0(I)/X(I))
70    CONTINUE
80    CONTINUE
      IF (M.LE.0) GO TO 180
      DO 170 K=1,M
      CALL GRAD1 (K)
      IF (RJ(K).GT.0.0) GO TO 110
C     ALL VIOLATED CONSTRAINT  GRADS ADDED TO OBJECTIVE FUNCTION
      DO 100 I=1,N
      IF (DEL(I)) 90,100,90
90    DELX0(I)=DELX0(I)-DEL(I)
100   CONTINUE
      GO TO 170
110   TT=RHO/RJ(K)
      DO 160 I=1,N
      IF (DEL(I)) 120,160,120
C     IF DEL(I)=3 SKIP ALL THE FOLLOWING COMPUTATION INVOLVING * BY
C     DEL(I)
120   T=TT*DEL(I)
      DELX0(I)=DELX0(I)-T
      GO TO (130,160), IS
130   T=T/RJ(K)
      DO 150 JJ=1,I
      IF (DEL(JJ)) 140,150,140
140   A(I,JJ)=A(I,JJ)+T*DEL(JJ)
150   CONTINUE
160   CONTINUE
170   CONTINUE
```

```
C        EQUALITY CHANGES FOR GRAD
180      IF (MZ.LE.0) GO TO 250
         GO TO (250,190,250), NPHASE
190      RQ=2./RHO
         DO 240 J=1,MZ
         K=M+J
         CALL GRAD1 (K)
         TT=RQ*RJ(K)
         DO 230 I=1,N
         IF (DEL(I).EQ.0.0) GO TO 230
         DELX0(I)=DELX0(I)+DEL(I)*TT
         GO TO (200,230), IS
200      T=RQ*DEL(I)
         DO 220 JJ=1,I
         IF (DEL(JJ)) 210,220,210
210      A(I,JJ)=A(I,JJ)+T*DEL(JJ)
220      CONTINUE
230      CONTINUE
240      CONTINUE
250      GO TO (260,280), IS
260      DO 270 I=1,N
270      DIAG(I)=A(I,I)
280      GO TO (290,330,290), NPHASE
C        LEAVES NEGATIVE GRADIENT IN DELP
290      DO 300 I=1,N
300      DELX0(I)=-DELX0(I)
310      ADELX=0.
         DO 320 I=1,N
320      ADELX=ADELX+DELX0(I)**2
         ADELX=SQRT(ADELX)
         RETURN
330      CALL GRAD1 (0)
         DO 340 I=1,N
340      DELX0(I)=-DELX0(I)-DEL(I)
C        LEAVES THE NEGATIVE GADIENT OF P IN DELX0
         GO TO 310
C
         END




         SUBROUTINE MATRIX (J,L)
C
C        THIS ROUTINE EVALUATES THE MATRIX OF SECOND PARTIAL DERIVATIVES
C        FOR THE OBJECTIVE FUNCTION AND CONSTRAINTS
C
         COMMON /SHARE/ X(100),DEL(100),A(100,100),N,M,MN,NP1,NM1
         COMMON /CONST/ AM, B, C, D, E
         DIMENSION AM(50,25), B(25), C(25,25), D(25), E(25)
C
         CALL DIFF2 (J)
C
         RETURN
         END
```

```
      SUBROUTINE INVERS (NSME)
C
C     INVERS SOLVES THE SET OF EQUATION FOR THE MOVE-VECTOR USING
C     THE CROUT PROCEDURE.  IF THE MASTRIX IS NOT POSITIVE DEFINITE, A
C     DIFFERENT METHOD IS USED.
C     PERFORMING A L-U DECOMPOSITION OF THE MATRIX A, TAKING ADVANTAGE
C     OF THE SYMMETRY OF THE A MATRIX.
C     IF A NON-POSITIVE PIVOT CANDIDATE IS GENERATED, THEN MCCORMICK,S
C     PRODECURE IS USED(SEE PP. 167-168 IN FIACCO AND MCCORMICK).
C     IF NSME =1 WORKING WITH A NEW A MATRIX  IF NSME= 2 USING PREVIOUS
C     A MATRIX, BUT HAVE A NEW RIGHT-HAND-SIDE.
C     NINV IS THE NUMBER OF NON-POSITIVE PIVOT CANDIDATES GENERATED.
C
      COMMON /SHARE/ X(100),DEL(100),A(100,100),N,M,MN,NP1,NM1
      COMMON /OPTNS/ NT1,NT2,NT3,NT4,NT5,NT6,NT7,NT8,NT9,NT10
      COMMON /CRST/ DELX(100),DELX0(100),RHOIN,RATIO,EPSI,THETA0,
     1 RSIG1, G1, X1(100), X2(100), X3(100), XR2(100), XR1(100), PR1,
     2 PR2, P1, F1, RJ1(200), DOTT, PGRAD(100), DIAG(100),
     3 PREV3, ADELX, NTCTR, NUMINI, NPHASE, NSATIS
      COMMON /EXPOPT/ NEXOP1, NEXOP2, XEP1, XEP2
      COMMON /DEVC/ NI, NO, NP
      DIMENSION B(100)
      EQUIVALENCE (B,DELX)
C
      GO TO(20, 170), NSME
20    NINV=0
      IF (A(1,1)) 40,30,50
30    NINV=1
      GO TO 70
40    NINV=1
50    A(1,1)=1./A(1,1)
      DO 60 I=2,N
60    A(1,I)=A(1,I)*A(1,1)
70    DO 160 J=2,N
      JM1=J-1
      T=0.
      DO 90 I=1,JM1
      IF (A(I,J)) 80,90,80
80    T=T+A(J,I)*A(I,J)
90    CONTINUE
      A(J,J)=A(J,J)-T
      IF (A(J,J)) 110,100,120
100   NINV=NINV+1
      GO TO 170
110   NINV=NINV+1
120   A(J,J)=1./A(J,J)
      IF (J.EQ.N) GO TO 170
      JP1=J+1
      DO 150 L=JP1,N
      T=0.
      DO 140 I=1,JM1
      IF (A(I,J)) 130,140,130
130   T=T+A(L,I)*A(I,J)
140   CONTINUE
      A(L,J)=A(L,J)-T
```

```
              A(J,L)=A(L,J)*A(J,J)
150     CONTINUE
160     CONTINUE
170     CONTINUE
        IF (NINV) 180,180,290
180     B(1)=B(1)*A(1,1)
        DO 210 J=2,N
        T=0.
        JM1=J-1
        DO 200 I=1,JM1
        IF (A(J,I)) 190,200,190
190     T=T+A(J,I)*B(I)
200     CONTINUE
        B(J)=(B(J)-T)*A(J,J)
210     CONTINUE
        DO 240 I=1,NM1
        NMK=N-I
        DO 230 J=1,I
        L=NP1-J
        IF (A(NMK,L)) 220,230,220
220     B(NMK)=B(NMK)-A(NMK,L)*B(L)
230     CONTINUE
240     CONTINUE
250     GO TO (280,260), NT3
260     WRITE (NO,430)
430     FORMAT (//,2X,12HDEL P VECTOR  )
        WRITE (NO,420) (I, DELX0(I), I=1,N)
420     FORMAT (/,3(2X,4HDEL(,I2,4H) = ,E16.8))
270     WRITE (NO,440)
440     FORMAT (//,2X,24HSECOND ORDER MOVE VECTOR  )
        WRITE (NO,420) (I, DELX(I), I=1,N)
280     RETURN
C       COMPUTE ORTHOGONAL MOVE
290     CONTINUE
        DO 350 II=1,N
        I=N-II+1
        IF (A(I,I)) 310,300,320
300     B(I)=0.0
        GO TO 350
310     B(I)=1.0
        GO TO 330
320     B(I)=0.0
330     IP1=I+1
        IF (IP1.GT.N) GO TO 350
        DO 340 J=IP1,N
340     B(I)=B(I)-A(I,J)*B(J)
350     CONTINUE
        GO TO 360
C       CHECK MAYBE DO DIFF FOR P.S.D.
360     ZC2=0.0
        DO 370 I=1,N
370     ZC2=ZC2+DELX0(I)*B(I)
        IF (ZC2) 380,400,400
380     DO 390 I=1,N
390     B(I)=-B(I)
```

```
  400     WRITE (NO,450)
  450     FORMAT (//,2X,23HORTHOGONAL MOVE VECTORS )
          IF (NEXOP2.NE.2) GO TO 250
          DO 410 K=1,N
  410     B(K)=B(K)+DELX0(K)
          GO TO 250
C
          END




          SUBROUTINE OUTPUT (K)
C
C     OUTPUT PRINTS OUT INFORMATION ON THE RESULTS OF EACH ITERATION AND
C     THE SOLUTION ESTIMATES AND THE ESTIMATES OF THE LAGRANGE
C     MULTIPLIERS
C
      COMMON /SHARE/ X(100),DEL(100),A(100,100),N,M,MN,NP1,NM1
      COMMON /EQAL/ H, H1, MZ
      COMMON /OPTNS/ NT1,NT2,NT3,NT4,NT5,NT6,NT7,NT8,NT9,NT10
      COMMON /VALUE/ F,G,P0,RSIGMA,RJ(100),RHO
      COMMON /CRST/ DELX(100),DELX0(100),RHOIN,RATIO,EPSI,THETA0,
     1 RSIG1, G1, X1(100), X2(100), X3(100), XR2(100), XR1(100), PR1,
     2 PR2, P1, F1, RJ1(200), DOTT, PGRAD(100), DIAG(100),
     3 PREV3, ADELX, NTCTR, NUMINI, NPHASE, NSATIS
      COMMON /DEVC/ NI, NO, NP
C
      NZ=M+MZ
      GO TO (10,20), K
C
   10 WRITE (NO,001) NTCTR
  001 FORMAT (//,2X,23HVALUES AT POINT NUMBER ,I2 )
      WRITE (NO,002) NPHASE, RHO, RSIGMA
  002 FORMAT (/,2X,9HNPHASE = ,I1,4X,6HRHO = ,E16.8,4X,9HRSIGMA = ,
     1 E16.8 )
C
   20 WRITE (NO,003) F, P0, G
  003 FORMAT  (/,2X,4HF = ,E16.8,4X,4HP = ,E16.8,4X,4HG = ,E16.8 )
      WRITE (NO,004)
  004 FORMAT (/,2X,18HVALUES OF X VECTOR )
      WRITE (NO,005) (J, X(J), J=1,N)
  005 FORMAT  (/,3(2X,2HX(,I2,4H) = ,E16.8))
      WRITE (NO,006)
  006 FORMAT (/,2X,25HVALUES OF THE CONSTRAINTS )
      GO TO (30,40), NT2
   30 WRITE (NO,007)
  007 FORMAT (/,2X,42HNOT INCLUDING THE NON-NEGATIVE CONSTRAINTS )
      WRITE (NO,008) (I, RJ(I), I=1,NZ)
  008 FORMAT (/,3(2X,2HG(,I2,4H) = ,E16.8))
      GO TO 50
   40 WRITE (NO,008) (I, RJ(I), I=1,NZ)
C
   50 RETURN
      END
```

```
      SUBROUTINE OPT
C
C     OPT LOOKS FOR A MINIMUM ALONG THE SEARCH VECTOR USING THE GOLDEN
C     SECTION SEARCH METHOD.
C
      COMMON /SHARE/ X(100),DEL(100),A(100,100),N,M,MN,NP1,NM1
      COMMON /VALUE/ F,G,P0,RSIGMA,RJ(100),RHO
      COMMON /CRST/ DELX(100),DELX0(100),RHOIN,RATIO,EPSI,THETA0,
     1 RSIG1, G1, X1(100), X2(100), X3(100), XR2(100), XR1(100), PR1,
     2 PR2, P1, F1, RJ1(200), DOTT, PGRAD(100), DIAG(100),
     3 PREV3, ADELX, NTCTR, NUMINI, NPHASE, NSATIS
      COMMON /DEVC/ NI, NO, NP
C
      KSW=1
      N405=1
      P31=P0
      P31=P0
      ISW=1
      DOTT=0.
      DO 10 J=1,N
10    DOTT=DOTT+DELX(J)*DELX0(J)
      GO TO 40
20    DO 30 I=1,N
30    DELX(I)=-DELX(I)
40    CONTINUE
      N404=0
      MN=MN+1
C     MN IS NOW NUMBERS OF POINTS AFTER MINIMUM ACHIEVED
      NTCTR=NTCTR+1
      DO 50 I=1,N
50    X2(I)=X(I)
      PX1=P0
      N401=0
60    N401=N401+1
      DO 70 I=1,N
70    X(I)=X2(I)+DELX(I)
      CALL EVALU
C     1 MEANS SATISFACTION CONSTRAINT NOT PREVIOUSLY SATISFIED.  2 MEANS
C     NO CHANGE, 3 MEANS VIOLATION.
C     IF POINT IS NOT FEASIBLE GIVE IT AN ARBRITRARILY HIGH VALUE.
      GO TO (540,90,80), NSATIS
80    PX2=10.E35
      P0=10.E35
      GO TO 100
90    CONTINUE
      PX2=P0
      IF (PX1-PX2) 100,100,150
100   IF (N401-2) 130,110,110
110   DO 120 I=1,N
120   X1(I)=X(I)
      P1=PX2
      GO TO 430
C     ONLY ONE POINT SO FAR COMPUTED
130   DO 140 I=1,N
140   X3(I)=X2(I)
```

```
              PREV3=PX1
              GO TO 180
150           DO 160 I=1,N
              X3(I)=X2(I)
              X2(I)=X(I)
160           DELX(I)=1.61803399*DELX(I)
              PREV3=PX1
              PX1=PX2
              GO TO 60
C       GOLDEN SECTION SEARCH METHOD.
C       B VECTOR GOES TO X1(I)
170           P0=1.E36
              N404=N404+1
180           DO 190 I=1,N
190           X1(I)=X(I)
              P1=P0
              DO 200 I=1,N
              X(I)=.38196601*(X1(I)-X3(I))+X3(I)
200           X2(I)=X(I)
              CALL EVALU
              GO TO (540,270,210), NSATIS
210           IF (N404.LT.30) GO TO 170
C       IT IS POSSIBLE NO FEASIBLE POINT EXISTS, IF NOT TRY MOVING ON
C       DELX0.  IF IT IS NOT POSSIBLE TO MOVE ON DELX0 THEN WE MUST BE
C       AT A SOLUTION OF THE NLP PROBLEM.
              IF (N404.GT.100) GO TO 240
220           DO 230 I=1,N
              IF (ABS(ABS(X3(I)/X1(I))-1.).GT.1.E-7) GO TO 170
230           CONTINUE
240           GO TO (250,260), N405
250           N405=2
C       TRY TO MOVE ON GRADIENT.
              NTCTR=NTCTR-1
              MN=MN-1
              GO TO 20
260           WRITE (NO,580)
580           FORMAT (//,2X,75HOPT CAN≠T FIND A FEASIBLE POINT THAT GIVES A LOWE
             1R VALUE OF THE P-FUNCTION  )
              CALL TIMEC
              CALL OUTPUT (1)
              CALL REJECT
              STOP 22042
C
270           CONTINUE
              N404=0
              PX1=P0
              DO 280 I=1,N
280           X(I)=0.38196601*(X1(I)-X2(I))+X2(I)
              CALL EVALU
              GO TO (540,290,220), NSATIS
290           PX2=P0
              N401=1
300           N401=N401+1
              IF (N401-25) 340,310,310
310           KSW=2
```

```
          IF (N401-40) 320,460,460
320       DO 330 I=1,N
          IF (ABS(X2(I)/X(I)-1.0).GE.1.E-7) GO TO 340
330       CONTINUE
          GO TO 460
340       IF (ABS(PX1/PX2-1.).LE.1.E-7) GO TO 460
          IF (PX1-PX2) 350,460,400
C         THROW AWAY RIGHT PART
350       DO 360 I=1,N
360       X1(I)=X(I)
          P1=PX2
          DO 370 I=1,N
C         POINT XP1 BECOMES XP2 TEMPORARILY IN X STORAGE
370       X(I)=.38196601*(X1(I)-X3(I))+X3(I)
          CALL EVALU
          GO TO (540,380,170), NSATIS
380       CONTINUE
          PX2=PX1
C         SWITCH VECTORS TO PROPER POSITION
          PX1=P0
          DO 390 I=1,N
          XX=X2(I)
          X2(I)=X(I)
390       X(I)=XX
          GO TO 300
C         LEFT SIDE TOSSED AWAY
C         CHANGES FOR NONUNIMODAL FUNCTION.  GO TO THROW AWAY RIGHT IN CASE
C         INITIAL VALUE LESS THAN FEASIBLE POINT.
400       IF (PREV3-PX2) 350,350,410
410       DO 420 I=1,N
          X3(I)=X2(I)
420       X2(I)=X(I)
          PREV3=PX1
          PX1=PX2
430       DO 440 I=1,N
440       X(I)=0.38196601*(X1(I)-X2(I))+X2(I)
          CALL EVALU
          GO TO (540,450,170), NSATIS
450       CONTINUE
          PX2=P0
          GO TO 300
C         THE INTERIOR POINTS NOW GIVE EQUAL VALUE FOR P.  COMPUTE MIDPOINT.
460       DO 470 I=1,N
          DELX0(I)=X(I)
          X(I)=(DELX0(I)+X2(I))*0.5
470       CONTINUE
          CALL EVALU
          GO TO (480,490), KSW
480       IF (ABS(P0/PX1-1.).GT.1.E-7) GO TO 520
490       GO TO (500,510), ISW
500       IF (P0.LT.P31) GO TO 510
          ISW=2
C         IF P-FUNCTION DIDN,T   GO DOWN, TRY NEGATIVE VECTOR.
          GO TO 20
510       RETURN
```

```
520     DO 530 I=1,N
530     X(I)=DELX0(I)
        GO TO 350
C       WE ARE NOW IN FEASIBILITY PHASE
540     DO 550 I=1,M
        IF (RJ(I)) 560,560,550
550     CONTINUE
        NSATIS=4
        RETURN
C       PROBLEM HAS BECOME FEASIBLE
C       P - FUNCTION CHANGES IF A CONSTRAINT BECOMES FEASIBLE
560     MN=0
        DO 570 I=1,M
570     RJ1(I)=RJ(I)
C
        RETURN
        END




        SUBROUTINE PUNCH
C
C       THIS SUBROUTINE PUNCHES THE STOPPING POINTS AND ASSOCIATED
C       PARAMETERS SO THAT ANOTHER RUN MAY BE MADE STARTING WHERE THE
C       CURRENT ONE STOPPED.
C       THIS ROUTINE IS CALLED IF NT6=2.
C
        COMMON /SHARE/ X(100),DEL(100),A(100,100),N,M,MN,NP1,NM1
        COMMON /EQAL/ H, H1, MZ
        COMMON /OPTNS/ NT1,NT2,NT3,NT4,NT5,NT6,NT7,NT8,NT9,NT10
        COMMON /VALUE/ F,G,P0,RSIGMA,RJ(100),RHO
        COMMON /CRST/ DELX(100),DELX0(100),RHOIN,RATIO,EPSI,THETA0,
       1 RSIG1, G1, X1(100), X2(100), X3(100), XR2(100), XR1(100), PR1,
       2 PR2, P1, F1, RJ1(200), DOTT, PGRAD(100), DIAG(100),
       3 PREV3, ADELX, NTCTR, NUMINI, NPHASE, NSATIS
        COMMON /EXPOPT/ NEXOP1, NEXOP2, XEP1, XEP2
        COMMON /DEVC/ NI, NO, NP
C
C       TIME LIMIT IS SET TO 60.0 SECONDS.
        T=60.0
        WRITE (NP,001) EPSI, RHO, THETA0, RATIO, T
001     FORMAT (8E10.4)
        WRITE (NP,002) M, N, MZ
002     FORMAT (8I10)
        WRITE (NP,001) (X(I),I=1,N)
C       SET RHO OPTION SO THIS VALUE OF RHO WILL BE USE FOR THE RESTART.
        NT1=3
        WRITE (NP,003) NT1,NT2,NT3,NT4,NT5,NT6,NT7,NT8,NT9,NT10
003     FORMAT (16I5)
        WRITE (NP,001) XEP1,XEP2
        WRITE (NP,002) NEXOP1,NEXOP2
C
        RETURN
        END
```

```
      SUBROUTINE PEVALU
C
C     PEVALU COMPUTES THE VALUE OF THE PENALTY FUNCTION AND THE VALUE
C     OF THE DUAL USING PREVIOUSLY COMPUTED VALUES FOR F, AND RJ.
C
      COMMON /SHARE/ X(100),DEL(100),A(100,100),N,M,MN,NP1,NM1
      COMMON /EQAL/ H, H1, MZ
      COMMON /OPTNS/ NT1,NT2,NT3,NT4,NT5,NT6,NT7,NT8,NT9,NT10
      COMMON /VALUE/ F,G,P0,RSIGMA,RJ(100),RHO
      COMMON /CRST/ DELX(100),DELX0(100),RHOIN,RATIO,EPSI,THETA0,
     1 RSIG1, G1, X1(100), X2(100), X3(100), XR2(100), XR1(100), PR1,
     2 PR2, P1, F1, RJ1(200), DOTT, PGRAD(100), DIAG(100),
     3 PREV3, ADELX, NTCTR, NUMINI, NPHASE, NSATIS
C
      H=0.0
      RSIGMA=0.0
C     NONNEGS IF INCLUDED ARE ADDED TO  P--ARE POSITIVE IN ALL PHASES
      GO TO (10,30), NT2
10    DO 20 I=1,N
20    RSIGMA=RSIGMA-RHO*ALOG(X(I))
30    GO TO (40,50,150), NPHASE
C     OBJECTIVE FUNCTION   - SIGMA VIOLATED CONSTRAINTS
40    F=0.0
50    IF (M) 100,100,60
60    DO 90 J=1,M
      IF (RJ(J)) 80,80,70
70    RSIGMA=RSIGMA-RHO*ALOG(RJ(J))
      GO TO 90
80    F=F-RJ(J)
90    CONTINUE
C     EQUALITIES NOT ADDED IN FEASIBILITY PHASE
100   CONTINUE
      IF (MZ) 140,140,110
110   GO TO (140,120,150), NPHASE
120   DO 130 I=1,MZ
      K=M+I
130   H=H+RJ(K)**2
      H=H/RHO
140   HS=H+RSIGMA
      P0=F+HS
      HMS=2.*H-RHO*FLOAT(M)
      G=F+HMS
      IF (NT2.EQ.1) G=G-RHO*FLOAT(N)
C
150   RETURN
      END




      SUBROUTINE RHOCOM
C
C     SUBROUTINE  TO COMPUTE INITIAL RHO VALUE
C
      COMMON /SHARE/ X(100),DEL(100),A(100,100),N,M,MN,NP1,NM1
      COMMON /OPTNS/ NT1,NT2,NT3,NT4,NT5,NT6,NT7,NT8,NT9,NT10
```

```
      COMMON /VALUE/ F,G,P0,RSIGMA,RJ(100),RHO
      COMMON /CRST/ DELX(100),DELX0(100),RHOIN,RATIO,EPSI,THETA0,
     1 RSIG1, G1, X1(100), X2(100), X3(100), XR2(100), XR1(100), PR1,
     2 PR2, P1, F1, RJ1(200), DOTT, PGRAD(100), DIAG(100),
     3 PREV3, ADELX, NTCTR, NUMINI, NPHASE, NSATIS
C
      GO TO (110,50,10,190), NT1
10    RHO=RHOIN
20    IF (RHO) 30,30,40
30    RHO=1.
40    RETURN
50    NPAR1=1
60    RHO=1.
C     2 MEANS RHO WHICH MINIMIZES GRADIENT MAGNITUDE.
      CALL GRAD (2)
      DO 70 I=1,N
70    PGRAD(I)=DELX0(I)
      RHO=2.
      CALL GRAD (2)
      DO 80 I=1,N
      DELX0(I)=DELX0(I)-PGRAD(I)
80    PGRAD(I)=PGRAD(I)-DELX0(I)
      GO TO (90,130), NPAR1
90    DOT1=0.
      DOT2=0.
      DO 100 I=1,N
      DOT1=DOT1+DELX0(I)*PGRAD(I)
100   DOT2=DOT2+DELX0(I)**2
      RHO=ABS(DOT1/DOT2)
      GO TO 20
C     3 MEANS COMPUTE RHO SO AS TO MINIMIZE DEL P(/DDP/1.)DEL P
110   NPAR2=1
120   NPAR1=2
      GO TO 60
130   RHO=1.
C     ASSUME SIGMA TERM IS CONSIDERABLY GREATER THAN F TERM.
      CALL SECORD (2)
      DO 140 I=1,N
140   DELX(I)=PGRAD(I)
      CALL INVERS (1)
      DO 150 I=1,N
      X1(I)=DELX(I)
150   DELX(I)=DELX0(I)
      CALL SECORD (2)
      CALL INVERS (1)
      DO 160 I=1,N
160   XR2(I)=DELX(I)
      GO TO (170,200), NPAR2
170   DOT1=0.
      DOT2=0.
      DO 180 I=1,N
      DOT1=DOT1+PGRAD(I)*X1(I)
180   DOT2=DOT2+DELX0(I)*XR2(I)
      RHO=SQRT(ABS(DOT1/DOT2))
      GO TO 20
190   NPAR2=2
C     RHO MINIMIZES  2ND ORDER MOVE
      GO TO 120
```

```
200   DOT1=0.0
      DOT2=0.0
      DO 210 I=1,N
      DOT1=X1(I)**2+DOT1
210   DOT2=X1(I)*XR2(I)+DOT2
      RHO=ABS(DOT1/DOT2)
      GO TO 20
C
      END




      SUBROUTINE SECORD (IS)
C
C     SECORD EVALUATES THE MATRIX OF SECOND PARTIALS OF THE PENALTY
C     FUNCTION.
C     (1) MEANS DON,T COMPUTE GRADIENT OUTER PRODUCT ( IN SECORD).
C
      COMMON /SHARE/ X(100),DEL(100),A(100,100),N,M,MN,NP1,NM1
      COMMON /EQAL/ H, H1, MZ
      COMMON /OPTNS/ NT1,NT2,NT3,NT4,NT5,NT6,NT7,NT8,NT9,NT10
      COMMON /VALUE/ F,G,P0,RSIGMA,RJ(100),RHO
      COMMON /CRST/ DELX(100),DELX0(100),RHOIN,RATIO,EPSI,THETA0,
     1 RSIG1, G1, X1(100), X2(100), X3(100), XR2(100), XR1(100), PR1,
     2 PR2, P1, F1, RJ1(200), DOTT, PGRAD(100), DIAG(100),
     3 PREV3, ADELX, NTCTR, NUMINI, NPHASE, NSATIS
C
      DO 10 I=1,N
      DO 10 J=I,N
10    A(I,J)=0.
      GO TO (230,20), IS
C     GRADIENT TERM NOT PREVIOUSLY COMPUTED.
20    DO 30 I=1,N
      DO 30 J=1,I
      A(I,J)=0.0
30    CONTINUE
      GO TO (40,60), NT2
40    DO 50 I=1,N
50    A(I,I)=RHO/X(I)**2
60    CONTINUE
      IF (M.LE.0) GO TO 130
      DO 120 IN=1,M
      IF (RJ(IN)) 120,120,70
70    CALL GRAD1 (IN)
      TT = RHO/RJ(IN)**2
      DO 110 I=1,N
      IF (DEL(I)) 80,110,80
80    T=TT*DEL(I)
      DO 100 J=1,I
      IF (DEL(J)) 90,100,90
90    A(I,J)=A(I,J)+T*DEL(J)
100   CONTINUE
110   CONTINUE
120   CONTINUE
C     EQUALTIY CONSTRAINTS
```

```
130       IF (MZ) 210,210,140
140       GO TO (210,150,230), NPHASE
150       RQ=2./RHO
          DO 200 JJ=1,MZ
          IN=M+JJ
          CALL GRAD1 (IN)
          DO 190 I=1,N
          IF (DEL(I)) 160,190,160
160       T=RQ*DEL(I)
          DO 180 J=1,I
          IF (DEL(J)) 170,180,170
170       A(I,J)=A(I,J)+T*DEL(J)
180       CONTINUE
190       CONTINUE
200       CONTINUE
210       DO 220 I=1,N
          DIAG(I)=A(I,I)
220       A(I,I)=0.
C         READY NOW FOR MATRIX OF 2ND PARTIALS OF RESTRAINTS
230       GO TO (240,510,520), NT10
240       IF (M.LE.0) GO TO 340
          DO 330 IN=1,M
          LORN=2
C         CONSTRAINT ASSUMED NONLINEAR
          CALL MATRIX (IN,LORN)
          IF (LORN.LT.2) GO TO 330
          IF (RJ(IN) .GT. 0.0) GO TO 280
          DO 260 I=2,N
          IM1=I-1
          DO 260 J=1,IM1
          IF (A(J,I)) 250,260,250
250       A(I,J)=A(I,J)-A(J,I)
          A(J,I)=0.
260       CONTINUE
          DO 270 I=1,N
          DIAG(I)=DIAG(I)-A(I,I)
270       A(I,I)=0.0
          GO TO 330
280       T=-RHO/RJ(IN)
          DO 300 I=2,N
          IM1=I-1
          DO 300 J=1,IM1
          IF (A(J,I)) 290,300,290
290       A(I,J)=A(I,J)+T*A(J,I)
          A(J,I)=0.
300       CONTINUE
          DO 320 I=1,N
          IF (A(I,I)) 310,320,310
310       DIAG(I)=DIAG(I)+T*A(I,I)
          A(I,I)=0.
320       CONTINUE
330       CONTINUE
340       CONTINUE
          GO TO (520,350,520), NPHASE
350       IF (MZ.EQ.0) GO TO 420
C         EQUALITY SECOND PARTIALS HERE
          IF (NT10.GE.2) GO TO 420
          DO 410 II=1,MZ
```

```
          IN=M+II
          LORN=2
          CALL MATRIX (IN,LORN)
          IF (LORN.LT.2) GO TO 410
          T=2.*RJ(IN)/RHO
          DO 380 I=2,N
          IM1=I-1
          DO 370 J=1,IM1
          IF (A(J,I)) 360,370,360
360       A(I,J)=A(I,J)+T*A(J,I)
          A(J,I)=0.0
370       CONTINUE
380       CONTINUE
          DO 400 I=1,N
          IF (A(I,I)) 390,400,390
390       DIAG(I)=DIAG(I)+T*A(I,I)
          A(I,I)=0.0
400       CONTINUE
410       CONTINUE
C         GET MATRIX OF 2ND PARTIALS OF OBJECTIVE FUNCTION
420       LLL=2
          CALL MATRIX (0,LLL)
          IF (LLL.LT.2) GO TO 490
          DO 440 I=2,N
          IM1=I-1
          DO 440 J=1,IM1
          IF (A(J,I)) 430,440,430
430       A(I,J)=A(I,J)+A(J,I)
440       A(J,I)=A(I,J)
          DO 470 I=1,N
          IF (A(I,I)) 450,460,450
450       A(I,I)=DIAG(I)+A(I,I)
          GO TO 470
460       A(I,I)=DIAG(I)
470       CONTINUE
480       RETURN
490       DO 500 I=1,N
          A(I,I)=DIAG(I)
          DO 500 J=I,N
500       A(I,J)=A(J,I)
          GO TO 480
510       GO TO (520,350,350), NPHASE
520       DO 530 I=2,N
          IM1=I-1
          DO 530 J=1,IM1
530       A(J,I)=A(I,J)
          DO 540 I=1,N
540       A(I,I)=DIAG(I)
          GO TO 480
C
          END
```

```
      SUBROUTINE XMOVE
C
C     XMOVE DETERMINES THE VECTOR ALONG WHICH THE SEARCH  FOR A MINIMUM
C     IS USING OPT.
C     NEXOP2 DETERMINES HOW MOVE IS TO BE MADE
C               1   USE MODIFIED  NEWTON RAPHSON METHOD.
C               2   USE MODIFIED  NEWTON RAPHSON METHOD, BUT ADD DELX0 TO
C                   ORTHOGONAL MOVE VECTOR IF HEXXIAN IS INDEFINITE.
C               3   USE STEEPEST DESCENT METHOD.
C               4   USE MCCORMICK,S MODIFICATION OF THE FLECTCHER-POWELL
C                   METHOD.
C
      COMMON /SHARE/ X(100),DEL(100),A(100,100),N,M,MN,NP1,NM1
      COMMON /CRST/ DELX(100),DELX0(100),RHOIN,RATIO,EPSI,THETA0,
     1 RSIG1, G1, X1(100), X2(100), X3(100), XR2(100), XR1(100), PR1,
     2 PR2, P1, F1, RJ1(200), DOTT, PGRAD(100), DIAG(100),
     3 PREV3, ADELX, NTCTR, NUMINI, NPHASE, NSATIS
      COMMON /EXPOPT/ NEXOP1, NEXOP2, XEP1, XEP2
      COMMON /XVE/ SIG(100),YY(100),XXX(100),DELL(100)
C
      GO TO (10,10,180,30), NEXOP2
C     NEWTON-RAPHSON WITH WHATEVER METHOD IS IN INVERSE
10    CALL GRAD (1)
C     ONE (1) MEANS ACCUMULATE MATRIX OF SECOND PARTIAL DERIVATIVES
      CALL SECORD (1)
      DO 20 I=1,N
20    DELX(I)=DELX0(I)
      CALL INVERS (1)
C     IF A NONPOSITIVE PIVOT IS ENCOUNTERED IN INVERSE AN ATTEMPT IS
C     MADE TO COMPUTE A VECTOR HAVING A POSITIVE DOT PRODUCT WITH A
C     NEGATIVE EIGENVECTOR AND THE NEGATIVE OF DEL P.
      CALL STORE
      CALL OPT
      RETURN
30    CALL GRAD (2)
C     MN IS NO. OF MOVES FOR THIS VALUE OF RHO
      IF (MN.NE.0) GO TO 70
40    IREP=0
      IT=0
C     SET INITIAL GUESS INVERS MATRIX OF SECOND PARTIAL DERIVATIVES
C     USE PARTIAL INVERSE IF KNOWN
      DO 50 I=1,N
      DO 50 J=1,N
50    A(I,J)=0.0
      DO 60 I=1,N
60    A(I,I)=1.0
70    DO 80 I=1,N
80    DELX(I)=DELX0(I)
      IF (IREP.GT.N) GO TO 40
      IF (IT.EQ.0) GO TO 130
      DO 90 I=1,N
      SIG(I)=X(I)-XXX(I)
90    YY(I)=DELL(I)-DELX0(I)
C     NEGATIVE GRADIENT STORED AND COMPUTED.  COMPUTE HY.
      DO 100 I=1,N
```

```
              DELX(I)=0.0
              DO 100 J=1,N
100           DELX(I)=DELX(I)+A(I,J)*YY(J)
C       COMPUTE  Y(SIG -HY) -1
              ZCON=0.0
              DO 110 I=1,N
110           ZCON=ZCON+YY(I)*(SIG(I)-DELX(I))
              IF (ZCON.EQ.0.0) GO TO 130
              IREP=IREP+1
              ZC=1./ZCON
C       UPDATE H MATRIX USING MCC FORMULA WHEN SCALAR NOT EQUAL TO ZERO.
              DO 120 I=1,N
              T1=ZC*(SIG(I)-DELX(I))
              DO 120 J=I,N
              A(I,J)=A(I,J)+T1*(-DELX(J)+SIG(J))
120           A(J,I)=A(I,J)
C       STORE CURRENT POINT AND CURRENT GRADIENT (NEG)
130           DO 140 I=1,N
              XXX(I)=X(I)
140           DELL(I)=DELX0(I)
              DO 150 I=1,N
              DELX(I)=0.0
              DO 150 J=1,N
150           DELX(I)=DELX(I)+A(I,J)*DELX0(J)
              ZC1=0.0
              DO 160 I=1,N
160           ZC1=DELX(I)**2+ZC1
              ZC1=SQRT(ZC1)
              DO 170 I=1,N
170           DELX(I)=DELX(I)/ZC1
              CALL STORE
              CALL OPT
              IT=IT+1
              RETURN
180           CONTINUE
C       STEEPEST DESCENT
              CALL GRAD (2)
              DO 190 I=1,N
190           DELX(I)=DELX0(I)
              CALL STORE
              CALL OPT
C
              RETURN
              END
```

```fortran
      SUBROUTINE   RESTNT (IN,VAL)
C
C     THIS ROUTINE EVALUATES THE OBJECTIVE FUNCTION AND CONSTRAINTS FOR
C     A GIVEN VALUE OF THE X VECTOR
C
      COMMON /SHARE/ X(100),DEL(100),A(100,100),N,M,MN,NP1,NM1
      COMMON /CONST/ AM, B, C, D, E
      DIMENSION AM(50,25), B(25), C(25,25), D(25), E(25)
C
      IF (IN) 10, 10, 20
C
   10 VAL = C(1,1)*X(1) * C(1,2)*X(2) * C(1,3)*X(3)
      VAL = - VAL
      GO TO 99
C
   20 GO TO (1,2,3,4), IN
    1 VAL = B(1) - AM(1,1) * X(1)
      GO TO 99
    2 VAL = B(2) - AM(2,2) * X(2)
      GO TO 99
    3 VAL = B(3) - AM(3,3) * X(3)
      GO TO 99
    4 VAL = AM(4,1)*X(1) + AM(4,2)*X(2) + AM(4,3)*X(3) - B(4)
      VAL = - VAL
C
   99 RETURN
      END




      SUBROUTINE TIMEC
C
C     TIMEC CHECKS THE NUMBER OF SECONDS THAT HAVE ELAPSED SINCE THE
C     START OF THE PROBLEM.  IT PRINTS THIS NUMBER.  IF THE SOLUTION
C     IS TAKING LONGER THAN THE ESTIMATES MAXIMUM TIME, A SWITCH IS
C     SET TO TERMINATE THE RUN.
C
      COMMON /TSW/ NSWW
      COMMON /TMX/ TMO,EXT,EXT90
      COMMON /DEVC/ NI, NO, NP
C
C     SECOND GIVES JOB CPU EXECUTION TIME IN 1/1000 OF A SECOND
C
      CALL SECOND (SECS)
      X=SECS-TMO
      WRITE (NO,001) X
  001 FORMAT (//,2X,7HTIME = ,F9.3,8H SECONDS )
      IF (SECS.LT.EXT) GO TO 10
      NSWW=2
C
   10 RETURN
      END
```

```
      SUBROUTINE STORE
C
C     STORE STORES THE VALUES OF THE CURRENT POINT AND THE ASSOCIATED
C     VALUES OF THE FUNCTION IN A TEMPORARY AREA.
C
      COMMON /SHARE/ X(100),DEL(100),A(100,100),N,M,MN,NP1,NM1
      COMMON /EQAL/ H, H1, MZ
      COMMON /VALUE/ F,G,P0,RSIGMA,RJ(100),RHO
      COMMON /CRST/ DELX(100),DELX0(100),RHOIN,RATIO,EPSI,THETA0,
     1 RSIG1, G1, X1(100), X2(100), X3(100), XR2(100), XR1(100), PR1,
     2 PR2, P1, F1, RJ1(200), DOTT, PGRAD(100), DIAG(100),
     3 PREV3, ADELX, NTCTR, NUMINI, NPHASE, NSATIS
C
      DO 10 I=1,N
   10 X1(I)=X(I)
      MMZ=M+MZ
      DO 20 J=1,MMZ
   20 RJ1(J)=RJ(J)
      P1=P0
      F1=F
      G1=G
      RSIG1=RSIGMA
      H1=H
C
      RETURN
      END




      SUBROUTINE TCHECK
C
C     TCHECK CHECKS THE NUMBER OF SECONDS THAT HAVE ELAPSED SINCE THE
C     START OF THE PROBLEM.  IF THE SOLUTION IS TAKING LONGER THAN 90
C     PER-CENT OF THE ESTIMATED MAXIMUM TIME, A SWITCH IS SET TO GIVE
C     MORE OUTPUT.
C
      COMMON /OPTNS/ NT1,NT2,NT3,NT4,NT5,NT6,NT7,NT8,NT9,NT10
      COMMON /TSW/ NSWW
      COMMON /TMX/ TM0,EXT,EXT90
      COMMON /DEVC/ NI, NO, NP
C
      CALL SECOND (SECS)
      IF (SECS.LT.EXT90) RETURN
C     GETTING CLOSE TO EXCEEDING THE TIME LIMIT SET OUTPUT OPTION TO
C     GIVE MORE OUTPUT.
      NT3=2
      X=SECS - TM0
      WRITE (NO,001) X
  001 FORMAT (//,2X,7HTIME = ,F9.3,8H SECONDS )
      IF(SECS .GT. EXT) NSWW=2
      CALL OUTPUT (1)
C
      RETURN
      END
```

```
      SUBROUTINE READIN
C
C     THIS ROUTINE READS THE USERS DATA
C     IF NEWOLD=1 READ IN DATA FOR NEW PROBLEM
C     IF NEWOLD=2 ONLY READ IN NEW B VECTOR
C     IF NEWOLD=3 USE SAME CONSTANTS AS USED IN PREVIOUS PROBLEM
C
      COMMON /CONST/ A, B, C, D, E
      COMMON /DEVC/ NI, NO, NP
      DIMENSION A(50,25), B(25), C(25,25), D(25), E(25)
C
      READ (NI,001) NEWOLD, NR, MR
  001 FORMAT (8I10)
C
      GO TO (10,20,30), NEWOLD
C
   10 READ (NI,002) (D(I), I=1,NR)
  002 FORMAT (8E10.4)
      READ (NI,002) (E(I), I=1,NR)
      DO 100 I=1,MR
  100 READ (NI,002) (A(I,K), K=1,NR)
      DO 200 J=1,NR
  200 READ (NI,002) (C(J,K), K=1,NR)
C
   20 READ (NI,002) (B(I), I=1,MR)
C
   30 RETURN
      END



      SUBROUTINE REJECT
C
C     REJECT RETURNS THE STORED VALUES OF THE OBJECTIVE FUNCTION, THE
C     CONSTRAINT FUNCTION AND THE PENALTY FUNCTION TO THEIR NORMAL
C     LOCATION.
C
      COMMON /SHARE/ X(100),DEL(100),A(100,100),N,M,MN,NP1,NM1
      COMMON /EQAL/ H, H1, MZ
      COMMON /VALUE/ F,G,P0,RSIGMA,RJ(100),RHO
      COMMON /CRST/ DELX(100),DELX0(100),RHOIN,RATIO,EPSI,THETAG,
     1 RSIG1, G1, X1(100), X2(100), X3(100), XR2(100), XR1(100), PR1,
     2 PR2, P1, F1, RJ1(200), DOTT, PGRAD(100), DIAG(100),
     3 PREV3, ADELX, NTCTR, NUMINI, NPHASE, NSATIS
C
      DO 10 I=1,N
   10 X(I)=X1(I)
      MMZ=M+MZ
      DO 20 J=1,MMZ
   20 RJ(J)=RJ1(J)
      P0=P1
      RSIGMA=RSIG1
      G=G1
      F=F1
      H=H1
C
      RETURN
      END
```

NONLINEAR PROGRAMMING ROUTINE -- SUMT VERSION 4

PARAMETERS

N = 3   M = 4   MZ = 0   TMMAX =   60.000
RHO = .1000E+01   RATIO = .4000E+01   EPSILON = .1000E-08   THETA = .1000E-05

OPTIONS SELECTED

NT1 = 1   NT2 = 1   NT3 = 1   NT4 = 1   NT5 = 2
NT6 = 2   NT7 = 1   NT8 = 1   NT9 = 1   NT10 = 2
NEXOP1 = 4   NEXOP2 = 4

TOLERANCES FOR DIFFERENCING AND MOVES

XEP1 = .1000E-03   XEP2 = 0.

TIME =   .058 SECONDS

F =   -.10000000E+04   P =   0.   G =   0.

VALUES OF X VECTOR

X(1) =   .10000000E+02   X(2) =   .10000000E+02   X(3) =   .10000000E+02

VALUES OF THE CONSTRAINTS

NOT INCLUDING THE NON-NEGATIVE CONSTRAINTS

G(1) =   .32000000E+02   G(2) =   .32000000E+02   G(3) =   .32000000E+02
G(4) =   .22000000E+02   G(

VALUES OF OBJECTIVE FUNCTION PARTIALS

ANALYTICAL FIRST PARTIALS

DEL(1) =   .1234568E+01   DEL(2) =   .1234568E+01   DEL(3) =   .1234568E+01

NUMERICAL FIRST PARTIALS

DEL(1) =   -.1000000E+03   DEL(2) =   -.1000000E+03   DEL(3) =   -.1000000E+03

VALUES FOR CONSTRAINT NUMBER 1

ANALYTICAL FIRST PARTIALS

DEL(1) =   -.1000000E+01   DEL(2) =   0.   DEL(3) =   0.

NUMERICAL FIRST PARTIALS

DEL(1) =   -.1000000E+01   DEL(2) =   0.   DEL(3) =   0.

VALUES FOR CONSTRAINT NUMBER 2

ANALYTICAL FIRST PARTIALS

DEL(1) =   0.   DEL(2) =   -.1000000E+01 DEL(3) =   0.

NUMERICAL FIRST PARTIALS

DEL(1) =   0.   DEL(2) =   -.1000000E+01 DEL(3) =   0.

VALUES FOR CONSTRAINT NUMBER 3

ANALYTICAL FIRST PARTIALS

DEL(1) =   0.   DEL(2) =   0.   DEL(3) =   -.1000000E+01

NUMERICAL FIRST PARTIALS

DEL(1) =   0.   DEL(2) =   0.   DEL(3) =   -.1000000E+01

VALUES FOR CONSTRAINT NUMBER 4

ANALYTICAL FIRST PARTIALS

DEL(1) =   -.1000000E+01   DEL(2) =   -.2000000E+01 DEL(3) =   -.2000000E+01

NUMERICAL FIRST PARTIALS

DEL(1) =   -.1000000E+01   DEL(2) =   -.2000000E+01 DEL(3) =   -.2000000E+01

VALUES OF OBJECTIVE FUNCTION PARTIALS

ANALYTICAL SECOND PARTIALS

A(1, 1) = -.909495E-04   A(1, 2) = -.999999E+01   A(1, 3) = -.100001E+02
A(2, 1) = 0.   A(2, 2) = -.909495E-04   A(2, 3) = -.100001E+02
A(3, 1) = 0.   A(3, 2) = 0.   A(3, 3) = 0.

NUMERICAL SECOND PARTIALS

A(1, 1) = -.909495E-04   A(1, 2) = -.999999E+01   A(1, 3) = -.100001E+02
A(2, 1) = 0.   A(2, 2) = -.909495E-04   A(2, 3) = -.100001E+02

VALUES FOR CONSTRAINT NUMBER 1

ANALYTICAL SECOND PARTIALS

A(1, 1) = 0.   A(1, 2) = 0.   A(1, 3) = 0.

A(2, 1) = 0.   A(2, 2) = 0.   A(2, 3) = 0.

A(3, 1) = 0.   A(3, 2) = 0.   A(3, 3) = 0.

NUMERICAL SECOND PARTIALS

VALUES FOR CONSTRAINT NUMBER 2

ANALYTICAL SECOND PARTIALS

A(1, 1) = 0.   A(1, 2) = 0.   A(1, 3) = 0.

A(2, 1) = 0.   A(2, 2) = 0.   A(2, 3) = 0.

A(3, 1) = 0.   A(3, 2) = 0.   A(3, 3) = 0.

NUMERICAL SECOND PARTIALS

VALUES FOR CONSTRAINT NUMBER 3

ANALYTICAL SECOND PARTIALS

A(1, 1) = 0.   A(1, 2) = 0.   A(1, 3) = 0.

A(2, 1) = 0.   A(2, 2) = 0.   A(2, 3) = 0.

A(3, 1) = 0.   A(3, 2) = 0.   A(3, 3) = 0.

NUMERICAL SECOND PARTIALS

VALUES FOR CONSTRAINT NUMBER 4

ANALYTICAL SECOND PARTIALS

A(1, 1) = 0.   A(1, 2) = 0.   A(1, 3) = 0.

A(2, 1) = 0.   A(2, 2) = .568434E-05   A(2, 3) = 0.

A(3, 1) = 0.   A(3, 2) = 0.   A(3, 3) = .568434E-05

NUMERICAL SECOND PARTIALS

A(2, 1) = 0.   A(2, 2) = .568434E-05   A(2, 3) = 0.

A(3, 1) = 0.   A(3, 2) = 0.   A(3, 3) = .568434E-05

TIME =   .343 SECONDS

THE FEASIBLE STARTING POINT AND VALUES

F =   -.10000000E+04   P =   0.   G =   0.

VALUES OF X VECTOR

X(1) =   .10000000E+02 X(2) =   .10000000E+02 X(3) =   .10000000E+02

VALUES OF THE CONSTRAINTS

NOT INCLUDING THE NON-NEGATIVE CONSTRAINTS

G(1) =   .32000000E+02 G(2) =   .32000000E+02 G(3) =   .32000000E+02
G(4) =   .22000000E+02 G(

ORTHOGONAL MOVE VECTORS

ORTHOGONAL MOVE VECTORS

ROUNDOFF ERRORS PREVENT MORE ACCURATE DETERMINATION
OF THE MINIMUM OF THIS SUBPROBLEM.

TIME =   .669 SECONDS

VALUES AT POINT NUMBER 7

NPHASE = 2   RHO =   .65601442E+02   RSIGMA =   -.11195674E+04

F =   -.33900152E+04   P = -.45095826E+04   G = -.38492253E+04

VALUES OF X VECTOR

X(1) =   .23576020E+02 X(2) =   .11991309E+02 X(3) =   .11991253E+02

VALUES OF THE CONSTRAINTS

NOT INCLUDING THE NON-NEGATIVE CONSTRAINTS

G(1) =   .18423980E+02 G(2) =   .30008691E+02 G(3) =   .30008747E+02
G(4) =   .45885655E+00 G(

SOLUTION OF THE SUBPROBLEM

F = -.33900152E+04   P = -.45095826E+04   G = -.38492253E+04

VALUES OF X VECTOR

X(1) =   .27825494E+01   X(2) =   .54707491E+01   X(3) =   .54707746E+01

VALUES OF THE CONSTRAINTS

NOT INCLUDING THE NON-NEGATIVE CONSTRAINTS

G(1) =   .35606554E+01   G(2) =   .21860814E+01   G(3) =   .21860773E+01
G(4) =   .14296721E+03   G(

ROUNDOFF ERRORS PREVENT MORE ACCURATE DETERMINATION
OF THE MINIMUM OF THIS SUBPROBLEM.

TIME =   .887 SECONDS

(Printouts at points number 12, 17, 25, 28, 36, 43, 49, 51, 66, and 73 are omitted.)

VALUES AT POINT NUMBER 22

NPHASE = 2   RHO =   .10250225E+01   RSIGMA =   -.13217600E+02

F = -.34549756E+04   P = -.34681932E+04   G = -.34621508E+04

VALUES OF X VECTOR

X(1) =   .23995572E+02   X(2) =   .12000944E+02   X(3) =   .11997713E+02

VALUES OF THE CONSTRAINTS

NOT INCLUDING THE NON-NEGATIVE CONSTRAINTS

G(1) =   .18004428E+02   G(2) =   .29999056E+02   G(3) =   .30002287E+02
G(4) =   .71139907E-02   G(

2ND ORDER ESTIMATES

F = -.34559829E+04   P = -.34544842E+04   G = -.34559882E+04

VALUES OF X VECTOR

X(1) =   .24002895E+02   X(2) =   .12001068E+02   X(3) =   .11997425E+02

VALUES OF THE CONSTRAINTS

NOT INCLUDING THE NON-NEGATIVE CONSTRAINTS

G(1) =   .17997105E+02   G(2) =   .29998932E+02   G(3) =   .30002575E+02
G(4) =   .11775323E-03   G(

1ST ORDER ESTIMATES

F =   -.34559881E+04   P =   -.34541049E+04   G =   -.34559882E+04

VALUES OF X VECTOR

X(1) =   .24002865E+02   X(2) =   .12001028E+02   X(3) =   .11997498E+02

VALUES OF THE CONSTRAINTS

NOT INCLUDING THE NON-NEGATIVE CONSTRAINTS

G(1) =   .17997135E+02   G(2) =   .29998972E+02   G(3) =   .30002502E+02
G(4) =   .81597562E-04   G(

SOLUTION OF THE SUBPROBLEM

F =   -.34559881E+04   P =   -.34541049E+04   G =   -.34559882E+04

VALUES OF X VECTOR

X(1) =   .42717154E-01   X(2) =   .85411825E-01   X(3) =   .85434825E-01

VALUES OF THE CONSTRAINTS

NOT INCLUDING THE NON-NEGATIVE CONSTRAINTS

G(1) =   .56931690E-01   G(2) =   .34168493E-01   G(3) =   .34164813E-01
G(4) =   .14408545E+03   G(

ROUNDOFF ERRORS PREVENT MORE ACCURATE DETERMINATION
OF THE MINIMUM OF THIS SUBPROBLEM.

TIME =   1.595 SECONDS

VALUES AT POINT NUMBER 39

NPHASE = 2   RHO =   .40039942E-02   RSIGMA =   -.30102997E-01

F =   -.34559951E+04   P =   -.34560252E+04   G =   -.34560231E+04

VALUES OF X VECTOR

X(1) =   .23993833E+02   X(2) =   .12002129E+02   X(3) =   .12000938E+02

VALUES OF THE CONSTRAINTS

NOT INCLUDING THE NON-NEGATIVE CONSTRAINTS

G(1) =   .18006167E+02   G(2) =   .29997871E+02   G(3) =   .29999062E+02
G(4) =   .32880503E-04   G(

2ND ORDER ESTIMATES

F =   -.34559992E+04   P =   -.34559938E+04   G =   -.34559993E+04

VALUES OF X VECTOR

X(1) =   .23992214E+02   X(2) =   .12002783E+02   X(3) =   .12001108E+02

VALUES OF THE CONSTRAINTS

NOT INCLUDING THE NON-NEGATIVE CONSTRAINTS

G(1) =   .18007786E+02   G(2) =   .29997217E+02   G(3) =   .29998892E+02
G(4) =   .32001499E-05   G(

1ST ORDER ESTIMATES

F =   -.34559992E+04   P =   -.34559923E+04   G =   -.34559993E+04

VALUES OF X VECTOR

X(1) =   .23992578E+02   X(2) =   .12002615E+02   X(3) =   .12001094E+02

VALUES OF THE CONSTRAINTS

NOT INCLUDING THE NON-NEGATIVE CONSTRAINTS

G(1) =   .18007422E+02   G(2) =   .29997385E+02   G(3) =   .29998906E+02
G(4) =   .35937543E-05   G(

SOLUTION OF THE SUBPROBLEM

F =   -.34559992E+04   P =   -.34559923E+04   G =   -.34559993E+04

VALUES OF X VECTOR

X(1) =   .16687597E-03   X(2) =   .33360699E-03   X(3) =   .33364012E-03

VALUES OF THE CONSTRAINTS

NOT INCLUDING THE NON-NEGATIVE CONSTRAINTS

G(1) =   .22236794E-03   G(2) =   .13347595E-03   G(3) =   .13347065E-03
G(4) =   .12177412E+03   G(

ROUNDOFF ERRORS PREVENT MORE ACCURATE DETERMINATION
OF THE MINIMUM OF THIS SUBPROBLEM.

TIME =   2.573 SECONDS

VALUES AT POINT NUMBER 59

NPHASE = 2   RHO =   .39101506E-05   RSIGMA =   -.16626139E-04

F =   -.34559998E+04   P =   -.34559998E+04   G =   -.34559998E+04

VALUES OF X VECTOR

X(1) =   .24000150E+02   X(2) =   .12001159E+02   X(3) =   .11998766E+02

VALUES OF THE CONSTRAINTS

NOT INCLUDING THE NON-NEGATIVE CONSTRAINTS

G(1) =   .17999850E+02   G(2) =   .29998841E+02   G(3) =   .30001234E+02
G(4) =   .12547399E-05   G(

2ND ORDER ESTIMATES

F =   -.34559998E+04   P =   -.34559999E+04   G =   -.34559999E+04

VALUES OF X VECTOR

X(1) =   .23998241E+02   X(2) =   .12000842E+02   X(3) =   .12000037E+02

VALUES OF THE CONSTRAINTS

NOT INCLUDING THE NON-NEGATIVE CONSTRAINTS

G(1) =   .18001759E+02   G(2) =   .29999158E+02   G(3) =   .29999963E+02
G(4) =   .13260837E-05   G(

1ST ORDER ESTIMATES

F =   -.34559998E+04   P =   -.34559999E+04   G =   -.34559999E+04

VALUES OF X VECTOR

X(1) =   .23998643E+02   X(2) =   .12000909E+02   X(3) =   .11999769E+02

VALUES OF THE CONSTRAINTS

NOT INCLUDING THE NON-NEGATIVE CONSTRAINTS

G(1) =   .18001357E+02   G(2) =   .29999091E+02   G(3) =   .30000231E+02
G(4) =   .12881778E-05   G(

SOLUTION OF THE SUBPROBLEM

F =   -.34559998E+04   P =   -.34559999E+04   G =   -.34559999E+04

VALUES OF X VECTOR

X(1) =   .16292193E-06   X(2) =   .32581443E-06   X(3) =   .32587940E-06

VALUES OF THE CONSTRAINTS

NOT INCLUDING THE NON-NEGATIVE CONSTRAINTS

G(1) =   .21723240E-06   G(2) =   .13034339E-06   G(3) =   .13033299E-06
G(4) =   .31163037E+01   G(

ROUNDOFF ERRORS PREVENT MORE ACCURATE DETERMINATION
OF THE MINIMUM OF THIS SUBPROBLEM.

TIME =   3.999 SECONDS

VALUES AT POINT NUMBER 80

NPHASE = 2   RHO =   .61096104E-07   RSIGMA =   -.26212086E-06

F = -.34559998E+04   P = -.34559998E+04   G = -.34559998E+04

VALUES OF X VECTOR

X(1) =   .23999970E+02   X(2) =   .12000430E+02   X(3) =   .11999584E+02

VALUES OF THE CONSTRAINTS

NOT INCLUDING THE NON-NEGATIVE CONSTRAINTS

G(1) =   .18000030E+02   G(2) =   .29999570E+02   G(3) =   .30000416E+02
G(4) =   .13036654E-05   G(

2ND ORDER ESTIMATES

F = -.34559998E+04   P = -.34559998E+04   G = -.34559998E+04

VALUES OF X VECTOR

X(1) =   .23999970E+02   X(2) =   .12000429E+02   X(3) =   .11999585E+02

VALUES OF THE CONSTRAINTS

NOT INCLUDING THE NON-NEGATIVE CONSTRAINTS

G(1) =   .18000030E+02   G(2) =   .29999571E+02   G(3) =   .30000415E+02
G(4) =   .13042286E-05   G(

1ST ORDER ESTIMATES

F =   -.34559998E+04   P =   -.34559998E+04   G =   -.34559998E+04

VALUES OF X VECTOR

X(1) =   .23999970E+02   X(2) =   .12000430E+02   X(3) =   .11999585E+02

VALUES OF THE CONSTRAINTS

NOT INCLUDING THE NON-NEGATIVE CONSTRAINTS

G(1) =   .18000030E+02   G(2) =   .29999570E+02   G(3) =   .30000415E+02
G(4) =   .13039731E-05   G(

SOLUTION OF THE SUBPROBLEM

F =   -.34559998E+04   P =   -.34559998E+04   G =   -.34559998E+04

VALUES OF X VECTOR

X(1) =   .25456742E-08   X(2) =   .50911594E-08   X(3) =   .50915184E-08

VALUES OF THE CONSTRAINTS

NOT INCLUDING THE NON-NEGATIVE CONSTRAINTS

G(1) =   .33942223E-08   G(2) =   .20365660E-08   G(3) =   .20365085E-08
G(4) =   .46864866E-01   G(

FINAL VALUE OF F = -3.45599981E+03

FINAL X VALUES

X(1) =   2.39999699E+01   X(2) =   1.20004303E+01   X(3) =   1.19995841E+01

V. CONSTRAINED FLETCHER-POWELL (CONMIN ALGORITHM)*

A. Purpose

This program finds the minimum of a multivariable, nonlinear function subject to nonlinear equality constraints:

Minimize $\quad F(X_1, X_2, \ldots, X_N)$

Subject to $\quad G_k(X_1, X_2, \ldots, X_N) = 0, \quad k = 1, 2, \ldots, M.$

B. Method

The procedure has been described by Haarhoff and Buys (23). The method incorporates the constraints into a modified, unconstrained objective function which is then optimized by the unconstrained minimization technique of Fletcher and Powell (see Chapter 9). Derivatives of the objective function with respect to the independent variables are thus required. Inequality constraints can be treated by use of slack variables and transformations (see Test Problem). The algorithm proceeds as follows:

1) A new unconstrained objective function is formulated from the original function and constraints,

$$\Phi = F - \sum_{k=1}^{M} \lambda_k G_k + B \sum_{k=1}^{M} G_k^2$$

where λ_k and B are constants.

2) A starting point is selected (feasible or nonfeasible) and the value and derivatives of F and values of G_k are determined. The derivatives can be either analytical or numerical approximations. The H matrix in the Fletcher and Powell procedure is set equal to the identity matrix for the first iteration. The λ_k values are determined from the following system of equations,

$$\sum_{i=1}^{N} \sum_{j=1}^{M} \left(\lambda_j \frac{\partial G_j}{\partial X_i} \frac{\partial G_k}{\partial X_i} \right) = \sum_{i=1}^{N} \left(\frac{\partial G_k}{\partial X_i} \frac{\partial F}{\partial X_i} \right), \quad k = 1, 2, \ldots, M.$$

---

*Computer code developed by P. C. Haarhoff, J. D. Buys, and H. von Molendorff, Atomic Energy Board, Pretoria, South Africa. Used by permission.

The B value is set equal to some positive number--the authors suggest a value of 30.

3) A series of search directions and one dimensional search steps are then determined per the unconstrained Fletcher and Powell general procedure for the modified objective function with updating of the λ_k values for each new iteration. When convergence is achieved, $G_k = 0$ and $F = \Phi$.

C. Program Description

1) <u>Usage</u>:

The program consists of a main program, a primary subroutine (CONMIN), seven additional general subroutines (MATUP, LAMB, LINMIN, DSIMQ, SCALER, LINK, GRADFG) and a user supplied subroutine (FUNC). The program is setup without data card input. Output is from the main program and subroutines FUNC and CONMIN. Format changes may be required.

2) <u>Subroutines Required</u>:

SUBROUTINE CONMIN(FNS, N, M, X, F, GRADF, G, GRADG, B, R, H, FLAM, DELX, EPS, DEFX, DELG, IT, IPR, IND, IER, S, GO, GN, W, A) main optimization subroutine--performs or coordinates all calculations.

SUBROUTINE MATUP(H, IND, N, S, GO, GN, ALFA, DELX, DELG, W) computes search directions.

SUBROUTINE LAMB(N, M, F, GM, GV, SL, FLAM, R, DELG, W, A, ICV) computes λ_k values.

SUBROUTINE LINMIN(IND, ND, NPT, BEGIN, S, ALFA, END, FN, DELX, DEFX, EPS) performs one dimensional search.

SUBROUTINE DSIMQ(A, B1, B2, N, KS) solves a system of simultaneous linear equations.

SUBROUTINE SCALER(N, M, X, DEL, TOP, BOT, B, GRADG1, GRADG2) calculates scaling factors for function and independent variables.

SUBROUTINE LINK(IF, X, F, GRADF, G, GRADG) applies scaling factors to function and independent variables.

SUBROUTINE GRADFG(IF, X, F, GRADF, G, GRADG) calculates derivatives (numerical as written--modify if analytical derivatives used).

SUBROUTINE FUNC(X, F, G) specifies objective function and constraints (user supplied).

3) Description of Parameters:

The parameters involved in the argument lists of the various subroutines and main program are as follows:

CONMIN

    FNS   Subroutine of the form FNS(IF,X,F,GRADF,G,GRADG), providing information about the function to be minimized and the constraining functions, with

        IF   Parameter indicating whether only F and G must be determined for the given X(IF=1) or GRADF and GRADG as well (IF=2)

        X   N-Dimensional vector containing the values of the independent variables at the point considered

        F   Value of the function to be minimized

        GRADF   N-Dimensional vector containing the gradient of F

        G   M-Dimensional vector containing the values of the constraining functions

        GRADG   N*M-Dimensional vector containing the gradient of each G(I) in sequence

    N   Number of independent variables

    M   Number of constraints

    X   N-Dimensional vector containing the initial values of the independent variables. On return, X corresponds to the constrained minimum

    F   Variable containing the minimum function value on return

    GRADF   N-Dimensional vector containing the gradient of F on return

    G   M-Dimensional vector containing on return the values of the constraining functions at the minimum. These should be zero

    GRADG   N*M-Dimensional vector containing the gradient of each G(I) in sequence on return

    B   constant used in the construction of the auxiliary function PHI

    R   Constant used in the determination of the direction of the first linear iteration after new values of the LAMBDA(I) have been found

    H $N*(N+1)/2$ dimensional vector containing the matrix H in triangular form

    FLAM   M-Dimensional vector containing the values of the parameters LAMBDA(I) on return

    DELX   Test value representing the absolute error in the position of a point in the N-dimensional variable space

EPS	Test value representing the absolute error in the magnitude of the auxiliary function PHI near the minimum
DEFX	Test value representing the greatest distance (in N-dimensional space) from the minimum corresponding to an increase of EPS in the value of PHI
DELG	Test value representing the absolute error in the magnitudes of the gradients of F and G
IT	Maximum number of linear iterations
IPR	Printout index consisting of two digits. IPR=10*K+L where K is the data set reference number for the printer, and the results of every L^{th} linear iteration should be presented. If K=0, no information is printed, and if L=0, no results for linear iterations are printed
IND	Index indicating the action to be taken by CONMIN IND=0 H is changed to the unit matrix before commencing minimization IND=1 H is used as given
IER	Error parameter IER=0 convergence obtained IER=1 no convergence in IT iterations IER=2 a minimum for the function PHI cannot be found (try a larger value of B)
S	Working storage of dimension N
GO	Working storage of dimension N
GN	Working storage of dimension N
W	Working storage of dimension 2*N
A	Working storage of dimension M**2

Additional symbols used in printout:

ICV	Convergence index for LAMBDA determinations ICV=0 convergence obtained ICV=1 probable that convergence was obtained but that DELG has been underestimated ICV=2 no convergence yet
INL	Convergence index for linear iterations INL=0 convergence obtained INL=1 no convergence yet
ALFA	Ratio of distance moved along line in linear minimization to distance recommended beforehand using gradient and matrix H

REMARKS

The subroutine replacing FNS(LINK) should be declared as EXTERNAL in the main program.

MATUP

H	Vector of dimension N(N+1)/2 containing the matrix H in triangular form
IND	Index indicating the action to be taken by MATUP (see below)
N	Number of independent variables in function to be minimized
S	N-Dimensional vector containing the direction of the line on which function should be minimized
GO	N-Dimensional vector containing gradient before minimization
GN	N-Dimensional vector containing gradient after minimization
ALFA	Displacement along line S from initial to final point, as a multiple of the vector S
DELX	Absolute error in magnitude of ALFA*S
DELG	Absolute error in magnitudes of GO and GN
W	Working storage of dimension 2*N

REMARKS

1) Action taken by MATUP
   IND=0,1   H is generated as the identity matrix
   IND=0,2   S is put equal to -GO
   IND=3,4   H is updated and GO is set equal to the initial value of GN, while the latter is destroyed
   IND=3,5   S is updated using the current values of H and GO

   The minimization process is normally initialized by IND=0 or 1 and continued by IND=3

2) When H is accurately known, the minimum on the line S will lie at ALFA=1

LAMB

N	Number of independent variables
M	Number of constraints
F	N-Dimensional vector giving the gradient of the function to be minimized, at a point X
GM	N*M-Dimensional vector giving the gradient G(I) of each of the constraining functions at X, in sequence
GV	M-Dimensional vector giving the values of the constraining functions at X
SL	N-Dimensional vector containing the direction of the line on which the function should be minimized when the new LAMDA(I) values are first used
FLAM	M-Dimensional vector giving the current values of LAMDA(I). These are improved by the subroutine

R	Constant used in the determination of the SL(I)
DELG	Absolute error in the magnitudes of F and G(I)
W	Working storage of dimension N
A	Working storage of dimension M**2
ICV	Convergence parameter

    ICV=0   the changes in the values of FLAM may be ascribed to errors in F and GM

    ICV=1   the history of FLAM indicates that DELG has been underestimated and that better values for FLAM than those given will not be obtained

    ICV=2   no convergence yet

REMARKS

When calling LAMB for the first time, ICV must be put equal to -1, and each FLAM(I) to zero.

LINMIN

IND	Index indicating progress of minimization (see below)
ND	Number of independent variables which determine FN
NPT	Total number of points used to obtain the minimum
BEGIN	ND-Dimensional vector containing the initial point for minimization
S	ND-Dimensional vector containing the direction of the line on which the function should be minimized
ALFA	Displacement along line S from initial point, as a multiple of the vector S. At the start of the minimization process, ALFA must give an estimate of the position of the minimum, and at the end of the process, ALFA corresponds to the minimum
END	ND-Dimensional vector containing the minimum point at the end of the process
FN	Minimum function value
DELX	Absolute error in the position of the initial point, as a distance in the ND-Dimensional space
DEFX	Spatial distance from the minimum point at which the increase in the function value will definitely be observable
EPS	Maximum error in function value at the minimum point

REMARKS

1) For functions which are quadratic at the minimum, DEFX may be chosen considerably greater than the distance from the minimum point corresponding to an increase of EPS in the function value. Otherwise, DEFX should not be much greater than this distance.

2) Values of the index IND

   IND=0   When LINMIN is called at the start of the minimization process, the index IND should be zero. (All parameters other than END and FN must be supplied)

   IND=1   LINMIN will repeatedly return with IND=1, in which case it should again be called, with FN equal to the function value corresponding to END, parameters other than FN remaining unchanged

   IND=2   A linear minimum cannot be found

   IND=-1 The minimization process is complete, and FN has been significantly improved

   IND=-2 The minimization process is complete, and FN has not been significantly improved

DSIMQ

   A   Matrix of coefficients stored columnwise. These are destroyed in the computation. The size of matrix A is N by N.

   B1&B2 Vectors of original constants (length N). These are replaced by final solution values, vectors X1 and X2

   N   Number of equations and variables for each set

   KS   Output digit
          0 for normal solutions
          1 for singular sets of solutions

REMARKS

Matrix A must be general. If matrix is singular, solution values are meaningless.

SCALER

   N   Number of independent variables

   M   Number of constraints

   X   N-Dimensional vector containing the unscaled values of the independent variables at the initial point. On return, X contains the scaled coordinates of the initial point

   DEL   Test value representing the maximum absolute error in the scaled values of the object and constraining functions

   TOP   Maximum value allowed for XSC(I)

   BOT   Minimum value allowed for XSC(I)

   B   Constant used in the construction of the auxiliary function PHI

   GRADG1 Working storage of dimension N*M

   GRADG2 Working storage of dimension N*M

Description of parameters evaluated by subroutine SCALER:

   XSC   N-Dimensional vector containing the scales for the independent variables. (An unscaled value is divided by its scale to obtain a scaled value)

FSC	Scale for the function F
GSC	M-Dimensional vector containing the scales for the constraining functions G

REMARKS

When M=0 the parameters B, GRADG1 and GRADG2 may be omitted from the argument list.

LINK

Description of parameters required by subroutine LINK:

IF	Parameter indicating whether only F and G must be determined for the given X (IF=1) or GRADF and GRADG as well (IF=2)
X	N-Dimensional vector containing the scaled values of the independent variables at the point considered
N	Number of independent variables
M	Number of constraints
XSC	N-Dimensional vector containing the scales for the independent variables. (An unscaled value is divided by its scale to obtain a scaled value)
FSC	Scale for the function F
GSC	M-Dimensional vector containing the scales for the constraining functions G

Description of parameters evaluated by subroutine LINK:

F	Scaled value of the function to be minimized
GRADF	N-Dimensional vector containing the scaled gradient of F
G	M-Dimensional vector containing the scaled values of the constraining functions
GRADG	N*M-Dimensional vector containing the scaled gradients of the constraining functions, in sequence

REMARKS

1) When M=0 the parameters G and GRADG in the argument list, as well as GSC in the common statements, may be omitted when LINK is called.
2) When used, subroutine LINK takes the place of the subroutine FNS called by 'CONMIN' and must be declared as 'EXTERNAL LINK' in the main program.

GRADFG

IF	Parameter indicating whether only F and G must be determined for the given X (IF=1) or GRADF and GRADG as well (IF=2)
X	N-Dimensional vector containing the values of the independent variables at the point considered
F	Value of the function to be minimized

GRADF	N-Dimensional vector containing the gradient of F
G	M-Dimensional vector containing the values of the constraining functions
GRADG	N*M-Dimensional vector containing the gradient of each G(I), in sequence
N	Number of independent variables
M	Number of constraints
DELDIF	N-Dimensional vector containing increments of X to be used when finding gradients by first differences

FUNC

X	Vector of independent variables (length N)
F	Objective function
G	Constraint vector (length M)
Y	Slack variables

Main Program

NI	Card reader unit number
NO	Printer unit number

4) DIMENSION Requirements:

Dimensioning is handled by DIMENSION and COMMON statements and should be modified according to the requirements of each particular problem. The appropriate dimension limits are given in the Description of Parameters section.

5) Input Formats:

The program as written does not require card input data. The main program is used to specify the starting values and system parameters.

6) Output:

Printing is from the main program and subroutines FUNC and CONMIN. Starting point information is printed first followed by intermediate printouts of current value information keyed by IPR. The final values are then printed.

7) Summary of User Requirements:

a) Determine values for the various control parameters. The following starting guesses are suggested for the listed subroutines (24):

CONMIN

$B = 30$

$R = 10$

$DELX = 10^{-12}$

$$EPS = 10^{-12}$$
$$DELG = 10^{-12} \text{ (analytical derivatives used)}$$
$$= 10^{-6} \text{ (numerical derivatives used)}$$
$$DEFX = 10^{-4}$$

IT = 5 times the number of independent variables

<u>SCALER</u>

$$DEL = 10^{-12}$$
$$TOP = 20$$
$$BOT = 0.05$$

b) Specify starting guesses and system parameters in the main program.

c) Specify objective function and constraints in FUNC.

d) Adjust DIMENSION statements and output FORMAT statements as required.

Further details on this program are available elsewhere (24).

D. <u>Test Problem</u>

The following problem was formulated by Rosenbrock (41). The calculations were performed on a CDC 6400 computer.

Function: $F = -X_1 X_2 X_3$

Constraints: $0 \leq X_i \leq 42, \quad i = 1, 2, 3$
$0 \leq (X_4 = X_1 + 2X_2 + 2X_3) \leq 72$

Starting Point: $X_1 = X_2 = X_3 = 10.0$

Parameters:
  IT = 20   DELX = 10^{-12}
  IPR = 61   EPS = 10^{-12}
  IND = 0   DEFX = 10^{-4}
  B = 100.   DELG = 10^{-6}
  R = 10.

Algorithm Answers: F = -3456.000
  X_1 = 24.000
  X_2 = 12.000
  X_3 = 12.000

Number of Iterations: 16

Central Processor Time: 12 seconds

Note on Reformulation of Test Problem

The method as presented will handle equality constraints only. Thus the problem was reformulated as follows:

a) Introduce slack variables (X_4, X_5, X_6, X_7) to the upper constraints to form equalities,

$$X_1 + X_4^2 = 42$$
$$X_2 + X_5^2 = 42$$
$$X_3 + X_6^2 = 42$$
$$X_4 + X_7^2 = 72 = X_1 + 2X_2 + 2X_3$$

b) Solving for X_1, X_2 and X_3 from the first three constraints and inserting the values into the fourth constraint, a single equality constraint is formed,

$$G_1 = 0 = (42 - X_4^2) + 2(42 - X_5^2) + 2(42 - X_6^2) - (72 - X_7^2)$$

At a starting point of (10, 10, 10), the slack variables will be initially set at

$$X_4 = X_5 = X_6 = \sqrt{42 - 10} = \sqrt{32}$$
$$X_7 = \sqrt{72 - 50} = \sqrt{22}$$

and updated as the algorithm progresses. When the equality constraint is satisfied, the slack variables will equal zero.

c) The lower constraints are handled by means of a penalty function, $U \times 10^{20}$, added to the objective function,

$$f = F - U \times 10^{20}$$

where U is set to the minimum current value of X_1, X_2, X_3, X_4 or 0. Thus if the lower constraints are violated ($X_i \leq 0$), the penalty function will drive the search back into the feasible region.

d) The reformulated problem is thus treated as consisting of 4 variables with one equality constraint (N=4, M=1).

The listing and output for this problem are contained in the following section.

E. Program Listings and Example Output

```
C
C      MAIN LINE PROGRAM FOR CONMIN ALGORITHM
C
       DIMENSION X(4),GRADF(4),G(2),GRADG(4),H(10),FLAM(2),S(4),GO(4),
      1GN(4),W(8),A(2)
       COMMON NI,NO,IFUNC,NN,MM,XSC(100),FSC,GSC(100),DELDIF(100)
       EXTERNAL LINK
C
       NI=50
       NO=66
       U=SQRT(3.2E1)
       DO 10 I=1,3
    10 X(I)=U
       X(4)=SQRT(2.2E1)
       IFUNC=0
C
       CALL SCALER(4,1,X,1.E-12,2.E1,5.E-2,1.E2,GO,GN)
C
       WRITE(NO,20)
    20 FORMAT (1H1,3X,22HRESULTS OF CALCULATION//4X,14HINITIAL VALUES)
       IFUNC=1
C
       CALL LINK (1,X,F,GRADF,G,GRADG)
C
       IFUNC=0
C
       CALL CONMIN(LINK,4,1,X,F,GRADF,G,GRADG,1.E2,1.E1,H,FLAM,1.E-12,
      11.E-12,1.E-4,1.E-6,20,61,0,IER,S,GO,GN,W,A)
C
       WRITE(NO,30)
    30 FORMAT (///4X,12HFINAL VALUES)
       IFUNC=1
C
       CALL LINK (1,X,F,GRADF,G,GRADG)
C
       END




       SUBROUTINE FUNC(X,F,G)
C
       DIMENSION X(1),G(1),Y(4)
       COMMON NI,NO,IFUNC,NN,MM,XSC(100),FSC,GSC(100),DELDIF(100)
C
C
C    Y(I)=42-X(I)**2, I=1,2,3    Y(4)=Y(1)+2*Y(2)+2*Y(3) =72-X(4)**2
C
       DO 10 I=1,3
```

```
   10 Y(I)=4.2E1-X(I)**2
      Y(4)=7.2E1-X(4)**2
      U=MIN1(Y(1),Y(2),Y(3),Y(4),0.E0)
      F=-Y(1)*Y(2)*Y(3)-1.E20*U
      G(1)=Y(1)+2.E0*(Y(2)+Y(3))-Y(4)
      IF(IFUNC) 40,40,20
   20 U=Y(1)*Y(2)*Y(3)
      C=G(1)+Y(4)
      WRITE(NO,30)(Y(I),I=1,3),U,C
   30 FORMAT (4X,4HX1 =,1PE13.6,4X,4HX2 =,1PE13.6,4X,4HX3 =,1PE13.6/
     14X,9HPRODUCT =,1PE13.5,4X,14HX1+2*X2+2*X3 =,1PE13.6)
C
   40 RETURN
      END




      SUBROUTINE CONMIN(FNS,N,M,X,F,GRADF,G,GRADG,B,R,H,FLAM,DELX,EPS,DE
     &FX,DELG,IT,IPR,IND,IER,S,GO,GN,W,A)
C
      DIMENSION X(25),GRADF(25),G(25),GRADG(25),H(50),FLAM(25),S(25),
     &GO(25),GN(25),W(50),A(25)
      COMMON NI,NO,IFUNC,NN,MM,XSC(100),FSC,GSC(100),DELDIF(100)
C
C     INITIALIZE MATRIX H
C
      IF (IND-1)10,20,20
   10 CALL MATUP(H,1,N,S,GO,GN,ALFA,DELX,DELG,W)
C     PRINT OUT INITIAL CONDITIONS
   20 IP=IPR/10
      IR=IPR-10*IP
      IF (IP)40,40,30
   30 WRITE (NO,420)N,M,IT,IPR,IND,B,R,DELX,EPS,DEFX,DELG
      WRITE (NO,430)(X(I),I=1,N)
C     FUNCTIONS AND GRADIENTS AT INITIAL POINT
   40 CALL FNS(2,X,F,GRADF,G,GRADG)
      IET=0
      ICV=-1
      IEL=0
      INL=1
C
C     OBTAIN NEW SET OF LAMBDA VALUES
      IF (1.GT.M)GO TO 5
      DO 50 I=1,M
   50 FLAM(I) = 0.0
    5 CONTINUE
   60 CALL LAMB(N,M,GRADF,GRADG,G,S,FLAM,R,DELG,W,A,ICV)
      IEL=IEL+1
C     OBTAIN THE AUXILIARY FUNCTION (PHI), ITS GRADIENT(GN), THE
C     MAGNITUDE(GNA) OF GN AND THE SUM(GS) OF B*G*(I)**2 FOR I=1 TO M
      L=1
   70 PHI=F
      IF (1.GT.M)GO TO 15
```

```
      DO 80 I=1,M
   80 PHI=PHI+G(I)*(B*G(I)-FLAM(I))
   15 CONTINUE
      ICKCGT=L
      IF (ICKCGT.LE.1)GO TO 90
      IF (ICKCGT.GE.2)GO TO 240
      GO TO (90,240),ICKCGT
   90 GNA = 0.
      IF (1.GT.N)GO TO 25
      DO 110 I=1,N
      T=GRADF(I)
      K=I
      IF (1.GT.M)GO TO 35
      DO 100 J=1,M
      T = T + GRADG(K)*(2.*B*G(J) - FLAM(J))
  100 K=K+N
   35 CONTINUE
      GN(I)=T
  110 GNA=GNA+T*T
   25 CONTINUE
      GNA = SQRT(GNA)
      GS = 0.
      IF (1.GT.M)GO TO 45
      DO 120 I=1,M
  120 GS=GS+B*G(I)*G(I)
   45 CONTINUE
      ICKCGT=L
      IF (ICKCGT.LE.1)GO TO 130
      IF (ICKCGT.GE.2)GO TO 280
      GO TO (130,280),ICKCGT
C     PRINT OUT DATA FOR LAMBDA DETERMINATION
  130 IF (IP)150,150,140
  140 WRITE (NO,450)IEL,IET,ICV
      WRITE (NO,430)(X(I),I=1,N)
      WRITE (NO,440)(FLAM(I),I=1,M)
      WRITE (NO,460)(G(I),I=1,M)
      WRITE (NO,470)PHI,F,GS,GNA
C     DETERMINE WHETHER CONVERGENCE HAS BEEN OBTAINED
  150 IF (ICV-1)160,170,180
  160 IF (INL)180,170,180
  170 IER=0
      GO TO 380
C
C     LINEAR ITERATION
  180 L=2
      INL=1
      ITN=0
      IF (1.GT.N)GO TO 55
      DO 190 I=1,N
  190 GO(I)=GN(I)
   55 CONTINUE
  200 IF (IET-IT)220,210,210
  210 IER=1
      GO TO 380
  220 IF (1.GT.N)GO TO 65
      DO 230 I=1,N
```

```
      230 W(I)=X(I)
       65 CONTINUE
          IET=IET+1
          ITN=ITN+1
          INLP=INL
          INL=0
          ALFA = 1.0
      240 CALL LINMIN(INL,N,NF,W,S,ALFA,X,PHI,DELX,DEFX,EPS)
          IF (INL-1)270,260,250
      250 IER=2
          GO TO 380
      260 CALL FNS(1,X,F,GRADF,G,GRADG)
          GO TO 70
C         OBTAIN GN,GNA, AND GS
      270 CALL FNS(2,X,F,GRADF,G,GRADG)
          GO TO 90
C         FIND INDEX INL FOR LINEAR ITERATIONS
      280 INL=INL+2
          IF (ALFA)300,300,290
      290 IF (GNA-DELG)300,300,310
      300 INL=0
C         PRINT OUT DATA FOR LINEAR ITERATION
      310 IF (IP*IR)340,340,320
      320 IF (MOD(IET,IR))340,330,340
      330 WRITE (NO,480)IET,INL,ALFA,NF
          WRITE (NO,430)(X(I),I=1,N)
          WRITE (NO,470)PHI,F,GS,GNA
C         UPDATE H AND OBTAIN NEW S
      340 CALL MATUP(H,3,N,S,GO,GN,ALFA,DELX,DELG,W)
C         CONVERGENCE OBTAINED IF INLP=0 AND INL=0 (ITN=2)
          IF (INLP+INL)350,350,360
      350 IER=0
          GO TO 380
C         OBTAIN NEW SET OF LAMBDA VALUES OR CARRY OUT A FURTHER ITERATION
      360 IF (ITN-1)200,200,370
      370 IF (INL*(N-M+1-ITN))60,60,200
C
C         CONCLUSION
      380 IF (IP)410,410,390
      390 WRITE (NO,490)IER
          K=1
          IF (1.GT.N)GO TO 75
          DO 400 I=1,N
          W(I)=H(K)
      400 K=K+N+1-I
       75 CONTINUE
          WRITE (NO,500)(W(I),I=1,N)
      410 RETURN
C
      420 FORMAT (42H1INITIAL DATA FOR CONSTRAINED MINIMIZATION,//3H N=,I3,3
         &X,2HM=,I3,3X,3HIT=,I4,3X,4HIPR=,I2,3X,4HIND=,I1/3H B=,1PE10.3,5X,2
         &HR=,E10.3,5X,5HDELX=,E10.3,5X,4HEPS=,E10.3/1X,5HDEFX=,E10.3,5X,5HD
         &ELG=,E10.3)
      430 FORMAT (/1X,2HX=,(1P5E16.7))
      440 FORMAT (12X,7HLAMBDA=,(1P5E16.7))
      450 FORMAT (///21H LAMBDA DETERMINATION,I4,16H AFTER ITERATION,I5,7X,4
         &HICV=,I1)
```

```
      460 FORMAT (17X,2HG=,(1P5E16.7))
      470 FORMAT (5H PHI=,1PE14.7,6X,2HF=,E14.7/1X,14HSUM OF B*G**2=,E14.7,6
     &X,23HMAGNITUDE OF GRAD(PHI)=,E14.7)
      480 FORMAT (/10H ITERATION,I5,7X,4HINL=,I1,6X,5HALFA=,1PE10.3,6X,14HNO
     &. OF POINTS=,I3)
      490 FORMAT (///35H CONSTRAINED MINIMIZATION COMPLETED,7X,4HIER=,I1)
      500 FORMAT (24H DIAGONAL ELEMENTS OF H=,(1P5E12.3))
C
      END



      SUBROUTINE MATUP(H,IND,N,S,GO,GN,ALFA,DELX,DELG,W)
C
      DIMENSION X(25),GRADF(25),G(25),GRADG(25),H(50),FLAM(25),S(25),
     &GO(25),GN(25),W(50),A(25)
C
C     GENERATE IDENTITY MATRIX
C     -(DIAGONAL ELEMENTS ARE H(1),H(N+1),H(2*N),H(3*N-2),----,H(N*N))
      IF (IND-2)10,60,90
   10 K=1
      IF (1.GT.N)GO TO 5
      DO 40 I=1,N
      H(K) = 1.0
      NJ=N-I
      IF (NJ)50,50,20
   20 IF (1.GT.NJ)GO TO 15
      DO 30 J=1,NJ
      KJ=K+J
   30 H(KJ) = 0.0
   15 CONTINUE
   40 K=KJ+1
    5 CONTINUE
   50 IF (IND)60,60,80
C
C     DIRECTION OF STEEPEST DESCENT
   60 IF (1.GT.N)GO TO 25
      DO 70 I=1,N
   70 S(I)=-GO(I)
   25 CONTINUE
   80 RETURN
C
C     UPDATE MATRIX H
C     -COMPUTE GRADIENT CHANGE Y AND STORE IN GN
   90 IF (IND-4)100,100,210
  100 IF (1.GT.N)GO TO 35
      DO 110 I=1,N
      T=GN(I)
      GN(I)=T-GO(I)
  110 GO(I)=T
C     -COMPUTE H.Y AND STORE IN G(1) TO G(N)
   35 CONTINUE
      IF (1.GT.N)GO TO 45
      DO 150 I=1,N
      T = 0.0
```

```
          K=I
          IF (1.GT.N)GO TO 55
          DO 140 J=1,N
          T=T+GN(J)*H(K)
          IF (J-I)120,130,130
    120   K=K+N-J
          GO TO 140
    130   K=K+1
    140   CONTINUE
     55   CONTINUE
    150   W(I)=T
     45   CONTINUE
C         -COMPUTE POSITION CHANGE Z AND STORE IN G(N+1) TO G(2*N)
          K=N
          IF (1.GT.N)GO TO 65
          DO 160 I=1,N
          K=K+1
    160   W(K)=ALFA*S(I)
     65   CONTINUE
C         -COMPUTE Q=Y.H.Y, P=Y.Z AND R=SQRT(Z*Z)
          R = 0.0
          Q = 0.0
          P = 0.0
          K=N
          IF (1.GT.N)GO TO 75
          DO 170 I=1,N
          K=K+1
          R=R+W(K)*W(K)
          Q=Q+GN(I)*W(I)
    170   P=P+GN(I)*W(K)
     75   CONTINUE
          R = SQRT(R)
C         -LEAVE H UNCHANGED IF P OR R ARE VERY SMALL
          IF (R-DELX)210,210,180
    180   IF (P-R*DELG)210,210,190
C         -COMPUTE NEW H
    190   K=1
          IF (1.GT.N)GO TO 85
          DO 200 I=1,N
          LI=N+I
          DO 200 J=I,N
          LJ=N+J
          H(K)=H(K)+W(LI)*W(LJ)/P-W(I)*W(J)/Q
    200   K=K+1
     85   CONTINUE
C
C         COMPUTE NEW DIRECTION
    210   IF (IND-4)220,270,220
    220   IF (1.GT.N)GO TO 95
          DO 260 I=1,N
          T = 0.0
          K=I
          IF (1.GT.N)GO TO 105
          DO 250 J=1,N
          T=T+GO(J)*H(K)
          IF (J-I)230,240,240
    230   K=K+N-J
```

```
          GO TO 250
  240 K=K+1
  250 CONTINUE
  105 CONTINUE
  260 S(I)=-T
   95 CONTINUE
  270 RETURN
C
      END




      SUBROUTINE LAMB(N,M,F,GM,GV,SL,FLAM,R,DELG,W,A,ICV)
C
      DIMENSION X(25),GRADF(25),G(25),GRADG(25),H(50),FLAM(25),S(25),
     &GO(25),GN(25),W(50),A(25),SL(50),GV(50),HIS(25),F(50),GM(50)
C
C
C     EVALUATE W(I)=F.G(I)
      AS = 0.0
      Q=0.0
      Z = 0.0
   10 K=1
      IF (1.GT.M)GO TO 5
      DO 30 I=1,M
      T = 0.0
      IF (1.GT.N)GO TO 15
      DO 20 J=1,N
      T=T+F(J)*GM(K)
   20 K=K+1
   15 CONTINUE
   30 W(I)=T
    5 CONTINUE
C
C     EVALUATE A(M*(I-1)+J)=G(I).G(J)
      K=1-N
      IF (1.GT.M)GO TO 25
      DO 50 I=1,M
      K=K+N
      L=K
      DO 50 J=I,M
      T = 0.0
      DO 40 IN=1,N
      T=T+GM(K)*GM(L)
      L=L+1
   40 K=K+1
   35 CONTINUE
      K=K-N
      LL=M*(I-1)+J
      A(LL)=T
      LL=M*(J-1)+I
   50 A(LL)=T
   25 CONTINUE
C
```

```
C       EVALUATE RD AND Q
C       -  S=F-SUM(I=1,M) OF FLAM(I)*G(I)
C       -  AGI=MAGNITUDE OF G(I)
C       -  SGI=ABS(S.G(I))
C       -  RD=MAXIMUM VALUE OF SGI/AGI/DELG WITH RESPECT TO I
C       -  Q =SUM OF DIAGONAL ELEMENTS OF A
        RD = 0.0
        K=1
        L=1
        IF (1.GT.M)GO TO 45
        DO 70 I=1,M
        SGI=W(I)
        IF (1.GT.M)GO TO 55
        DO 60 J=1,M
        SGI=SGI-FLAM(J)*A(K)
     60 K=K+1
     55 CONTINUE
        SGI = ABS(SGI)
        AGI = SQRT(A(L))
        Q=Q+A(L)
        L=L+M+1
        T=AGI*DELG
        T=SGI/T/DELG
     70 RD = AMAX1(RD,T)
     45 CONTINUE
C
C       COMPUTE SL(I) AND NEW VALUES FOR FLAM(I)
C       -  D=MAGNITUDE OF CHANGE IN FLAM
        K=1
        Q=Q*Z
        IF (1.GT.M)GO TO 65
        DO 80 I=1,M
        A(K)=A(K)+Q
        SL(I)=-GV(I)
     80 K=K+M+1
     65 CONTINUE
        CALL DSIMQ(A,W,SL,M,IER)
        IF (IER)100,100,90
     90 Z = 10.*Z + 1.E-12
        GO TO 10
    100 D = 0.0
        IF (1.GT.M)GO TO 75
        DO 110 I=1,M
        Z=W(I)-FLAM(I)
        FLAM(I)=W(I)
        W(I)=SL(I)
    110 D=D+Z*Z
     75 CONTINUE
        D = SQRT(D)
        IF (1.GT.N)GO TO 85
        DO 120 I=1,N
        K=I
        SL(I) = 0.0
        DO 120 J=1,M
        SL(I)=SL(I)+W(J)*GM(K)
    120 K=K+N
     85 CONTINUE
C
```

```
C     MODIFY SL(I) IF NECESSARY
C      - W=NORMALISED GRADIENT OF SUM OF GV(I)**2(I=1 TO M)
C      - SW=W.SL
C      - SP=MAGNITUDE OF COMPONENT OF SL PERPENDICULAR TO W
C      - R=MAXIMUM ALLOWED VALUE OF SP/ABS(SW)
      T = 0.0
      IF (1.GT.N)GO TO 95
      DO 140 I=1,N
      W(I) = 0.0
      K=I
      IF (1.GT.M)GO TO 105
      DO 130 J=1,M
      W(I)=W(I)+GV(J)*GM(K)
  130 K=K+N
  105 CONTINUE
      Z = W(I) + SIGN(1.E-20,W(I))
      W(I)=Z
  140 T=T+Z*Z
   95 CONTINUE
      T = SQRT(T)
      SW = 0.0
      IF (1.GT.N)GO TO 115
      DO 150 I=1,N
      W(I)=W(I)/T
  150 SW=SW+W(I)*SL(I)
  115 CONTINUE
      SW = SW + SIGN(1.E-20,SW)
      SP = 0.0
      IF (1.GT.N)GO TO 125
      DO 160 I=1,N
      Z=SL(I)-SW*W(I)
      SL(I)=Z
  160 SP=SP+Z*Z
  125 CONTINUE
      SP = SQRT(SP) + 1.E-20
      T = AMIN1(R*ABS(SW)/SP,1.)
      IF (1.GT.N)GO TO 135
      DO 170 I=1,N
  170 SL(I)=T*SL(I)+SW*W(I)
  135 CONTINUE
C
C     CONVERGENCE PARAMETER
      IF (ICV)180,200,200
  180 I100=0
      DO 190 I=1,4
  190 HIS(I) = 1.E20
  200 ICV=2
      T=HIS(1)
      DO 210 I=1,3
      T=T+HIS(I+1)
  210 HIS(I)=HIS(I+1)
      HIS(4)=D
      T = T/4.
      IF (RD-100.) 230,230,220
  220 I100=0
      RETURN
  230 I100=I100+1
```

```
          IF (RD-1.) 240,240,250
      240 ICV=0
          RETURN
      250 IF (I100-5)280,260,260
      260 IF (D-T)280,270,270
      270 ICV=1
C
      280 RETURN
          END




          SUBROUTINE LINMIN(IND,ND,NPT,BEGIN,S,ALFA,END,FN,DELX,DEFX,EPS)
C
          DIMENSION S(25),F(50),T(50),BEGIN(50),END(50)
C
          IF (IND)40,40,10
       10 F(KK)=FN
          NPT=NPT+1
          ICKCGT=NGOTO
          IF (ICKCGT.LE.1)GO TO 60
          IF (ICKCGT.GE.6)GO TO 360
          GO TO (60,70,110,150,290,360),ICKCGT
       20 IF (1.GT.ND)GO TO 5
C
C         GET FUNCTION VALUES
          DO 30 I=1,ND
       30 END(I)=BEGIN(I)+T(KK)*S(I)
        5 CONTINUE
          RETURN
C
C         CHANGE IN ALFA CORRESPONDING TO DELX AND DEFX
       40 IND=1
          NPT=0
          Z = 0.0
          IF (1.GT.ND)GO TO 15
          DO 50 I=1,ND
       50 Z=Z+S(I)*S(I)
       15 CONTINUE
          Z = SQRT(Z)
          Z = AMAX1(Z,1.E-20)
          DEL=DELX/Z
          DEF=DEFX/Z
C
C         OBTAIN THREE POINTS, L, M AND N, WITH L AND N ON OPPOSITE SIDES OF
C         M, F(L) AND F(N) NOT LESS THAN F(M), AND THE DISTANCES (L TO M)
C         AND (M TO N) AT LEAST DEFX
C         -INITIAL AND RECOMMENDED POINTS
          T(1) = 0.0
          KK=1
          NGOTO=1
          GO TO 20
       60 FKEEP=F(1)
          T(2)=ALFA
```

```
          KK=2
          NGOTO=2
          GO TO 20
C         -POINTS M AND L
   70 L=1
      M=2
      IF (F(1)-F(2))80,90,90
   80 M=1
      L=2
   90 IF (ABS(T(M) - T(L)) - DEF) 100,130,130
  100 T(L) = T(M) + SIGN(DEF,T(L)-T(M))
      KK=L
      NGOTO=3
      GO TO 20
  110 IF (F(L)-F(M))120,130,130
  120 I=L
      L=M
      M=I
C         -POINT N (DOUBLE STEPLENGTH EACH TIME)
  130 Z = 1.0
      N=3
  140 T(N)=T(M)+Z*(T(M)-T(L))
      KK=N
      NGOTO=4
      GO TO 20
  150 IF (F(N)-F(M))160,180,180
  160 I=L
      L=M
      M=N
      N=I
      Z = 2.0
      IF (ABS(T(M) - T(L)) - 1.E20) 140,170,170
  170 IND=2
      RETURN
C
C         DECREASE THE DISTANCE (L TO N) TO LESS THAN 4*DEFX, KEEPING THE
C         DISTANCES (L TO M) AND (M TO N) AT LEAST DEFX
  180 NEW=4
      NBAD=0
C         -LET L BE CLOSER TO M THAN IS N
  190 IF (ABS(T(M)-T(L)) - ABS(T(M)-T(N))) 210,210,200
  200 I=L
      L=N
      N=I
C         -ESTIMATE THE POSITION OF THE MINIMUM POINT (NEW) FROM A PARABOLIC
C         FIT F=A+B*T+C*T**2
  210 T1=T(L)-T(M)
      T2=T(N)-T(M)
      H1 = ABS(F(L) - F(M))/T1
      H2 = ABS(F(N) - F(M))/T2
      C=(H2-H1)/(T2-T1)
      B=(H1*T2-H2*T1)/(T2-T1)
      T(NEW) = T(M) - B/2./(C+SIGN(1.E-40,C))
C         -END CYCLE WHEN DISTANCE (L TO N) IS LESS THAN 4*DEFX
      IF (ABS(T(N) - T(L)) - 4.*DEF) 350,350,220
C         -IF NBAD=2, CHANGE T(NEW) SO THAT THE DISTANCE (L TO NEW) IS THE
C         GEOMETRIC AVERAGE OF THE DISTANCES (L TO M) AND (L TO N)
```

```
      220 IF (NBAD-2)240,230,230
      230 T(NEW) = SQRT(T1*(T1-T2))
          T(NEW) = T(L) + SIGN(T(NEW),T2)
          NBAD=0
C         -NEW WILL BE CLOSER TO M THAN TO N OR L.   IF NEW LIES WITHIN
C         DEFX OF M, CHANGE T(NEW) SO THAT NEW LIES BETWEEN M AND N, AT A
C         DISTANCE DEFX FROM M
      240 IF(ABS(T(NEW)-T(M)) - DEF) 250,260,260
      250 T(NEW) = T(M) + SIGN(DEF,T2)
C         -LET NEW LIE BETWEEN M AND N
      260 IF ((T(NEW)-T(M))*T2)270,280,280
      270 I=L
          L=N
          N=I
C         -IMPROVE L,M AND N
      280 KK=NEW
          NGOTO=5
          GO TO 20
      290 Z = ABS(T(N)-T(L))
          IF (F(NEW)-F(M))300,310,310
      300 I=L
          L=M
          M=NEW
          NEW=I
          GO TO 320
      310 I=N
          N=NEW
          NEW=I
C         -TEST THAT THE DISTANCE (L TO N) DECREASED BY AT LAST TEN PER CENT
      320 IF (ABS(T(N)-T(L))/Z - .9) 330,340,340
      330 NBAD=0
          GO TO 190
      340 NBAD=NBAD+1
          GO TO 190
C
C         OBTAIN THE FUNCTION VALUE AT THE ESTIMATED MINIMUM POINT
      350 KK=NEW
          NGOTO=6
          GO TO 20
      360 IF (F(NEW)-F(M))380,380,370
      370 NEW=M
      380 ALFA=T(NEW)
          FN=F(NEW)
          IF (1.GT.ND)GO TO 25
          DO 390 I=1,ND
      390 END(I)=BEGIN(I)+ALFA*S(I)
       25 CONTINUE
C
C         TEST WHETHER IMPROVEMENT IS SIGNIFICANT
          IF (FKEEP-FN-EPS)400,410,410
      400 IND=-2
          RETURN
      410 IF (ABS(ALFA) - DEL) 400,420,420
      420 IND=-1
C
          RETURN
          END
```

```
      SUBROUTINE DSIMQ(A,B1,B2,N,KS)
C
      DIMENSION A(25),B1(50),B2(50)
C
C        FORWARD SOLUTION
C
      TOL = 1.E-20
      KS=0
      JJ=-N
      IF (1.GT.N)GO TO 5
      DO 80 J=1,N
      JY=J+1
      JJ=JJ+N+1
      BIGA=0
      IT=JJ-J
      IF (J.GT.N)GO TO 15
      DO 20 I=J,N
C
C        SEARCH FOR MAXIMUM COEFICIENT IN COLUMN
C
      IJ=IT+I
      IF (ABS(BIGA) - ABS(A(IJ))) 10,20,20
   10 BIGA=A(IJ)
      IMAX=I
   20 CONTINUE
   15 CONTINUE
C
C        TEST FOR PIVOT LESS THAN TOLERANCE (SINGULAR MATRIX)
C
      IF (ABS(BIGA) - TOL) 30,30,40
   30 KS=1
      RETURN
C
C        INTERCHANGE ROWS IF NECESSARY
C
   40 I1=J+N*(J-2)
      IT=IMAX-J
      IF (J.GT.N)GO TO 25
      DO 50 K=J,N
      I1=I1+N
      I2=I1+IT
      SAVE=A(I1)
      A(I1)=A(I2)
      A(I2)=SAVE
C
C        DIVIDE EQUATION BY LEADING COEFFICIENT
C
   50 A(I1)=A(I1)/BIGA
   25 CONTINUE
      SAVE=B1(IMAX)
      B1(IMAX)=B1(J)
      B1(J)=SAVE/BIGA
      SAVE=B2(IMAX)
```

```
      B2(IMAX)=B2(J)
      B2(J)=SAVE/BIGA
C
C         ELIMINATE NEXT VARIABLE
C
      IF (J-N)60,90,60
   60 IQS=N*(J-1)
      DO 80 IX=JY,N
      IXJ=IQS+IX
      IT=J-IX
      IF (JY.GT.N)GO TO 35
      DO 70 JX=JY,N
      IXJX=N*(JX-1)+IX
      JJX=IXJX+IT
   70 A(IXJX)=A(IXJX)-(A(IXJ)*A(JJX))
   35 CONTINUE
      B1(IX)=B1(IX)-(B1(J)*A(IXJ))
   80 B2(IX)=B2(IX)-(B2(J)*A(IXJ))
    5 CONTINUE
C
C         BACK SOLUTION
C
   90 IF (N-1)120,120,100
  100 NY=N-1
      IT=N*N
      IF (1.GT.NY)GO TO 45
      DO 110 J=1,NY
      IA=IT-J
      IB=N-J
      IC=N
      DO 110 K=1,J
      B1(IB)=B1(IB)-A(IA)*B1(IC)
      B2(IB)=B2(IB)-A(IA)*B2(IC)
      IA=IA-N
  110 IC=IC-1
   45 CONTINUE
C
  120 RETURN
      END




      SUBROUTINE SCALER(N,M,X,DEL,TOP,BOT,B,GRADG1,GRADG2)
C
      DIMENSION X(25),GRADG1(25),GRADG2(25),G(100),GP(100),GM(100),
     &GRADF1(100),GRADF2(100),ABGD(100)
      COMMON NI,NO,IFUNC,NN,MM,XSC(100),FSC,GSC(100),DELDIF(100)
C
      NN=N
      MM=M
C
C FIND FIRST AND SECOND DERIVATIVES
      DEL1 = SQRT(DEL)
      DEL3 = DEL**(1./3.)
      FSC = 1.0
```

```
      IF (1.GT.N)GO TO 5
      DO 10 I=1,N
   10 XSC(I) = 1.0
    5 CONTINUE
      IF (M)40,40,20
   20 IF (1.GT.M)GO TO 15
      DO 30 J=1,M
   30 GSC(J) = 1.0
   15 CONTINUE
   40 CALL LINK(1,X,F,GRADF1,G,GRADG1)
      IF (1.GT.N)GO TO 25
      DO 70 I=1,N
      X(I)=X(I)+DEL3
      CALL LINK(1,X,FP,GRADF1,GP,GRADG1)
      X(I) = X(I) - 2.*DEL3
      CALL LINK(1,X,FM,GRADF1,GM,GRADG1)
      X(I)=X(I)+DEL3
      GRADF1(I) = (FP-FM)/2./DEL3
      GRADF2(I) = (FP+ FM - 2.*F)/DEL3/DEL3
      IF (M)70,70,50
   50 K=I
      IF (1.GT.M)GO TO 35
      DO 60 J=1,M
      GRADG1(K) = (GP(J) - GM(J))/2./DEL3
      GRADG2(K) = (GP(J) + GM(J) - 2.*G(J))/DEL3/DEL3
   60 K=K+N
   35 CONTINUE
   70 CONTINUE
   25 CONTINUE
C
C  FIND FSC AND FIRST APPROXIMATION TO XSC
      FSC = ABS(F)
      IF (M)160,160,80
   80 K=1
      IF (1.GT.M)GO TO 45
      DO 100 J=1,M
      S = 0.0
      IF (1.GT.N)GO TO 55
      DO 90 I=1,N
      S=S+GRADG1(K)**2
   90 K=K+1
   55 CONTINUE
  100 ABGD(J) = SQRT(S)
   45 CONTINUE
      RN = 10.*SQRT(FLOAT(N))
      IF (1.GT.N)GO TO 65
      DO 130 I=1,N
      FS1 = ABS(GRADF1(I)/F)
      FS2 = SQRT(ABS(GRADF2(I)/F))
      S = 0.0
      K=I
      IF (1.GT.M)GO TO 75
      DO 120 J=1,M
      GRAT = ABS(GRADG1(K))/ABGD(J)*TOP*RN/BOT
      IF (GRAT-1.) 120,120,110
  110 T = ABS(GRADG2(K)/GRADG1(K))
      S = AMAX1(S,T)
```

```
  120 K=K+N
   75 CONTINUE
      S = AMAX1(FS1,FS2,S,1./TOP)
      S = AMIN1(S,1./BOT)
  130 XSC(I) = 1./S
   65 CONTINUE
C
C  FIND GSC
      K=1
      IF (1.GT.M)GO TO 85
      DO 150 J=1,M
      S = 0.0
      IF (1.GT.N)GO TO 95
      DO 140 I=1,N
      T=GRADG1(K)*XSC(I)
      S=S+T*T
  140 K=K+1
   95 CONTINUE
  150 GSC(J) = SQRT(S)
   85 CONTINUE
  160 IF (1.GT.N)GO TO 105
C
C  FIND FINAL VALUES OF XSC AND DELDIF
      DO 200 I=1,N
      S = 0.0
      IF (M)190,190,170
  170 K=I
      IF (1.GT.M)GO TO 115
      DO 180 J=1,M
      T=GRADG1(K)/GSC(J)
      S=S+T*T
  180 K=K+N
  115 CONTINUE
      S = 2.*B*S
  190 S=S+GRADF2(I)/FSC
      S = AMAX1(S,1./TOP/TOP)
      S = AMAX1(SQRT(1./S),BOT)
      XSC(I)=S
      DELDIF(I)=XSC(I)*DEL1
  200 X(I)=X(I)/XSC(I)
  105 CONTINUE
      RETURN
C
      END




      SUBROUTINE LINK(IF,X,F,GRADF,G,GRADG)
C
      DIMENSION X(25),GRADF(25),G(25),GRADG(25),XRAW(100)
      COMMON NI,NO,IFUNC,NN,MM,XSC(100),FSC,GSC(100),DELDIF(100)
      N=NN
      M=MM
C
```

```
      IF (1.GT.N)GO TO 5
      DO 10 I=1,N
   10 XRAW(I)=X(I)*XSC(I)
    5 CONTINUE
      IF (M)40,40,20
   20 CALL GRADFG(IF,XRAW,F,GRADF,G,GRADG)
      IF (1.GT.M)GO TO 15
      DO 30 J=1,M
   30 G(J)=G(J)/GSC(J)
   15 CONTINUE
      GO TO 50
   40 CALL GRADFG(IF,XRAW,F,GRADF,G,GRADG)
   50 F=F/FSC
      IF (IF-1)100,100,60
   60 IF (1.GT.N)GO TO 25
      DO 90 I=1,N
      GRADF(I)=GRADF(I)/FSC*XSC(I)
      IF (M)90,90,70
   70 K=I
      IF (1.GT.M)GO TO 35
      DO 80 J=1,M
      GRADG(K)=GRADG(K)/GSC(J)*XSC(I)
   80 K=K+N
   35 CONTINUE
   90 CONTINUE
   25 CONTINUE
  100 RETURN
C
      END




      SUBROUTINE GRADFG(IF,X,F,GRADF,G,GRADG)
C
      DIMENSION X(25),GRADF(25),G(25),GRADG(25),GD(100)
      COMMON NI,NO,IFUNC,NN,MM,XSC(100),FSC,GSC(100),DELDIF(100)
      N=NN
      M=MM
C
C  FIND VALUES
      IF (M)20,20,10
   10 CALL FUNC(X,F,G)
      GO TO 30
   20 CALL FUNC(X,F,G)
   30 IF (IF-1)40,40,50
   40 RETURN
C  FIND GRADIENTS
   50 IF (1.GT.N)GO TO 5
      DO 100 I=1,N
      DT=DELDIF(I)
      X(I)=X(I)+DT
      IF (M)80,80,60
   60 CALL FUNC(X,FD,GD)
```

```
      K=I
      IF (1.GT.M)GO TO 15
      DO 70 J=1,M
      GRADG(K)=(GD(J)-G(J))/DT
   70 K=K+N
   15 CONTINUE
      GO TO 90
   80 CALL FUNC(X,FD,G)
   90 X(I)=X(I)-DT
  100 GRADF(I)=(FD-F)/DT
    5 CONTINUE
      RETURN
C
      END
```

RESULTS OF CALCULATION

INITIAL VALUES
X1 = 1.00000E+01   X2 = 1.00000E+01   X3 = 1.00000E+01
PRODUCT = 1.0000E+03   X1+2*X2+2*X3 = 5.00000E+01

INITIAL DATA FOR CONSTRAINED MINIMIZATION

N= 4   M= 1   IT= 20   IPR=61   IND=0
B= 1.00E+02   R= 1.00E+01   DELX= 1.00E-12   EPS= 1.00E-12
DEFX= 1.00E-04   DELG= 1.00E-06

X=   1.979220E+01   3.934456E+01   3.934002E+01   1.347090E+01

LAMBDA DETERMINATION   1 AFTER ITERATION   0   ICV=2

X=   1.979220E+01   3.934456E+01   3.934002E+01   1.347090E+01
         LAMBDA=   -2.292765E+00
              G=   1.060738E-10
PHI= -1.000000E+00   F= -1.000000E+00
SUM OF B*G**2= 1.125166E-18   MAGNITUDE OF GRAD(PHI)= 2.297747E-01

ITERATION   1   INL=0   ALFA= -3.58E+03   NO. OF POINTS= 5

X=   1.979220E+01   3.934456E+01   3.934001E+01   1.347090E+01
PHI= -1.000000E+00   F= -1.000000E+00
SUM OF B*G**2= 1.441444E-11   MAGNITUDE OF GRAD(PHI)= 2.297744E-01

ITERATION   2   INL=1   ALFA= 5.72E+01   NO. OF POINTS= 19

X=   1.062773E+01   3.950164E+01   3.941706E+01   4.071455E+00
PHI= -2.860826E+00   F= -3.153362E+00
SUM OF B*G**2= 1.920576E-01   MAGNITUDE OF GRAD(PHI)= 5.017533E-01

ITERATION   3   INL=1   ALFA= 1.01E+01   NO. OF POINTS= 15

X=   1.014707E+01   3.936420E+01   3.990518E+01   2.705558E+00
PHI= -2.995451E+00   F= -3.038120E+00
SUM OF B*G**2= 1.478784E-02   MAGNITUDE OF GRAD(PHI)= 2.777388E-01

ITERATION   4   INL=1   ALFA= 2.07E+00   NO. OF POINTS= 12

X=   9.565203E+00   3.932290E+01   3.992335E+01   1.868844E+00
PHI= -3.020653E+00   F= -3.133523E+00
SUM OF B*G**2= 5.776439E-02   MAGNITUDE OF GRAD(PHI)= 1.368713E-01

```
LAMBDA DETERMINATION    3 AFTER ITERATION    8        ICV=2

X=    1.508011E+01    3.761002E+01    3.840694E+01    2.648116E-01
         LAMBDA=  -6.603675E+00
              G=  -1.079783E-03
PHI= -3.443935E+00        F= -3.436921E+00
SUM OF B*G**2= 1.165932E-04     MAGNITUDE OF GRAD(PHI)=  4.267085E-02


ITERATION    9        INL=1     ALFA= 1.04E+00     NO. OF POINTS=  7

X=    1.507517E+01    3.760378E+01    3.840058E+01    2.649432E-01
PHI= -3.444056E+00        F= -3.444345E+00
SUM OF B*G**2= 1.920534E-07     MAGNITUDE OF GRAD(PHI)=  3.544594E-02


ITERATION   10        INL=1     ALFA= 2.37E+00     NO. OF POINTS=  9

X=    1.478020E+01    3.775722E+01    3.847837E+01    2.970640E-01
PHI= -3.446448E+00        F= -3.446192E+00
SUM OF B*G**2= 1.502933E-07     MAGNITUDE OF GRAD(PHI)=  3.312300E-02


ITERATION   11        INL=1     ALFA= 7.26E+00     NO. OF POINTS= 12

X=    1.475627E+01    3.783156E+01    3.840151E+01   -1.404170E-01
PHI= -3.450411E+00        F= -3.459131E+00
SUM OF B*G**2= 1.677326E-04     MAGNITUDE OF GRAD(PHI)=  3.389957E-02


ITERATION   12        INL=1     ALFA= 1.20E+01     NO. OF POINTS= 12

X=    1.479290E+01    3.805516E+01    3.820883E+01   -1.067561E-01
PHI= -3.454558E+00        F= -3.437932E+00
SUM OF B*G**2= 6.873988E-04     MAGNITUDE OF GRAD(PHI)=  6.033008E-02
```

```
LAMBDA DETERMINATION   5 AFTER ITERATION   16      ICV=2

X=     1.484396E+01     3.809480E+01     3.809020E+01     -8.405724E-04
            LAMBDA=    -6.638399E+00
                 G=     7.941852E-05
PHI= -3.455999E+00      F= -3.456527E+00
SUM OF B*G**2=  6.307301E-07      MAGNITUDE OF GRAD(PHI)=  1.748406E-03

ITERATION   17      INL=1     ALFA=  1.05E+00      NO. OF POINTS=  6

X=     1.484432E+01     3.809527E+01     3.809067E+01     -8.407814E-04
PHI= -3.456000E+00      F= -3.455973E+00
SUM OF B*G**2=  1.644524E-09      MAGNITUDE OF GRAD(PHI)=  7.966904E-05

ITERATION   18      INL=1     ALFA=  5.08E-01      NO. OF POINTS=  7

X=     1.484404E+01     3.809531E+01     3.809091E+01     -4.644115E-04
PHI= -3.456000E+00      F= -3.455946E+00
SUM OF B*G**2=  6.645917E-09      MAGNITUDE OF GRAD(PHI)=  1.317112E-04

ITERATION   19      INL=0     ALFA=  0.00E+00      NO. OF POINTS=  6

X=     1.484404E+01     3.809531E+01     3.809091E+01     -4.644115E-04
PHI= -3.456000E+00      F= -3.455946E+00
SUM OF B*G**2=  6.645917E-09      MAGNITUDE OF GRAD(PHI)=  1.317112E-04


LAMBDA DETERMINATION   6 AFTER ITERATION   19      ICV=2

X=     1.484404E+01     3.809531E+01     3.809091E+01     -4.644115E-04
            LAMBDA=    -6.637885E+00
                 G=    -8.152250E-06
PHI= -3.456000E+00      F= -3.455946E+00
SUM OF B*G**2=  6.645917E-09      MAGNITUDE OF GRAD(PHI)=  1.855744E-04

ITERATION   20      INL=1     ALFA=  8.53E-01      NO. OF POINTS=  5

X=     1.484401E+01     3.809527E+01     3.809087E+01     -4.643941E-04
PHI= -3.456000E+00      F= -3.455992E+00
SUM OF B*G**2=  1.423146E-10      MAGNITUDE OF GRAD(PHI)=  1.150402E-04


CONSTRAINED MINIMIZATION COMPLETED      IER=1
DIAGONAL ELEMENTS OF H=    2.12E+01    1.11E+01    1.59E+01    7.52E+00


   FINAL VALUES
   X1 =  2.40003E+01    X2 =  1.19999E+01    X3 =  1.19999E+01
      PRODUCT =  3.4560E+03    X1+2*X2+2*X3 =  7.19999E+01
```

BIBLIOGRAPHY

(1) Balas, E. "An Additive Algorithm for Solving Linear Programs with Zero-One Variables," *Operations Research*, 13, 517-546, 1965.

(2) Bates, H. *Computer Code for Wolfe Algorithm*. Kansas State University.

(3) Bellman, R. *Dynamic Programming*. Princeton University Press, Princeton, N. J., 1957.

(4) Bellman, R., and S. Dreyfus. *Applied Dynamic Programming*. Princeton University Press, Princeton, N. J., 1962.

(5) Beveridge, G. S. G., and R. S. Schechter. *Optimization: Theory and Practice*. McGraw-Hill Book Co., New York, 1970.

(6) Blau, G. E. *Generalized Polynomial Programming: Extensions and Applications*. Ph.D. Dissertation, Stanford University, 1968.

(7) Box, M. J. "A New Method of Constrained Optimization and a Comparison with Other Methods," *Computer J.*, 8, 42-52, 1965.

(8) Box, M. J., D. Davies, and W. H. Swann. *Non-Linear Optimization Technique*. Imperial Chemical Industries Monograph No. 5, Oliver and Boyd, Edinburgh, 1969.

(9) Converse, A. O. *Optimization*. Holt, Rinehardt, and Winston, Inc., New York, 1970.

(10) Cooper, L., and D. Steinberg. *Introduction to Methods of Optimization*. W. B. Saunders Co., Philadelphia, 1970.

(11) Cutler, L., and P. Wolfe. "Experiments in Linear Programming," in Graves, R. L., and P. Wolfe, eds. *Recent Advances in Mathematical Programming*. McGraw-Hill Book Company, Inc., New York, 1963.

(12) Dantzig, G. B. *Linear Programming and Extensions*. Princeton University Press, Princeton, N. J., 1963.

(13) Derringer, G. C. "Sequential Method for Estimating Response Surface," *Ind. and Eng. Chem.*, 61 (12), 6-13, 1969.

(14) Dixon, L. C. W. *Nonlinear Optimization*. English Universities Press Ltd., London, 1972.

(15) Draper, N. R., and H. Smith. *Applied Regression Analysis*. John Wiley and Sons, Inc., New York, 1966.

(16) Fiacco, A. V., and G. P. McCormick. *Nonlinear Sequential Unconstrained Minimization Techniques.* John Wiley and Sons, Inc., New York, 1968.

(17) Fletcher, R., and M. J. D. Powell. "A Rapidly Convergent Descent Method for Minimization," *Computer J.*, 6, 1963-168, 1963.

(18) Fletcher, R., and C. M. Reeves. "Function Minimization by Conjugate Gradients," *Computer J.*, 7, 149-154, 1964.

(19) Fletcher, R., ed. *Optimization.* Academic Press, New York, 1969.

(20) Garcia, A., and G. Hogg. *Computer Code for Blau Algorithm.* University of Illinois at Urbana-Champaign.

(21) Geoffrion, A. "Integer Programming by Implicit Enumeration and Balas' Method," *The Rand Corporation.* RM-4783-PR, February, 1966.

(22) Glover, F. "A Multiphase-Dual Algorithm for the Zero-One Integer Programming Problems," *Operations Research*, 13, 879-919, 1965.

(23) Haarhoff, P. C., and J. D. Buys. "A New Method for the Optimization of a Nonlinear Function Subject to Nonlinear Constraints," *Computer J.*, 13, 178-184, 1970.

(24) Haarhoff, P. C., J. D. Buys, and H. von Molendorff. *CONMIN: A Computer Programme for the Minimization of a Non-Linear Function Subject to Non-Linear Constraints.* PEL190, Atomic Energy Board, Pretoria, South Africa.

(25) Hadley, G. *Linear Programming.* Addison-Wesley Publishing Co., Inc., Reading, Mass., 1963.

(26) Hartley, H. O. "The Modified Gauss-Newton Method for the Fitting of Non-Linear Regression Problems by Least Squares," *Technometrics* 3, 269-280, 1961.

(27) Henley, E. J., and E. M. Rosen. *Material and Energy Balance Computations.* John Wiley and Sons, Inc., New York, 1969.

(28) Hillier, F., and G. Lieberman. *Introduction to Operations Research.* Holden-Day, Inc., San Francisco, 1967.

(29) Himmelblau, D. M. *Applied Nonlinear Programming.* McGraw-Hill Book Co., New York, 1972.

(30) Hooke, R., and T. A. Jeeves. "Direct Search Solution of Numerical and Statistical Problems," *J. Assoc. Comp. Mach.*, 8, 212-229, 1961.

(31) Jacoby, S. L. S., J. S. Kowalik, and J. T. Pizzo. *Iterative Methods for Nonlinear Optimization Problems.* Prentice-Hall, Inc., Englewood Cliffs, N. J., 1972.

(32) Land, A. H., and A. G. Doig. "An Automatic Method of Solving Discrete Programming Problems," *Econometrica*, 28, 497-520, 1960.

(33) Marquardt, D. M. "An Algorithm for Least-Squares Estimation of Nonlinear Parameters," *J. Soc. Indust. Appl. Math.*, 11, 431-441, 1963.

(34) Mylander, W. C., R. L. Holmes, and G. P. McCormick. *A Guide to SUMT-Version 4: The Computer Program Implementing the Sequential Unconstrained Minimization Technique for Nonlinear Programming.* Research Analysis Corporation, McLean, Virginia, 1971.

(35) Nelder, J. A., and R. Mead. "A Simplex Method for Function Minimization," *Computer J.*, 7, 308-313, 1964.

(36) Peterson, C. "Computational Experience with Variants of the Balas Algorithm Applied to the Selection of R & D Projects," *Management Science*, 13, 736-784, 1967.

(37) Powell, M. J. D. "An Efficient Method for Finding the Minimum of a Function of Several Variables Without Calculating Derivatives," *Computer J.*, 7, 155-162, 1964.

(38) Powell, M. J. D. "A Method for Minimizing a Sum of Squares of Non-Linear Functions Without Calculating Derivatives," *Computer J.*, 7, 303-307, 1965.

(39) Rider, E. *General Computer Solution of Dynamic Programming Problems with Integer Restrictions.* M. S. Thesis, Arizona State University, 1971.

(40) Rosen, J. B. "The Gradient Projection Method for Nonlinear Programming. Part I, Linear Constraints," *J. Soc. Indust. Appl. Math.*, 8, 181-217, 1960.

(41) Rosenbrock, H. H. "An Automatic Method for Finding the Greatest or Least Value of a Function," *Computer J.*, 3, 175-184, 1960.

(42) Rosenbrock, H. H., and C. Storey. *Computational Techniques for Chemical Engineers.* Pergamon Press, New York, 1966.

(43) Spendley, W., G. R. Hext, and F. R. Himsworth. "Sequential Applications of Simplex Designs in Optimization and Evolutionary Operation," *Technometrics*, 4, 441-461, 1962.

(44) Terrell, M., and V. Sumaria. *Computer Code for Differential Algorithm.* Oklahoma State University.

(45) Tucker, A. W. "Combinatorial Theory Underlying Linear Programs," *Recent Advances in Mathematical Programming.* McGraw-Hill Book Company, Inc., New York, 1963.

(46) Wagner, H. M. *Principles of Operations Research.* Prentice-Hall, Inc., Englewood Cliffs, N. J., 1969.

(47) Wilde, D. J., and C. S. Beightler. *Foundations of Optimization.* Prentice-Hall, Inc., Englewood Cliffs, N. J., 1967.

(48) Wolfe, P. "The Simplex Method for Quadratic Programming," *Econometrica*, 27, 382-398, 1959.

(49) Wolfe, P. "The Composite Simplex Algorithm," *SIAM Review*, 7, No. 1, 1965.

(50) Yates, T. L. *Computer Code for Balas Algorithm*. Oregon State University.

(51) Zener, C. "A Mathematical Aid in Optimizing Engineering Design," *Proc. Nat'l. Academy of Science*, 47, 537-539, 1961.

(52) "Nonlinear Parameter Estimation and Programming," *Catalog of Programs for IBM System 360 Models 25 and Above*. GC 20-1619-8, Program Number 360.D-13.6.003. International Business Machines, White Plains, N. Y.

(53) "Branch and Bound Mixed Integer Programming," *Catalog of Programs for IBM System/360 Models 25 and Above*. GC 30-1619-8, Program Number 360 D-15.2.005. International Business Machines, White Plains, N. Y.

(54) "MFOR 360 Linear Programming Code," *Catalog of Programs for IBM System/360 Models 25 and Above*. GC 20-1619-8, Program Number 360 D-15.2.007. International Business Machines, White Plains, N. Y.

(55) "SYSTEM/360 FORTRAN Linear Programming System," *Catalog of Programs for IBM System/360 Models 25 and Above*. GC 20-1619-8, Program Number 360 D-15.2.006. International Business Machines, White Plains, N. Y.

(56) "Program FMFP (Fletcher and Powell Unconstrained Search Technique)," *System/360 Scientific Subroutine Package*. H20-0205-3, International Business Machines, White Plains, N. Y.

(57) "Program FMCG (Fletcher and Reeves Unconstrained Search Technique)," *System/360 Scientific Subroutine Package*. H20-0205-3, International Business Machines, White Plains, N. Y.